T0329596

Pressure Oscillation in Biomedical Diagnostics and Therapy

Wiley-ASME Press Series

Analysis of ASME Boiler, Pressure Vessel, and Nuclear Components in the Creep Range
Maan H. Jawad, Robert I. Jetter

Robust Control: Youla Parameterization Approach
Farhad Assadian, Kevin R. Mallon

Fabrication of Process Equipment
Owen Greulich, Maan H. Jawad

Engineering Practice with Oilfield and Drilling Applications
Donald W. Dareing

Metrology and Instrumentation: Practical Applications for Engineering and Manufacturing
Samir Mekid

Fabrication of Metallic Pressure Vessels
Owen R. Greulich, Maan H. Jawad

Flow-Induced Vibration Handbook for Nuclear and Process Equipment
Michel J. Pettigrew, Colette E. Taylor, Nigel J. Fisher

Vibrations of Linear Piezostructures
Andrew J. Kurdila, Pablo A. Tarazaga

Bearing Dynamic Coefficients in Rotordynamics: Computation Methods and Practical Applications
Lukasz Brenkacz

Advanced Multifunctional Lightweight Aerostructures: Design, Development, and Implementation
Kamran Behdinan, Rasool Moradi-Dastjerdi

Vibration Assisted Machining: Theory, Modelling and Applications
Li-Rong Zheng, Dr. Wanqun Chen, Dehong Huo

Two-Phase Heat Transfer
Mirza Mohammed Shah

Computer Vision for Structural Dynamics and Health Monitoring
Dongming Feng, Maria Q Feng

Theory of Solid-Propellant Nonsteady Combustion
Vasily B. Novozhilov, Boris V. Novozhilov

Introduction to Plastics Engineering
Vijay K. Stokes

Fundamentals of Heat Engines: Reciprocating and Gas Turbine Internal Combustion Engines
Jamil Ghojel

Offshore Compliant Platforms: Analysis, Design, and Experimental Studies
Srinivasan Chandrasekaran, R. Nagavinothini

Computer Aided Design and Manufacturing
Zhuming Bi, Xiaoqin Wang

Pumps and Compressors
Marc Borremans

Corrosion and Materials in Hydrocarbon Production: A Compendium of Operational and Engineering Aspects
Bijan Kermani and Don Harrop

Design and Analysis of Centrifugal Compressors
Rene Van den Braembussche

Case Studies in Fluid Mechanics with Sensitivities to Governing Variables
M. Kemal Atesmen

The Monte Carlo Ray-Trace Method in Radiation Heat Transfer and Applied Optics
J. Robert Mahan

Dynamics of Particles and Rigid Bodies: A Self-Learning Approach
Mohammed F. Daqaq

Primer on Engineering Standards, Expanded Textbook Edition
Maan H. Jawad and Owen R. Greulich

Engineering Optimization: Applications, Methods and Analysis
R. Russell Rhinehart

Compact Heat Exchangers: Analysis, Design and Optimization using FEM and CFD Approach
C. Ranganayakulu and Kankanhalli N. Seetharamu

Robust Adaptive Control for Fractional-Order Systems with Disturbance and Saturation
Mou Chen, Shuyi Shao, and Peng Shi

Robot Manipulator Redundancy Resolution
Yunong Zhang and Long Jin

Stress in ASME Pressure Vessels, Boilers, and Nuclear Components
Maan H. Jawad

Combined Cooling, Heating, and Power Systems: Modeling, Optimization, and Operation
Yang Shi, Mingxi Liu, and Fang Fang

Applications of Mathematical Heat Transfer and Fluid Flow Models in Engineering and Medicine
Abram S. Dorfman

Bioprocessing Piping and Equipment Design: A Companion Guide for the ASME BPE Standard
William M. (Bill) Huitt

Nonlinear Regression Modeling for Engineering Applications: Modeling, Model Validation, and Enabling Design of Experiments
R. Russell Rhinehart

Geothermal Heat Pump and Heat Engine Systems: Theory and Practice
Andrew D. Chiasson

Fundamentals of Mechanical Vibrations
Liang-Wu Cai

Introduction to Dynamics and Control in Mechanical Engineering Systems
Cho W.S. To

Pressure Oscillation in Biomedical Diagnostics and Therapy

Ahmed Al-Jumaily
Auckland University of Technology
Auckland, New Zealand

Lulu Wang
Shenzhen Technology University
Shenzhen, Guangdong, China

This Work is a co-publication between ASME Press and John Wiley & Sons Ltd

This edition is first published in 2022

This Work is a co-publication between ASME Press and John Wiley & Sons Ltd.

Registered Offices
John Wiley & Sons, Inc., 111 River Street, Hoboken, NJ 07030, USA
John Wiley & Sons Ltd, The Atrium, Southern Gate, Chichester, West Sussex, PO19 8SQ, UK

Editorial Office
111 River Street, Hoboken, NJ 07030, USA

For details of our global editorial offices, customer services, and more information about Wiley products, visit us at www.wiley.com.

Wiley also publishes its books in a variety of electronic formats and by print-on-demand. Some content that appears in standard print versions of this book may not be available in other formats.

Library of Congress Cataloging-in-Publication Data
Names: Al-Jumaily, Ahmed, author. | Wang, Lulu (Ph.D) author.
Title: Pressure oscillations in biomedical diagnostics & therapy / Ahmed Al-Jumaily, Lulu Wang.
Other titles: Pressure oscillations in biomedical diagnostics and therapy | Wiley-ASME Press series
Description: Hoboken, NJ : John Wiley & Sons, Ltd., 2023. | Series: Wiley-ASME press series | Includes bibliographical references and index.
Identifiers: LCCN 2021062923 (print) | LCCN 2021062924 (ebook) | ISBN 9781119265849 (cloth) | ISBN 9781119265900 (adobe pdf) | ISBN 9781119265917 (epub)
Subjects: MESH: Blood Pressure Determination–methods | Pulse Wave Analysis–methods | Elasticity Imaging Techniques–methods | Diagnostic Techniques, Cardiovascular | Respiratory Tract Diseases–therapy
Classification: LCC RC685.H8 (print) | LCC RC685.H8 (ebook) | NLM WB 280 | DDC 616.1/32075–dc23/eng/20220131
LC record available at https://lccn.loc.gov/2021062923
LC ebook record available at https://lccn.loc.gov/2021062924

Cover design: Wiley
Cover image: © Magic mine/Shutterstock

Set in 9.5/12.5pt STIXTwoText by Straive, Pondicherry, India
Printed and bound by CPI Group (UK) Ltd, Croydon, CR0 4YY

C9781119265849_200622

In the name of the CREATOR who fashioned humankind as the best of his creations.
To my wife Thana who stands by me in light and dark times.
To my kids, candles of my life.
To my postgraduate students from whom I have humbly learnt.

Contents

About the Author

Ahmed M. Al-Jumaily is currently a Professor of Biomechanical Engineering and the Founder of the Institute of Biomedical Technologies at the Auckland University of Technology, Auckland, New Zealand. He holds a PhD and MSc from the Ohio State University, USA, and a BSc from the University of Baghdad, Iraq. He is a Fellow member of the American Society of Mechanical Engineers (ASME) and the Acoustical Society of America in addition, to being a member of 11 more international professional societies. He is the Editor-in-Chief of the ASME Journal of Engineering and Science in Medical Diagnostics and Therapy, the Editor for the ASME monograph series-Biomedical and Nanomedical Technologies, and has been on the editorial and refereeing boards for several international journals. He has published more than 360 papers in international journals and conference proceedings including two ASME books on Vibration and Acoustics in Biomedical Applications and a third one on CPAP devices. He has supervised more than 100 postgraduate students from 35 countries in biomedical applications, vibrations, biomechanics, and electroactive polymers. During his academic career, he has forged strong alliances between academia and industries, in particular, in the medical devices area, which has resulted in many successful grants and contracts with companies and research organizations. Al-Jumaily's current research focuses on biomedical applications, particular interest in the applications of vibration and acoustics to airways constriction therapies, and artery noninvasive diagnostics.

Lulu Wang is currently a Distinguished Professor of Biomedical Device Innovation Center at Shenzhen Technology University in China. She received the ME (First class Hons.) and PhD degrees from the Auckland University of Technology, New Zealand, in 2009 and 2013, respectively. From 2013 to 2015, she was a Research Fellow with the Institute of Biomedical Technologies, Auckland University of Technology, New Zealand. In June 2015, Dr. Wang became an Associate Professor of biomedical engineering with the Hefei University of Technology. In June 2019, she became a Distinguished Professor of Biomedical Engineering at Shenzhen Technology University. Her research interests include medical devices, electromagnetic sensing and imaging, and computational mechanics. Over the past 5 years, Dr. Wang has authored more than 70 peer-reviewed publications, 2 ASME books, 7 book chapters, and 10 issued patents. Dr. Wang is a member of ASME, IEEE, MRSNZ, AAAS, PSNZ, and IPENZ. She is an active reviewer of numerous journals, books, and conferences. She has edited four books and three special issues of international journals. She has received multiple National and International Awards from various professional societies and organizations. She is an active organizer of several international conferences include ASME, IMECE, and ICCES.

Preface

This book compiles over 20 years of research and development in applying physiology and engineering principles to designing, modeling, and improving diagnostic and therapy devices and methods to serve the medical community. The fundamental and frontier theories and techniques of low- and high-frequency pressure oscillation are presented as the foundation for these principles. This area is evolving very fast, and documentation of such schemes is essential for various industries and clinical applications. Currently, there is no book with this title available in the open literature.

The book consists of eight chapters. Each chapter has a stand-alone content starting with an introductory material on the subject matter, followed by a review of available technologies to diagnose or treat the diseases with intensive literature survey, and then introducing the significance of pressure oscillation in the diagnostic or therapeutical applications. Each chapter finishes with some clinical applications and voluminous references and bibliography.

The first chapter is an introductory chapter which presents the foundation materials for the rest of the book. It introduces the basic principles of pressure oscillation and how they can be formulated into mathematical equations. The chapter explains how these equations can be converted to practical applications for biomedical diagnostics or therapy. The book is then split into two parts as detailed below.

Part I of the book consists of three chapters focusing on diagnostics, imaging, and characterization. Chapter 2 presents an application of pressure oscillation to develop a diagnostic technique. It briefly describes the basic concepts of arteries, arterial stiffness, arterial blood pressure, and pulse wave. It reviews several methods for measuring these variables. Various new contributions are discussed, including physics-based waveform measurements and human systemic arterial numerical models. Several medical applications of pulse wave analysis are also presented. The basic principles of converting pressure oscillation to a tool in biomedical imaging will be clearly explained in Chapter 3. Two of the methods will be explained in detail: (i) radiation force (RF), which is normally generated by high-frequency pressure oscillation. Several new radiation force-based elasticity imaging (EI) methods were proposed in the past two decades, and these techniques generate radiation force to an object and measure its dynamic displacement response in order to estimate the object's mechanical properties. (ii) Vibro-acoustography (VA) is a speckle-free acoustic radiation force (ARF)-based EI technique, which can visualize healthy and abnormal soft tissue through mapping the acoustic response of the object to a harmonic RF induced by ultrasound. This chapter briefly describes the history of ARF and its applications and provides an overview of ARF-based EI approaches. Examples of ARF-based EI–VA imaging and multifrequency VA techniques and their applications in medical and material evaluation areas are discussed. Several advantages and disadvantages of VA, comparison between VA and pulse–echo systems, as well as future directions are presented.

Respiratory diagnostic and characterization technique are summarized in Chapter 4. The forced oscillation technique (FOT) is a low-frequency-pressure wave technique. It has been used in the measurement of respiratory impedance and has evolved into powerful tools for the assessment of various mechanical phenomena in the mammalian healthy and diseased lung. This chapter briefly reviews the human respiratory system (RS) and its functions, mechanics, and various developed models and measurement methods. An example of the RS measurement method FOT is detailed, including working principles, instrumentation, measurements arrangement, and impedance measurement methods. Finally, clinical applications of FOT are also described.

Part II of the book focuses on respiratory therapies. To be equipped with sufficient knowledge on lung ailments in the upper, central, and lower airways covered in this section, an introductory chapter on the respiratory system is needed. Chapter 5 presents lung mechanics, how each part of the lung is associated with various diseases, and how pressure oscillation can target these parts and help in treating these diseases. In Part II, the following three chapters deal with specific diseases, namely, obstructive sleep apnea (OSA), asthma and respiratory distress syndromes (RDSs).

Chapter 6 briefly describes OSA syndromes, diagnostic methods based on the combined evaluation of clinical manifestations and objective sleep study findings, and currently available treatment methods. Polysomnography represents the gold standard to confirm the clinical suspicion of OSA syndrome, to assess its severity and to guide therapeutic choices. Continuous positive airway pressure (CPAP) is currently the most recommended method for OSA treatments. Some basic working principles of CPAP techniques and their types are discussed. Finally, clinical applications of these devices for the treatment of OSA and benefits are also highlighted. A new technique is introduced and detailed on how pressure oscillation can be used to improve the use of CPAP.

The importance of airway smooth muscle (ASM) in asthma was realized almost 150 years ago. Breathing has a strong relaxing and protective effect on ASM, inhibiting airway constriction. Understanding the behavior of ASM is crucial to understanding the reversible airway obstruction central to asthma. Chapter 7 briefly describes the basic concepts of asthma, ASM, and dynamic behavior of ASM through both experimental data and modeling results. A fading memory model is given to further describe the behavior of contracted ASM for finite duration length steps and longitudinal sinusoidal oscillations. Finally, the potential of pressure waves superimposed on breathing patterns in treating asthma is investigated using experimental research and animal models.

Chapter 8 briefly introduces the common neonatal respiratory diseases with traditional surfactant therapies and respiratory support devices currently used in practice. Various pressure oscillation techniques including high-frequency ventilation (HFV), continuous positive airway pressure (CPAP), and "noisy" ventilation as effective and cheaper methods for respiratory support are discussed. Clinical trials that describe the effectiveness of using such treatments are presented. The concept of stochastic resonance and its application to "noisy" ventilation are introduced. The potential advances in the use of pressure oscillations and "noisy" ventilation to treat both neonatal and adult diseases are presented.

This book can teach students how to turn mathematical equations into medical devices or methodologies, and it also offers an excellent reference for undergraduate and postgraduate students in Physiology, Radiology, Applied Mathematics, Physics, and Biomedical, Mechanical, and Electrical Engineering. The book will also appeal to fellow researchers, practitioners, lecturers, and professionals such as Biomedical Engineers, Clinicians, Medical Doctors, Radiologists, and Researchers in Biomedical Imaging, Diagnostics and Therapies, and medical device industry personnel. It is a helpful compilation that familiarizes the reader with practical modeling approaches to enhance the design process.

Introduction

The primary objectives of this book are to present recent developments, discoveries, and progress made in the implementation of pressure waves in biomedical diagnostics and therapies, with a focus on the arterial and respiratory systems. Based on engineering principles and physiology, the fundamental and frontier theories and techniques of low- and high-frequency pressure waves are applied to develop medical devices and technologies for biological systems imaging, diagnostics, and therapies. It is an interdisciplinary area which utilizes Mathematics, Physics, Chemistry, Engineering, Computer Sciences, Physiology, and other fields for Clinical Applications. As biomedical technologies are evolving very fast, documenting of such schemes are essential for medical industries and clinical applications.

The book is compiled of learning and findings gained in more than 20 years of research and development that I have conducted with my postgraduate students at the Auckland University of Technology. Each chapter summarizes a complete project, which is further detailed in theses cited as references. In this way, I would like to acknowledge the contributions of all my postgraduate students whose works are used as the main reference material for this book. I would also like to acknowledge my appreciation to other authors in the field, whose contributions are evidenced in the voluminous references and bibliography. Further, I would also like to acknowledge the effort of my previous student and postdoc, and current colleague, co-author Dr. Lulu Wang who has helped to compile some of the materials from my team's work during her postdoc position. Final reading of the manuscripts by Yelena Dumanovic is much appreciated.

1

Pressure Waves for Diagnostics and Therapy

1.1 Introduction

To understand the role of mathematics in an engineering discipline, we may talk about the capability of an engineer. Inspired by the laws of physics, an engineer uses mathematics as a tool to convert nature's resources to a product. This definition may be considered out of date now. A better definition may take the form: inspired by the laws of physics, using mathematics as a translator, to convert nature's resources to a product or to study a phenomenon (or phenomena) or a criterion (criteria). This latter definition takes engineers to go far into space and deep into the human body microstructure in order to investigate, analyze, and apply this knowledge to innovations for the betterment of the humankind. The process modeling could be done in the form of mathematical modeling or/and computer simulation.

1.2 Significance of Biological System Modeling

Various physical and biological systems can be modeled in the form of mathematical equations. The question often raised is why we put so much emphasis on mathematical modeling? The answer to this is the fact that mathematical modeling has so many advantages including but are not limited to:

1) It is a tool that helps to convert basic laws of nature (physics, chemistry, biology, etc.) to industrial application.
2) It converts real systems of interest into models which can be analyzed and tested on a piece of paper or a screen.
3) Nowadays, computers and new technologies have made math the most valuable tool to convert complicated real-life system into a virtual environment on the screen.

Of course, talking about math here implies all types of mathematical approaches including but not limited to differential mathematics, computational, statistical, and others. Statistical models applied to experimental data, as an example, may help in understanding the results for the purpose of assurance or development of a useful future formulation.

The two "Engineering" definitions stated earlier may sound the same; however, the second one is more general, and it represents the current methodology for engineering research. No more the engineer only designs and develops. Engineers go beyond the scope

Pressure Oscillation in Biomedical Diagnostics and Therapy, First Edition. Ahmed Al-Jumaily and Lulu Wang.

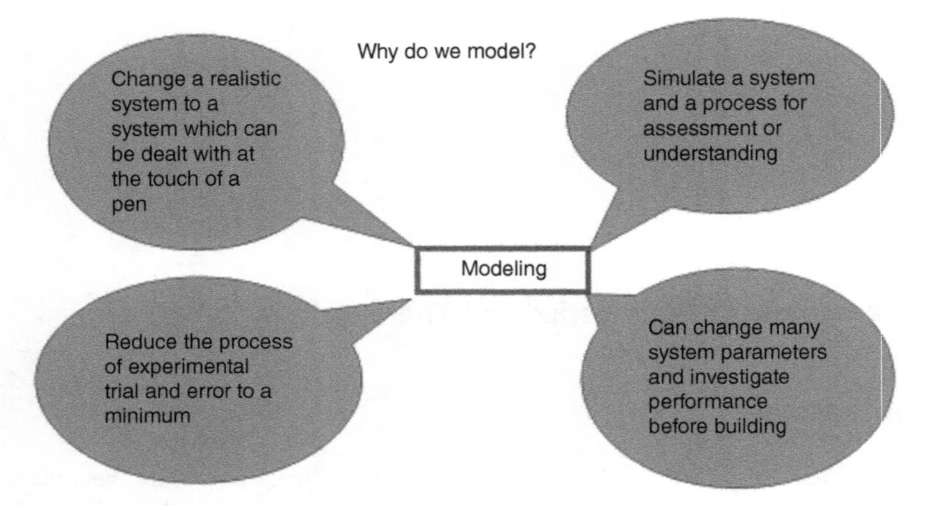

Figure 1.1 Advantages of mathematical modeling.

of developing a product to help humankind, by studying and investigating various systems and behaviors whether natural or synthetic. A biomedical engineer is not only responsible to build a medical device, a biomedical tool, or an artificial body part. Many biomedical engineers focus on investigating biological behaviors or responses in an optimum goal of serving the human body needs. Thus, this aligns with the second definition of "Engineers" stated earlier. The question which is normally raised is "what do we do with this modeling?" The response to this is summarized in Figure 1.1. While changing a realistic system, whether extremely large or at a nanoscale, to a system which can be represented virtually on a screen and reducing the process of experimental trial and error to a minimum time and cost without jeopardizing integrity are typical processes of mathematical modeling. Modeling could be used to simulate a system or a process to understand and assess, to change system parameters, and to investigate performance before building the device while reducing the time and costs of experimental investigations of the complicated very large or very small systems to investigate them. With this in mind, the main outcomes of the modeling process are

1) To turn unreachable systems to accessible ones.
2) To allow studying the real system without affecting or damaging it.
3) To keep a system model ready on the screen to optimize, design, analyze, etc.
4) To change the parameters and see how the system behaves.
5) To create specifications for the system to be ready for manufacturing.

Accessing unreachable systems and converting them to reachable has a very wide range of applications, particularly in the biomedical field. A study of an organ or a cell behavior is difficult if not impossible within the living human body. Thus, having a model on the screen helps to introduce many variations without affecting the human body as a living system itself. Further, if a model for a device to support the human body such as an artificial kidney is developed and needs optimization. A mathematical model for this device is extremely helpful in the process of designing and optimizing without jeopardizing the health of the person. Multiple variations of parameters and its effects could be studied affecting the human body itself. Obviously, out of the last design iteration and optimization process, new specifications could be generated for optimal operation and performance. However, the question which is always raised is how reliable modeling is? The answer is summarized by the following facts:

- Pace of technological change drives industries to implement modeling.
- Aircraft, underwater, space, and other industries have produced good and reliable models for the development of their products.
- Most research groups have developed reliable models in their fields of research and have worked with confidence to move toward innovation.
- Experience plays an important role in reliability.
- If a model is not perfect, its value at least is to show the trend of the behavior or variations.

Over the years and with the development of the computational capabilities, mathematical modeling has demonstrated advancement and success in all modern technologies, from space engineering to nano cells within the human body. All systems generated from modeling have worked reliably and with improved performance. Of course, like any other discipline, experience plays an important role in developing the best model for a particular device or a process. In general, if the model is not perfect, it will show the general trend of variation for the process or operation.

Without going into details of modeling, in this chapter, we will focus on the development of a single equation, as an example of modeling of one of the biological systems such as the respiratory or cardiovascular system and for the purpose of developing a medical diagnostic tool or therapy device/method or to investigate the behavior and measure performance. This simple equation is the wave equation. The nature of several systems within the human body conveys fluid for the purpose of living, particularly, the cardiovascular and the respiratory systems. In spite of the large differences between the functions of these two systems, wave propagation through them can be tuned into a diagnostic or a therapy approach. The significance of this equation will be clarified in Chapters 2, 3 and 5, but at this stage we will derive the equation and outline its significance for biomedical diagnostics and therapies.

1.3 Wave Equation

In this book, we focus on how mathematics can be applied as a modeling tool to convert basic laws of physics, biology, and chemistry sciences to a process, protocol, method, or/and a medical device, which can deliver an outcome for diagnostics or treatment for some critical lung, heart, or arterial diseases. Our focus will be on the simple wave equation, its basic principle, application, and how it can be used to investigate various biological processes and how this powerful equation could be used to develop a medical device for diagnostic purposes as well as various tools for treatments. There are many ways of deriving this equation from a simple string to a rod (bar) and to the flow in pipes. In this chapter, we focus on equations defining the flow in the pipes as it resembles an artery or an airway passage in the context of the human body.

Various transmission lines in the respiratory and cardiovascular systems may be treated as branching compliant tubes conveying fluid. Blood flow in the arterial system and airflow in the respiratory system induce forces and stresses in the arterial and respiratory walls, due to complex fluid–structure interactions. These forces and stresses play an important role in the onset and progression of many acquired and congenital cardiovascular diseases, such as arterial atherosclerosis and aneurysms, and airway ailments such as obstruction in obstructive sleep apnea and narrowing in asthma.

The blood flow in the arterial system and the airflow in the respiratory system diseases have a common physical problem which is change in the fluid flow passage as "restriction" or "enlargement" of fluid flow passage. Atherosclerosis, for example, involves the accumulation of plaque in the *intima* of the arterial wall, which reduces arterial lumen and increases local arterial stiffness. There is substantial evidence on the localization of these plaque deposits at sites with hemodynamic conditions commonly characterized by low wall shear stress (Caro et al. 1971; Taylor 1959). Aneurysms, on the other hand, involve the degradation of local arterial wall tissues, resulting in lowering of local arterial stiffness and enlargement of local vessel cross section. If, in extreme cases, the wall stress due to the transient fluid–structure interactions exceeds the strength limit of the dilated artery wall, it causes vessel rupture leading to death from internal hemorrhage which has been reported to be between 80 and 90% of the cases (Scotti et al. 2005). Aneurysms are common in locations with secondary flow and flow recirculation even in normal resting conditions (Peattie et al. 2004; Lasheras 2007).

In the respiratory system, on the other hand, the change in the air passages is due to either an "obstruction" or a "narrowing." The former is typical in obstructive sleep apnea where the upper airway tissue loses their capability to respond and may collapse and introduce obstruction. However, in asthma, the airway smooth muscles are considered as the main mechanism to introduce airway narrowing by a physiological process called cross-bridge cycling.

Considering the correlations between the various hemodynamic and aerodynamic conditions, and the onset and progression of different cardiovascular and respiratory diseases, there is worldwide consensus on the need for enhancements in the current understanding of cardiovascular mechanopathobiology (O'Rourke and Hashimoto 2007) and respiratory airflow dynamics. Since blood flow, airflow, and pressure are dependent on several factors in addition to physical phenomena, experimental studies demand rigorous screening, which make them expensive and time-consuming. A preferred mode of investigation is by computer simulation using mathematical models which in principle are based on the wave equation.

Principles of conservation of mass, momentum, and energy form the theoretical bases of mathematical models that have been developed for the study of the physical aspects of blood flow and airflow. However, an analytical solution of the full form of these equations has not been developed where only solutions for special cases are available today. Proper application of numerical techniques for solving these equations demands familiarity with assumptions and approximations that can be made to achieve a reasonable solution. This chapter develops a mathematical model capable of representing pressure propagation in human blood and respiratory air passages and use of this model to investigate the significance of different physical terms within this model on pressure propagation in any of the passages.

Although the development of one-dimensional mathematical models for studying pressure propagation in biological cylindrical tube-type passages and the effects of some phenomena have been previously discussed in the literature (Reuderink et al. 1989; Sherwin et al. 2001; Nardinocchi et al. 2005), the same equation can be derived for a wave traveling in a medium which could be used in the diagnostic process for cancer detection; see Chapter 3. Further, previous publications assume the special approximation to these passages such as to be straight and either linearly or exponentially tapered which is not anatomically representative. In this chapter, the general form of the wave equation is derived and discussed for the purpose of simplifying the process to develop a diagnostic or therapy device, method, or protocol.

The general form of deriving the governing equations for the wave propagation in a compliant tube conveying fluid leads to the partial differential equations which have a complex mathematical involvement without important physical significance. Therefore, a number of assumptions are made and justified to produce a practical model for the wave propagation.

In this chapter, it is assumed that the fluid conveyed by the tube is homogenous and non-viscous. This is obviously not true as the blood is both nonhomogenous and viscous. However, work by Taylor has shown that the effects of nonhomogeneity and viscosity of the fluid are only significant in tubes with very small diameters such as arterioles (Taylor 1959). In addition, it is assumed that the tubes are of cylindrical shape and elastic. The tube walls are nonlinearly viscoelastic, but work by Fung (1997) showed that the effect of nonlinear viscoelasticity on wave propagation is not important.

1.4 Governing Equation

The two main reasons we are focusing on the wave equation in this introductory chapter are as follows:

1) The main elements considered in this book are compliant tubes to represent the arterial and the respiratory system passages. Both the diagnostic and therapy techniques developed in this book can be easily explained by using the one-dimensional wave equation.
2) Although this equation will be used mainly for the two systems stated earlier, the equation could be used to explain the transmission and reflection of waves in any medium.

1.4.1 Assumptions

To derive this equation and to avoid nonlinear complexity, the following assumptions are made:

1) It is assumed that the length of the tube is long compared to its diameter. This assumption surely holds for passages in the conductive airways as well as for the central systemic arteries; thus, it can be hypothesized that the flow is one-dimensional (Lasheras 2007).
2) No friction, work, or heat transfer.
3) No action at a distance force.
4) Constant area except for radial elastic deflection.
5) Assume the density ρ of the conveyed fluid, blood or air, is constant.
6) The tube wall Young's modulus of elasticity E is assumed constant.

7) It is also assumed that the tube is thin-walled and obeys Hooke's law. Hence, any small change in transmural pressure dp is balanced by a change in the tube circumferential tension.

1.4.2 Derivation

By considering the free body diagram shown in Figure 1.2, the basic equations of conservation of mass and momentum can be derived by considering mass conservation in the segment of a passage dx as shown in Figure 1.2a which leads to

$$\frac{\partial A}{\partial t} + \frac{\partial}{\partial x}(Au) = 0 \tag{1.1}$$

Balancing the forces acting in the axial direction on a fluid element dx and cross-sectional area A as shown in Figure 1.2b leads to,

$$\frac{\partial u}{\partial t} + u\frac{\partial u}{\partial x} + \frac{1}{\rho}\frac{\partial p}{\partial x} = 0 \tag{1.2}$$

where u is the velocity of the fluid.

Considering assumption 5 and 6 and from equilibrium of the forces acting on a free body shown in Figure 1.3, one can write that the stress–strain relation can be written as:

$$\frac{Ehdr}{r} = rdp \tag{1.3}$$

Equations (1.1)–(1.3) can now be used to derive an equation governing steady-state flow through an elastic artery, and with further simplifications the theoretical artery wave speed can be determined.

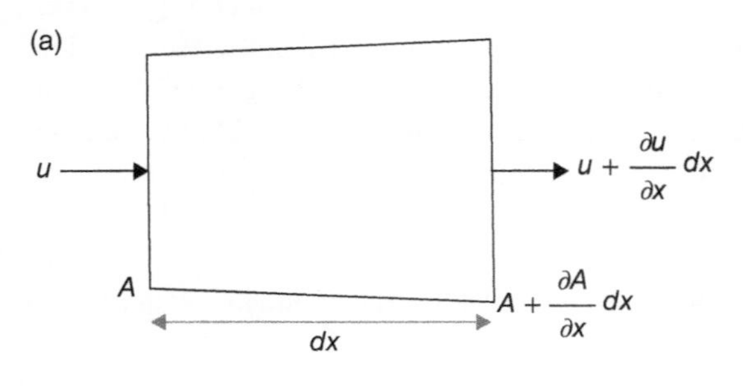

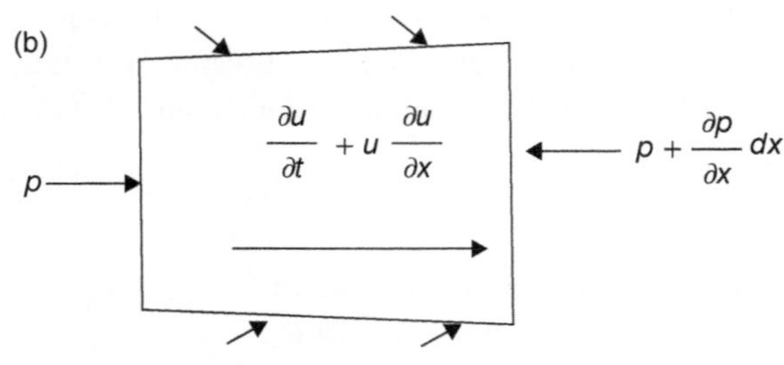

Figure 1.2 Free body diagram of a tube segment showing mass (a) and momentum (b) conservation.

Figure 1.3 Forces equilibrium tube segment free body diagram.

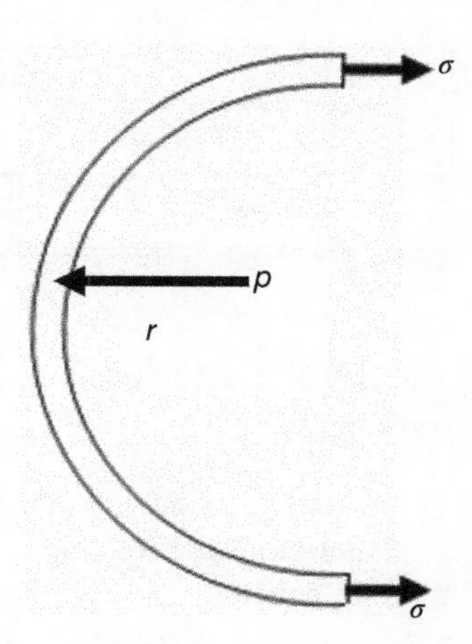

Equation (1.1) is linearized by substituting $A = \pi r^2$ and by assuming that the wave amplitude is much smaller than the wavelength. Neglecting the second-order terms reduces the equation to:

$$\frac{\partial u}{\partial x} + \frac{2}{r}\frac{\partial r}{\partial t} = 0 \tag{1.4}$$

Equation (1.2) can also be linearized by taking into account small disturbances in a motionless tube filled with fluid. Hence, the $u\frac{du}{dx}$ term is not significant and equation (1.2) can be written as:

$$\frac{\partial u}{\partial t} + \frac{1}{p}\frac{\partial p}{\partial x} = 0 \tag{1.5}$$

Combining equations (1.3) and (1.5) results in:

$$\frac{\partial u}{\partial x} + \frac{2r}{Eh}\frac{\partial p}{dt} = 0 \tag{1.6}$$

Differentiating equation (1.4) with respect to x and equation (1.6) with respect to t, then neglecting the second-order terms and substituting gives

$$\frac{\partial^2 p}{\partial x^2} + \frac{1}{c^2}\frac{\partial^2 p}{\partial t^2} = 0 \tag{1.7}$$

where

$$c^2 = \frac{Eh}{2\rho r} \tag{1.8}$$

1.4.3 Solution

Equation (1.7) is the wave equation which governs the pressure wave, Figure 1.3, where c is the wave speed:

$$c = \sqrt{\frac{Eh}{2\rho r}} \tag{1.9}$$

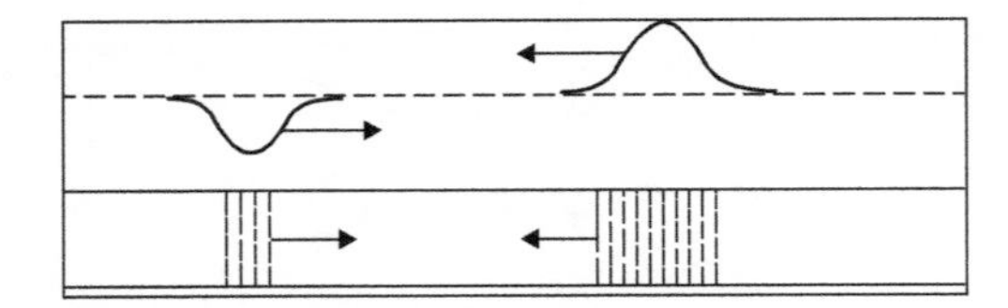

Figure 1.4 Pressure traveling waves: compressed and expanded.

The speed of sound c is a function of three main variables which significantly affect the traveling wave:

1) Geometry, namely, h and r.
2) material properties, and
3) fluid density.

If any of these properties changes, it will reflect on the shape of the traveling wave. This will influence the wavelength, frequency, and the shape of the wave, Figure 1.4, which may be implemented in a diagnostic procedure. However, we may use the pressure to forcibly change those parameters which can be implemented in a therapy scenario. These two scenarios can be used to develop a diagnostic device/method or a therapy device/method. Based on this fact, this is a powerful equation which could be used for diagnostics and/or therapy in many biomedical applications. The main parameter c plays an important role in the application. Clearly, it is a function of the tube wall properties E, h, and r and the fluid density ρ.

The solution of equation (1.7) can be written as:

$$P(x,t) = P_0 f(x - ct) + P_0' f(x + ct) \tag{1.10}$$

where $P_0 f(x - ct)$ represents an incident pressure wave traveling in the positive x direction and $P_0' f(x + ct)$ represents a pressure wave traveling in the negative x direction. Those traveling waves can reflect at any point due to a change in any of the properties. At any section along the length, if h, r, E, or ρ changes the wave will split into two, where one will travel through as an incident wave $P_0 f(x - ct)$ and the other $P_0' f(x + ct)$ will reflect back as a reflected wave. The values of these waves depend on the transmission and reflection parameters explained later in this chapter. Thus, at any section, the net wave will be the superposition of the two.

1.5 Bifurcation

Pressure waves are partially reflected when they experience a sudden change in the medium of transmission such as a bifurcation, change in the properties, or a sudden change in the arteries or respiratory airway's geometry or material properties. Central arteries such as the aorta bifurcate into a number of smaller arteries and in some diseases such as aneurysms where discrete batches of the arteries degenerate and hence there is a large change in

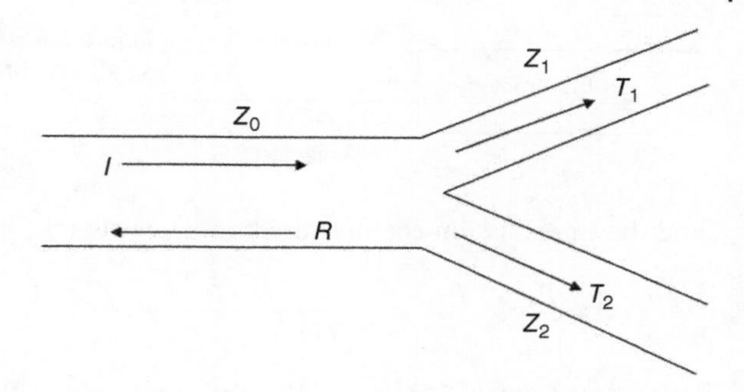

Figure 1.5 A schematic figure showing conditions at a bifurcation.

material properties and geometry. The respiratory system also branches from the main trachea to the bronchi, bronchioles, and smaller passages. When a pressure wave arrives at a bifurcation, part of that wave is transmitted while the other part is reflected.

Consider a tube that branches into two daughter tubes, T_1 and T_2, as shown in Figure 1.5. At the junction, it is assumed that pressure p is a single-valued function which can be written as:

$$p_I + p_R = p_{T_1} = p_{T_2} \tag{1.11}$$

where p_I is the incident pressure wave, p_R is the reflected pressure wave in the parent tube, p_{T_1} is the pressure transmitted into the first daughter tube T_1, and p_{T_2} is the pressure transmitted into the second daughter tube T_2. Also, it can be assumed that the flow is continuous:

$$Q_I - Q_R = Q_{T_1} + Q_{T_2} \tag{1.12}$$

where Q_I is the incident flow wave, Q_R is the reflected flow wave, Q_{T_1} is the flow into the first daughter tube T_1, and Q_{T_2} is the flow wave into the second daughter tube T_2. Equations (1.11) and (1.12) describe the pressure and flow conditions at the junction. Now a relationship between the pressure and flow is required. The characteristic impedance of the artery is a very important characteristic and is defined by:

$$Z = \frac{\rho c}{A} \tag{1.13}$$

where Z is also defined by the ratio of oscillatory pressure to oscillatory flow when the wave travels in the positive x direction:

$$Z = \frac{p}{Q} \tag{1.14}$$

Using equation (1.14), equation (1.12) can be written as:

$$\frac{p_I - p_R}{Z_0} = \frac{p_{T_1}}{Z_1} = \frac{p_{T_2}}{Z_2} \tag{1.15}$$

Using equations (1.11) and (1.15), the reflection coefficient R can be expressed as:

$$R = \frac{p_R}{p_I} = \frac{Z_0^{-1} - (Z_1^{-1} + Z_2^{-1})}{Z_0^{-1} + (Z_1^{-1} + Z_2^{-1})} \tag{1.16}$$

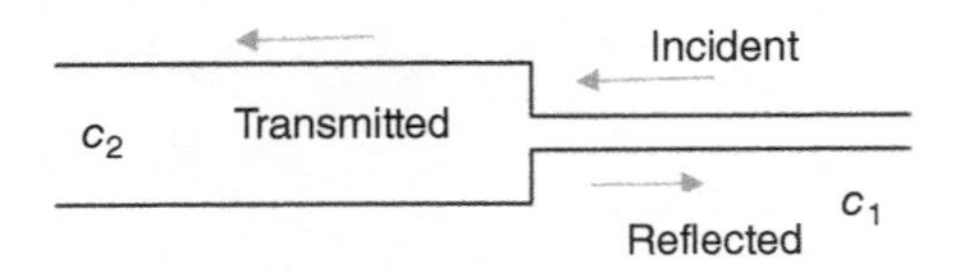

Figure 1.6 A schematic showing reflection from a small side branch.

and the transmission coefficient T can be written as:

$$T = \frac{p_{T_1}}{p_I} = \frac{p_{T_2}}{p_I} = \frac{2Z_0^{-1}}{Z_0^{-1} + \left(Z_1^{-1} + Z_2^{-1}\right)} \tag{1.17}$$

Using equations (1.16) and (1.17), the amplitude of the reflected pressure wave and the transmitted pressure wave at a junction can be calculated from the knowledge of the impedance of the parent and daughter arteries which is determined by the geometry and material properties of the respective passages.

It is also important to model the condition where pressure waves travel from a small tube into a larger tube as in reflected waves from any of the aorta's or airway's side branches.

Looking at Figure 1.6, similar simplifying assumptions can be made. The pressure at the junction can be assumed to be a single-valued function which can be written as:

$$p_I + p_R = p_T \tag{1.18}$$

where p_T is the pressure transmitted into the large artery. Furthermore, for fluid flow continuity,

$$Q_I - Q_R = Q_T \tag{1.19}$$

where Q_T is the flow transmitted into the large artery. Using equation (1.14), equation (1.19) can be written as:

$$\frac{p_I - p_R}{Z_0} = \frac{p_T}{Z_1} \tag{1.20}$$

Using equations (1.18) and (1.20), the reflection coefficient R can be expressed as:

$$R = \frac{p_R}{p_I} = \frac{Z_1^{-1} - Z_0^{-1}}{Z_1^{-1} + Z_0^{-1}} \tag{1.21}$$

and the transmission coefficient can be written as:

$$T = \frac{p_T}{p_I} = \frac{2Z_0^{-1}}{Z_0^{-1} + Z_1^{-1}} \tag{1.22}$$

From the preceding formulation, wave propagation phenomenon in branching tubes such as arteries or airway passages can be defined. The parent tube contains both the incident pressure wave and reflected waves, while the daughter tubes contain only transmitted waves.

The propagation speed of these waves is determined by the arterial and the airway geometry and material properties. The pressure in the parent artery can then be defined as:

$$p = p_I + p_R = p_I(t - x/c_0) + Rp_I(t + x/c_0) \tag{1.23}$$

and in the daughter tube or artery, the pressure can be defined as:

$$p_T = Tp_I(t - x/c_{1,2}) \tag{1.24}$$

where $c_{1,2}$ is the wave speed in either one of the daughters.

1.6 Diagnostics and Therapy

As explained earlier, the main parameter that plays an important role in the shape of the final pressure wave is c. It is a function of geometrical parameters, material properties, and fluid density. Any changes in these parameters will result in changes in c. Consequently, the wave shape will assume different shape parameters such as the periods of oscillation, peaks and valleys, and other critical parameters which are attributed to changes in the biological system under consideration. These changes could be considered in two scenarios; see Figure 1.7.

1.6.1 Diagnostic Applications

If we intend to use the wave equation for a diagnostic purpose, we may consider a realistic application. The pressure pulses coming from the heart pass through a series of arteries. If those arteries are healthy with no accumulation of plaque on the wall, a certain shape of a wave can be detected. Let us focus on the aorta as an example. If the aorta has a healthy diameter d_H and a wall thickness h_H with blood density ρ, then the speed of sound under healthy conditions is c_H, namely

$$C_H = \sqrt{\frac{E_H h_H}{2\rho r_H}} \tag{1.25}$$

with a period of $1/f_H$. The subscript H refers to healthy conditions. However, if there is a change say in the aorta thickness h such as thickening of the wall by plaque and/or other

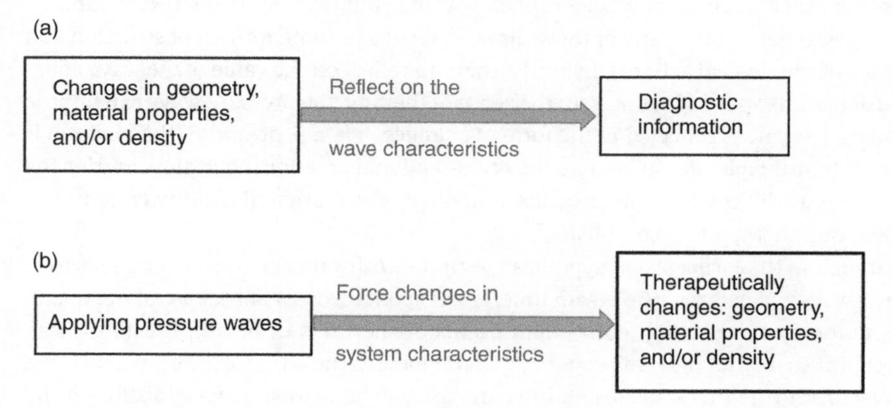

Figure 1.7 Diagnostics and therapy outcomes from a wave equation. (a) Diagnostics and imaging; (b) Therapy.

means, this will reflect on c and consequently on the wave characteristics determined by the wave variables.

$$C_u = \sqrt{\frac{E_u h_u}{2\rho r_u}} \tag{1.26}$$

where the subscript u describes unhealthy conditions. Studying the wave shape before and after the changes and putting this into a form of parameters such as a wavelength, amplitude ratio, etc., may lead to a diagnostic tool.

One of the applications of this diagnostic approach is detecting the pulse wave from the heart at a site which is easy to access noninvasively such as the brachial artery. Figure 1.8 shows a scenario on how the wave propagation in the arterial system could be converted to a diagnostic tool. The heart generates a pulse wave which travels in the arterial system. Let us follow one of the paths, the pulse wave coming from the heart and detected at the brachial artery. The pulse travels from the heart goes through several branches and at the same time at each bifurcation point the wave will split into two waves, to an incident and a reflected wave. Thus, at any location, the total wave will be the superposition of the traveling waves in both directions, namely the incident and the reflected waves. Figure 1.8a shows the actual physical system and Figure 1.8b shows the corresponding incident and reflected waves from the heart to the brachial artery including waves reflected at iliac bifurcation, brachial joint, and subclavian artery. Obviously, there are several other points for reflection and transmission, but here we are focusing on large arteries only. The final reading is recorded at the brachial artery, and it will be the superposition of all transmitted and reflected waves shown in Figure 1.8b. The detail of this technology is given in Chapter 2.

1.6.2 Therapy Applications

Considering the wave equation for a therapy scenario depends on the nature of the disease and its required therapy. In this book, we are focusing on diseases associated with the respiratory system such as obstructive sleep apnea, asthma, and respiratory distress syndrome. From biophysical perspective, any of these diseases results in some form of obstruction as a consequence of a biological action. Obviously, this will reflect on the value of the wave equation in particular the speed of sound c. In the sense of therapy, the pressure wave can impose some changes to c values reflected in the form of changes in h, r, ρ, or/and E (the medium is viscoelastic). In principle, it will change the cross-sectional area for the airflow and/or the elasticity of the medium where the pressure is applied. Thus, original healthy condition is approached, and therapy is accomplished.

As an example of the earlier stated hypothesis, asthma is a disorder characterized by narrowing of the airways which is reversible with time, either spontaneously or because of treatment. Classic symptoms are recurring intermittent events of shortness of breath, breathlessness, wheezing, tightness in the chest, and coughing. Consequently, the lining of the bronchial tubes swells in response to the attack and results in a narrowing of the airways, reduced airflow in and out of lungs in addition to excess mucus. Bronchospasm and hyperreactivity account for the contraction of airway smooth muscles (largely reversible). Logically, we can drive in a wave which can stimulate the tube to move from constricted to relaxed position. See Chapter 7 for more details.

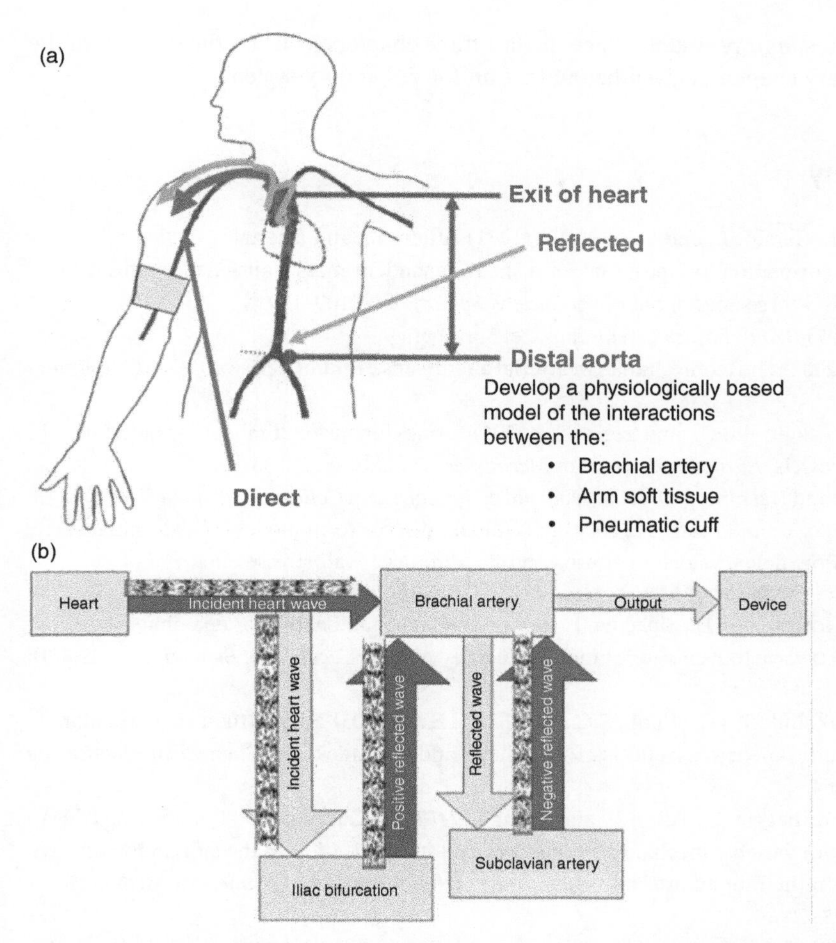

Figure 1.8 (a) Pulse wave traveling through the arterial system; (b) summation of transmitted and reflected waves.

Another example is obstructive sleep apnea. With this ailment the upper airways collapse due to the loose tissue and results in airway obstructions. A particular wave with certain frequencies and modes of vibration are applied to modulate the airways and push the obstruction away allowing for better breathing. See Chapter 6 for details.

1.7 Closure

In this chapter, we covered the basic principles of how a wave equation can be used for diagnostic and therapy applications. In the following chapters, more details are given for various practical applications in respiratory and cardiovascular systems. Both diagnostics and therapy techniques will be explained in more detail. The book is split into two main parts. Part I covers diagnostic applications, while Part II focuses on therapy applications, mainly in the respiratory system. Since the last three chapters focus on the diseases of the lung, a summary chapter is given before that on the respiratory system.

Bibliography

Caro, C.G., Fritz-Gerald, J., and Schroter, R. (1971) Atheroma and arterial wall shear: observation, correlation and proposal for a shear dependent mass transfer mechanism for atherogenesis. *Proceedings of the Royal Society, Series B* 177, 109–159.

Fung, Y.C. (1997) *Biomechanics Circulation*, 2e. Springer.

Lasheras, J.C. (2007) The biomechanics of arterial aneurysms. *Annual Reviews of Fluid Mechanics* 39, 293–319.

Nardinocchi, P., Pontrelli, G., and Teresi, L. (2005) A one-dimensional model for blood flow in prestressed vessels. *European Journal of Mechanics A/Soilds* 24, 23–33.

O'Rourke M. F. and Hashimoto, J. (2007) *Journal of the American College of Cardiology* 50(1) 1–13.

Peattie, R., Riehle, T., and Bluth, E. (2004) Pulsatile flow in fusiform models of abdominal aortic aneurysms: flow fields, velocity patterns and flow-induced wall stresses. *Journal of Biomechanical Engineering* 126(4): 438–446.

Reuderink, P., Hoogstraten, H., Sipkema, P. et al. (1989) Linear and nonlinear one-dimensional models of pulse wave transmission at high Womersley numbers. *Journal of Biomechanics* 22(8/9): 819–827.

Scotti, C.M., Shkolnik, A.D., Muluk, S.C., and Finol, E.A. (2005) Fluid-structure interaction in abdominal aortic aneurysms: effects of asymmetry and wall thickness. *Biomedical Engineering Online* 4(1): 64.

Sherwin, S.J., Formaggia, L., Peiro, J., and Franke, V. (2001) Computational modelling of 1D blood flow with variable mechanical properties and its application to the simulation of wave propagation in the human arterial system. *International Journal of Numerical Methods in Fluids* 43, 673–700.

Taylor, M. (1959) The influence of the anomalous viscosity of blood upon its oscillatory flow. *Physics of Medicine and Biology* 3, 273–290.

Part I

Diagnostics and Imaging

2

Pulse Wave for Arterial Diagnostics

2.1 Introduction

This chapter focuses on using pulse waves as a form of pressure oscillation for arterial diagnostics. The chapter starts with a brief description of the cardiovascular system, arterial stiffness, arterial blood pressure (BP), and pulse waves. Detailed descriptions of arterial system parameters used for system assessment such as the augmentation index, central *BP*, and others are presented. The chapter also reviews several methods for measuring these parameters. Methods for noninvasive detection of arterial stiffness are explained in detail. A physical-based waveform measurement method based on a systemic arterial numerical model is fully described. The mathematical modeling approach is based on the wave equation developed in Chapter 1 and used to develop a systematic diagnostic technique. Detailed descriptions of blood measurement methods are presented including an artificial neural network (ANN) approach. A full section is on pulse wave, its history and types, and how it is used for early detection of cardiovascular diseases (CVDs). The chapter finishes with a section on the various medical applications of pulse waves, including some clinical trial outcomes. All the sections in the chapter are supported by a comprehensive literature survey of related materials.

2.2 Cardiovascular System

The cardiovascular system consists of the heart, vasculature, and the blood constituents (cells and plasma). The blood vessels of the body include veins, arteries, and capillaries. Both veins and arteries contain smooth muscles (Tortora 2018). Veins have large lumens, relatively thin muscular walls, and less elastic fibers. They transport deoxygenated blood to the heart at low *BP* flow. Further, they have valves to prevent the backflow of the deoxygenated blood. Compared to veins, arteries are generally larger and have thick muscular walls and many elastic fibers. They carry blood away from the heart at high blood flow pressure. Capillaries are the smallest vessels that connect the arteries and the veins, and they take nutrients and substances to cells and take away the waste from the cells.

Pressure Oscillation in Biomedical Diagnostics and Therapy, First Edition. Ahmed Al-Jumaily and Lulu Wang.

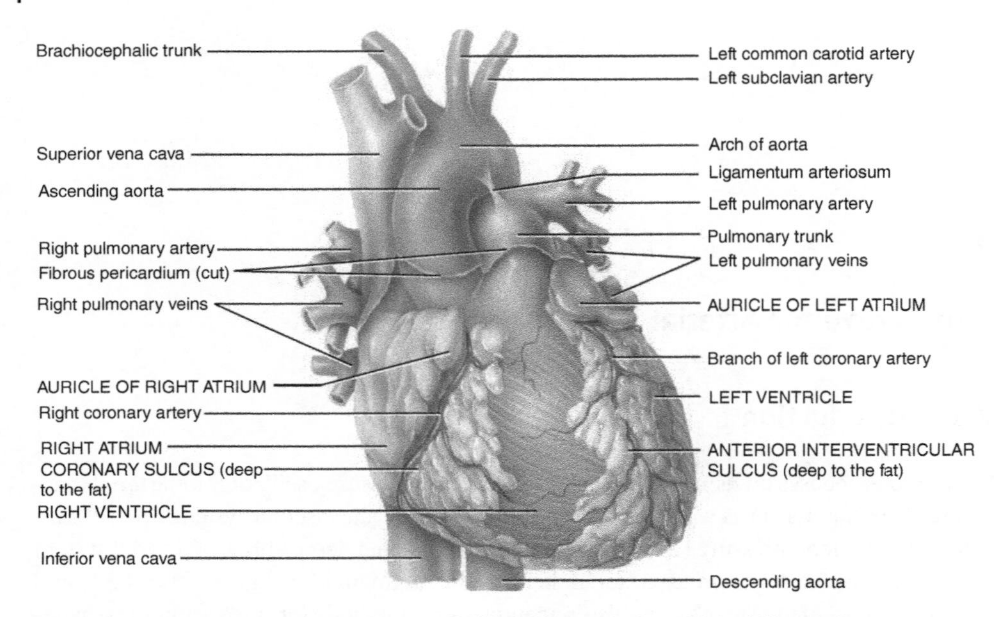

Figure 2.1 Human heart. *Source:* Tortora (2018)/with permission of John Wiley & Sons.

2.2.1 Arterial System

Arteries are blood vessels that contain a high percentage of smooth muscle, which can be controlled by hormones and unique signals from the nervous system. The heart first pumps the deoxygenated blood returning from the veins of the body into the lungs to oxygenate it and then passes it into the left ventricle, where it gets pumped to the ascending aorta. The aorta branches into smaller arteries that ultimately feed the capillary beds in the tissue, where the oxygen is exchanged for carbon dioxide, and nutrients are exchanged with tissue waste. The blood returns through veins to the right ventricle, where the cycle is repeated. A schematic diagram of the human heart illustrating its chambers is illustrated in Figure 2.1.

The aorta is divided into four main parts: the ascending aorta, arch of the aorta, thoracic aorta, and abdominal aorta. Also, all the systemic arteries branch from the aorta. Figure 2.2 illustrates the anatomy of all the major systemic arteries.

The arterial pressure includes the **systolic pressure** (*SP*) and the **diastolic pressure** (*DP*). *SP* varies between the peak pressure during heart contraction, while *DP* is the minimum pressure between contractions when the heart expands and refills. The pulse occurs due to the pressure changes within the artery, and it is observable in any artery to reflect the heart activity.

2.2.2 Properties of Arteries

The mechanical properties of the arteries determine the propagation of energy from the heart to the periphery (Dobrin 1978). These properties have been the subject of an intensive research of the coronary arteries in both animals (Patel and Janicki 1970; Gow et al. 1974; De Mey and Brutsaert 1984; Bund et al. 1996; Szekeres et al. 1998; Veress et al. 2000; Van

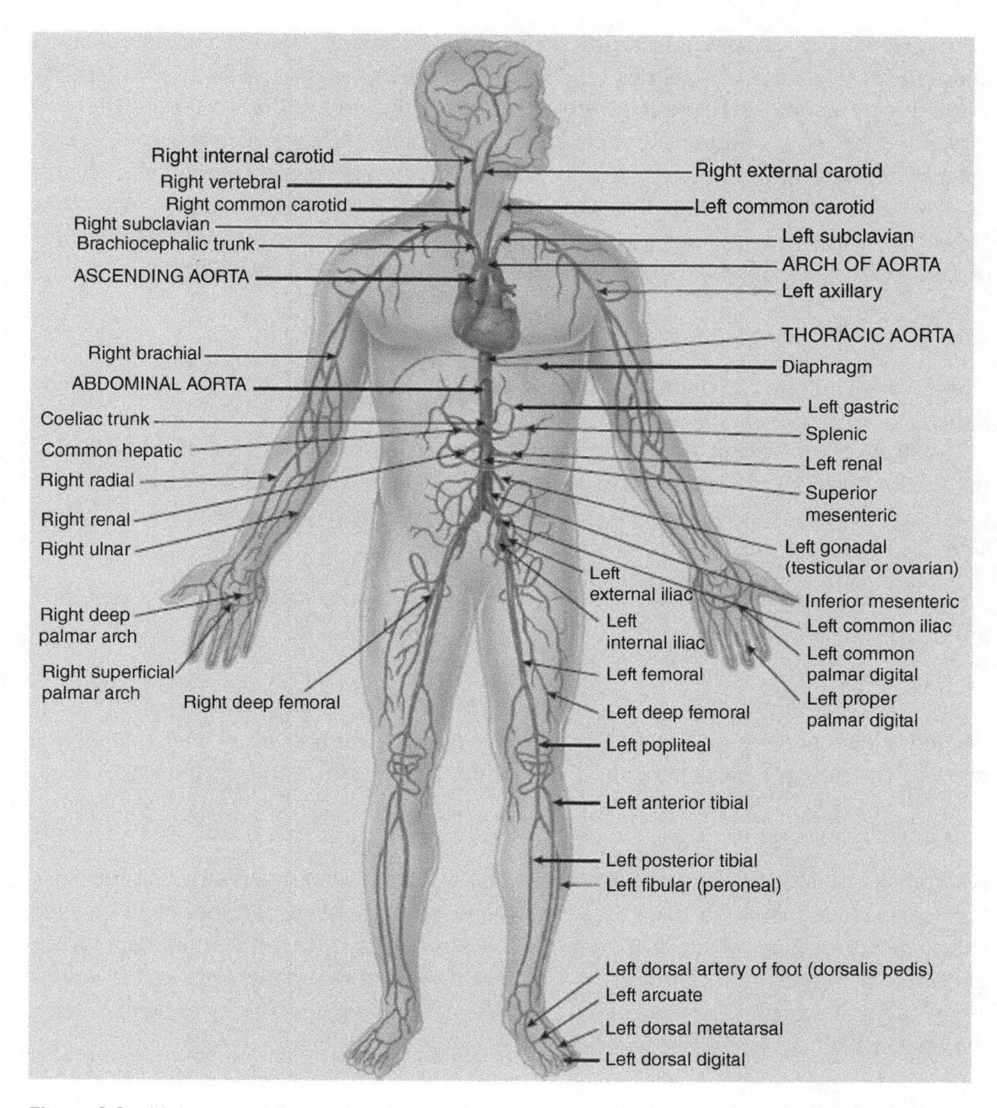

Figure 2.2 Major systemic arteries. *Source:* Tortora (2018)/with permission of John Wiley & Sons.

Andel et al. 2001; Lally et al. 2004; Wang et al. 2008) and human subjects (Gow and Hadfield 1979; Carmines et al. 1991; Opgaard and Edvinsson 1997; Ozolanta et al. 1998; Kasjanovs et al. 1999; Williams et al. 1999; van Andel et al. 2003; Holzapfel et al. 2005; Tajaddini et al. 2005). The focus has been on finding the relationship between the pressure and volume in segments of the vascular tree. Using segments of isolated vessels from the vascular tree, in vitro experiments were conducted to measure the properties of localized arteries and emphasize the role of elastin, collagen, and smooth muscle as determinants of artery wall properties. Both passive and active tissue components are used to determine the mechanical characteristics of blood vessels. Elastin and collagen are the most significant passive components of the fibrous connective tissues (Dobrin 1978).

The arteries have elastic characteristics, making them expand to accept a volume of blood and contract to squeeze back to their original size after the pressure is released. The elasticity of the arteries maintains the pressure on the blood when the heart relaxes and keeps it flowing forward. Elastin is a fibrous protein with remarkable elastomeric properties, i.e. it behaves mechanically as though it were composed of long, largely independent chains. Load–extension data indicate that elastin, like other elastomers, is highly extensible. Load–extension data are readily quantitated by describing stress–strain relationships or stress–extension ratio relationships. Strain (ϵ) is defined as (Dobrin 1978):

$$\epsilon = \frac{\Delta L}{L_0} = \frac{L - L_0}{L_0} \tag{2.1}$$

where L is the observed length of the tissue and L_0 is the original length of the unloaded tissue.

The strain is the fractional increase in tissue dimension relative to that of the unloaded tissue. However, the **extension ratio** is the observed length of tissue expressed as a fraction of the original length of the unloaded tissue. The relationship between the strain and extension ratio λ can be obtained by (Dobrin 1978):

$$\lambda = \epsilon + 1 \tag{2.2}$$

Stress (σ) can be calculated as:

$$\sigma = \frac{F}{A} \tag{2.3}$$

where F is the force exerted by the extended tissue and A is the area over which that force is exerted. The Young's elastic modulus (E) quantitatively measures the stiffness as:

$$E = \frac{\sigma}{\epsilon} \tag{2.4}$$

A stiffer material has larger values of E. Figure 2.3 shows a schematic stress–strain curve of a uniaxial extended tissue. If the stress–strain curve is linear, then the slopes will give the values of E. However, if the stress–strain curve is nonlinear, then it is more appropriate to obtain a "tangent" or "incremental" modulus by determining the slope of the stress–strain curve at designated points along the curve, see Figure 2.3. The incremental elastic modulus is defined by:

$$E_{inc} = \frac{\Delta \sigma}{\Delta \epsilon} \tag{2.5}$$

The elastic modulus for healthy tissue was observed to be $1.5\text{–}4.1 \times 10^6$ dynes/cm^2 in several studies (Krafka 1939; Wood 1954; Remington 1957; Ayer et al. 1958).

Mammalian studies have shown that the number of elastic lamellae in the abdominal aortic media is proportional to vessel diameter and to mean circumferential tension (Wolinsky and Glagov 1969). Human subjects have relatively fewer elastic lamellae compared to mammals (Wolinsky and Glagov 1967; Lustig and Glagov 1969; Bessette and Glagov 1970).

2.2.3 Arterial Stiffness

Arterial stiffness (AS) describes the reduced capability of an artery to expand and contract in response to pressure changes (Cecelja and Chowienczyk 2012). Changes in aortic stiffness

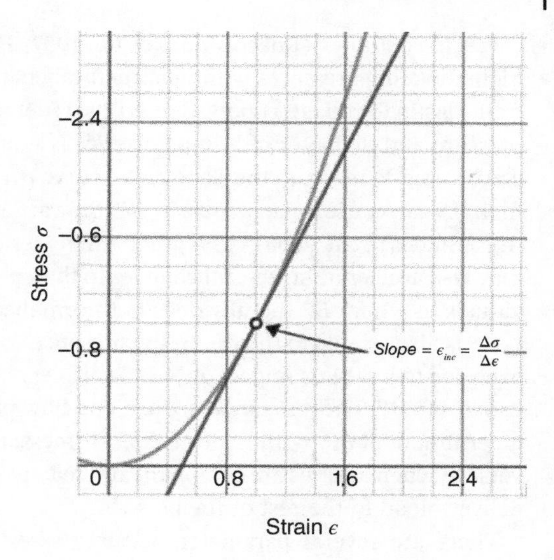

Figure 2.3 Stress–strain curve of a uniaxial extended tissue.

can increase aortic pulse pressure (Benetos 1999; Nichols and Edwards 2001) and cardiac pressure afterload, which can cause left ventricular hypertrophy (Nussbacher et al. 1999). Arterial stiffness is a significant risk factor in coronary heart disease (Ohtsuka et al. 1994; Franklin et al. 1999) and is considered an independent predictor of cardiovascular mortality (Laurent et al. 2001; Meaume et al. 2001; Boutouyrie et al. 2002; Safar et al. 2002; Blacher et al. 2003). Compliance (*C*) and distensibility (*D*) are two main parameters used to describe vessel stiffness. Compliance measures the volume change in response to a change in *BP*. The distensibility describes the ability of the artery to expand during systole, and it is related to initial volume and therefore relates more closely to wall stiffness (Cecelja and Chowienczyk 2012):

$$D = \frac{\Delta V}{\Delta P} \times V \tag{2.6}$$

where ΔV is the volume change and ΔP is the *BP* variation.

The most used variables of characterizing the *AS* are distensibility and pulse wave velocity (*PWV*) (McDonald 1968; Mohiaddin et al. 1993; Groenink et al. 1998; Vulliemoz et al. 2002; Laffon et al. 2005). *AS* information can be obtained using pulse contour to analyze peripheral pressure waveforms by early researchers (Szekeres et al. 1998; Lally et al. 2004); however, the results are mainly qualitative, and pulse contour analysis has been largely abandoned by practicing clinicians in favor of conventional sphygmomanometer.

Thomas Young first predicted the relationship between *AS* and *PWV* (1809), which is usually described by the Moens–Korteweg equation (O'Rourke 1982b) or the Bramwell–Hill equation (Bramwell and Hill 1922). Studies among East Asian populations with a low prevalence of atherosclerosis showed that *AS* increases with age by approximately 0.1 m/s/year (Avolio et al. 1983). A linear relationship between *AS* and age is reported (Avolio et al. 1983), and accelerated stiffening in the age group between 50 and 60 years is also reported (McEniery et al. 2005). In contrast, the stiffness of the peripheral arteries increases less (Avolio et al. 1983) or not at all with increasing age (Mitchell et al. 2004).

Previous studies (Roach and Burton 1957; Bergel 1961) showed that *AS* increases at higher loading pressures without structural changes. This is due to stress being transferred from elastin to collagen fibers. Bergel (1961) demonstrated that the relationship between the pressure and diameter of human arteries is nonlinear, and Young's incremental modulus can be used to measure the slope of the curve per increment in pressure change. In reality, a high *AS* can cause an acute rise in *BP* (Stewart et al. 2003). Compared to age-matched controls (Stewart et al. 2006), *AS* is greater in hypertensive individuals. Sustained hypertension may also accelerate structural changes to the arterial wall, particularly in hypertensive individuals in whom *BP* therapy does not normalize *AS* (Benetos et al. 2002).

AS is also related to the *BP*. In the presence of structural changes, *BP* reduction might be expected to have less impact on *AS* (Guerin et al. 2001). The strength of each heartbeat contributes to *BP*. The resistance to the blood flow provided by the arteries is also an important determinant of *BP*. With large arterial stiffness, resistance is greater and *BP* increases. Conversely, when arteries are compliant and reactive, the heart does not need to work so hard to deliver blood to the rest of the body.

There are several parameters used to assess *AS* including augmentation index (*AI*), central blood pressure (*CBP*), *PWV*, and carotid. Obviously, large *AS* is not desirable and there are many methods to reduce its value, including exercise, some types of drugs, and diet control. *AS* parameters can be measured by several commercial devices, including Arteriograph (Horvath et al. 2010; Nemes et al. 2010; Gavallér et al. 2011; Nemes et al. 2011; Rezai et al. 2011; Gaszner et al. 2012), BPLab, CVProfilor, Complior (Baulmann et al. 2008; Jatoiet al. 2009), Meditech, Mobil-O-Graph NG, Pulsecor, PulsePen (Salvi et al. 2010; Palombo et al. 2011) PeriScope, Hanbyul, Vasotens, and Sphygmocor (Millasseau et al. 2002; Hope et al. 2008; Wassertheurer et al. 2010; Kracht et al. 2011).

2.3 Noninvasive Arterial Stiffness Detection

There are three main noninvasive *AS* detection approaches:

2.3.1 Local Methods

These methods make use of the pressure area analysis of a specific region of the central or peripheral arteries. These are based on the principle that the distension of a region in the arteries to a given pressure is dependent on the stiffness/compliance at this specific location. The compliance (*C*) of the artery can be described by (Bund et al. 1996):

$$C = \frac{\Delta A}{A \cdot PP} \tag{2.7}$$

where ΔA is the change in the cross-sectional area between the *SP* and *DP*, *A* is the diastolic cross-sectional area, and *PP* is the pulse pressure.

The advantages of the local-based methods are that no circulatory models are required and they can directly measure the local *AS*. However, the limitations of these methods include high cost, requirement of a high level of expertise, inability to directly measure the *BP* at the central arteries, and inaccurate measurement of the arterial diameter

(Verbeke et al. 2005). Ultrasonic devices are usually used to measure the artery's diameter, while applanation tonometry is used to measure the local *BP*. Recently, researchers were also able to measure the stiffness of central arteries such as the aorta using magnetic resonance imaging (MRI).

2.3.2 Regional Methods

These methods use the *PWV* in a segment of the arterial tree. These methods are based on the principle that *PWV* is directly dependent on the stiffness of the artery where the wave is propagating. *PWV*-based methods are simple, noninvasive, cheap, and reasonably accurate in determining arterial stiffness. The regional stiffness of the aorta is usually of significant interest for several reasons. First, the aorta is the most significant contributor to the arterial buffering function, and also, aortic *PWV* is an independent predictor of CVD in several populations as it has the largest effect on the left ventricle function.

The aortic *PWV* is usually determined by measuring the pressure, artery destination, or Doppler waveforms at the carotid artery and the femoral artery. The distance traveled by the waves is usually estimated by measuring the body's surface distance between the two measuring sites. The *PWV* can be estimated as:

$$PWV = \frac{Distance}{Time\ Lag} \tag{2.8}$$

The foot-to-foot method is commonly used to estimate the transient time which is defined as the time it takes the foot (end of diastole) of the wave to travel over a defined distance. Even though the *PWV* technique is based on an accepted propagative model of the arterial system, there are inaccuracies introduced by the difficulty in estimating the actual distance traveled by the wave and the difficulty in accurately locating the wave's foot. Furthermore, the femoral artery pressure wave can be difficult to accurately measure in individuals suffering from obesity, diabetes, and femoral artery disease.

2.3.3 Waveform Analysis Methods

These methods are based on the fact that the arterial pressure waveforms are a composite of forwarding traveling waves produced by the contraction of the left ventricle and backward traveling reflected waves from peripheral sites of impedance mismatch, see Figure 2.4. The timing of the arrival of the reflected wave to the ascending aorta depends on both the reflection site and the *PWV*. In elastic arteries, *PWV* is relatively low, and hence the reflected wave arrives in the ascending aorta during diastole. As the stiffness of the central arteries increases, the *PWV* increases, and reflected waves arrive in the ascending aorta earlier, where it augments and adds to the *systolic pressure*. This phenomenon is used as an indicator of arterial stiffness, where the *AI* is used to quantify its magnitude, as shown in Figure 2.4. In the central arteries, the *AI* is defined as:

$$AI = \frac{P_s - P_i}{P_s - P_d} \tag{2.9}$$

where P_i is the pressure at the inflection point corresponding to the arrival of the reflected wave, P_s is the systolic pressure, and P_d is the diastolic pressure.

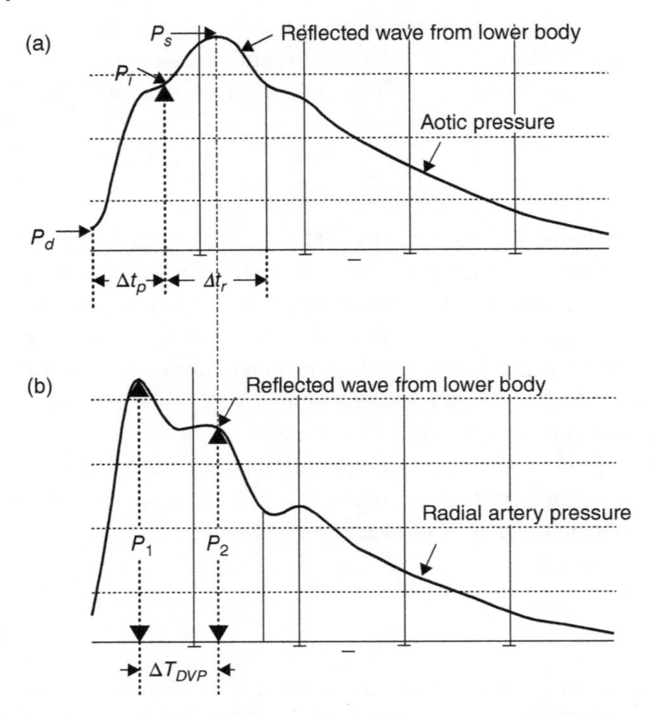

Figure 2.4 (a) Central pressure waveform showing wave reflection from the lower body; (b) radial artery waveform showing wave reflection from the lower body.

The ascending aorta pressure waves reveal the real load that the left ventricle has to pump against; therefore, the ascending aorta pressure waveforms should be analyzed. Direct noninvasive measurement of ascending aortic pressure is impossible. Hence, ascending aortic pressure is either estimated using the standard carotid pressure waveform or a transfer function that is used to estimate the aortic pressure from the radial artery pressure. These pressures are usually obtained using applanation tonometry. The transfer functions used to assess aortic waveforms can be either individualized or general. The accuracy of these transfer functions has been the topic of much debate lately, and its application, especially at high frequencies, is broadly disputed. Even though the common carotid artery tonometry does not require the use of a transfer function, it requires a high degree of technical expertise and cannot be applied in obese patients and patients with an advanced degree of atherosclerosis.

2.4 Arterial Model Development

In this work, we intend to develop a diagnostic method that is noninvasive, physiologically based, and quantitative. The model should not require high level of expertise and it should make use of the easily accessible brachial artery pressure waveform.

This method is based on the development of a model for the wave propagations in the arterial system which is used to accurately investigate the *AS* values (El-Aklouk 2007). The main model is developed with one idea in mind, is it possible to predict the central blood pulse wave from measurements at the brachial artery. Thus, we need to trace the pressure waves from the heart to various parts of the arterial system. However, any actual non-invasive measurement has to be conducted at a site that is externally accessible such as the brachial artery.

Tracing a pulse wave from the heart to the aorta and then to the brachial artery requires consideration of each reflection point and summing the transmitted and reflected waves at the site of interest. From Chapter 1, assuming a thin circular tube model for an artery, the wave equation can be written as:

$$\frac{\partial^2 p}{\partial x^2} = \frac{1}{c^2}\frac{\partial^2 p}{\partial t^2} = 0 \tag{2.10}$$

where the speed of sound c can take the form:

$$c^2 = \frac{Eh}{2\rho r}$$

where E, h, and r are the modulus of elasticity, thickness, and radius of the artery, respectively, and ρ is the blood density. The solution for equation (2.10) may take the form:

$$P(x,t) = P_0 f(x - ct) + P'_0 f(x + ct) \tag{2.11}$$

where $P_0 f(x - ct)$ is an incident wave traveling in the positive x direction and $P'_0 f(x + ct)$ presents a pressure wave traveling in the negative x direction. The coefficients P_0 and P'_0 are calculated using impedance mismatch (Figure 2.5).

Heart contractions produce pressure waves which travel through the arterial system. Waves reflect at points of impedance mismatch such as bifurcations, vascular beds, and arterial branching. The iliac bifurcation is a major reflection site. *BP* cuffs cause the brachial artery to narrow and lead to reflection waves. Any measured wave is a combination of

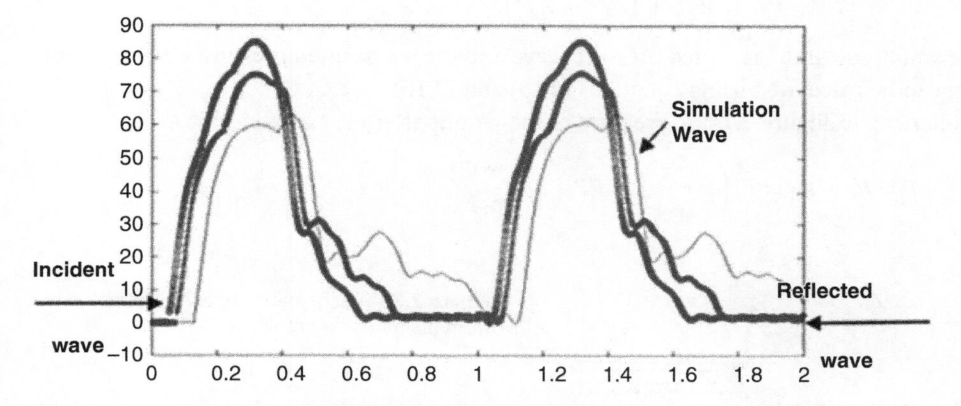

Figure 2.5 Incident and reflected waves with summation of waves at a point.

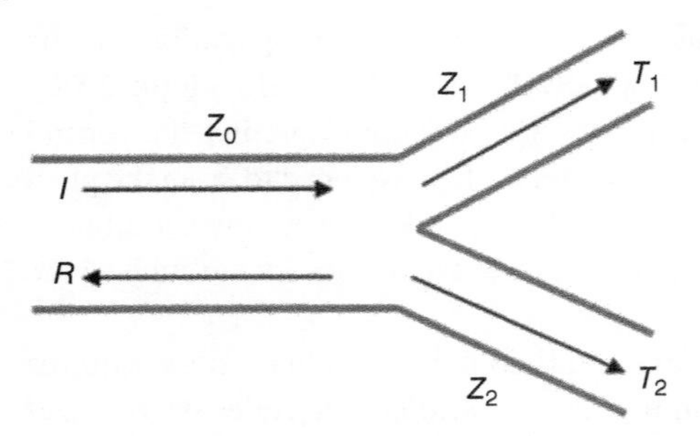

Figure 2.6 A schematic figure showing conditions at a bifurcation.

incident and reflected waves. Figure 2.6 shows a parent artery and two daughter arteries, and pressure at the junction is obtained:

$$P_I + P_R = P_{T_1} = P_{T_2} \tag{2.12}$$

where P_I and P_R are the incident and reflected pressure waves in the parent tube, P_{T_1} and P_{T_2} are the pressure transmitted into the first daughter tube T_1 and the second daughter tube T_2, respectively. The flow condition at the junction is defined by:

$$Q_I - Q_R = Q_{T_1} + Q_{T_2} \tag{2.13}$$

where Q_I and Q_R are the incident and reflected flow waves, and Q_{T_1} and Q_{T_2} are the flow into the first daughter tube T_1 and the second daughter tube T_2, respectively.

The characteristic impedance of the artery (Z) is defined by:

$$Z = \frac{P}{Q} \tag{2.14}$$

The reflection coefficient R can be expressed as:

$$R = \frac{P_R}{P_I} = \frac{Z_0^{-1} - (Z_1^{-1} + Z_2^{-1})}{Z_0^{-1} + (Z_1^{-1} + Z_2^{-1})} \tag{2.15}$$

The transmission coefficient T is written as:

$$T = \frac{P_{T_1}}{P_I} = \frac{P_{T_2}}{P_I} = \frac{2Z_0^{-1}}{Z_0^{-1} + (Z_1^{-1} + Z_2^{-1})} \tag{2.16}$$

The amplitude of the reflected pressure wave and the transmitted pressure wave at a junction can be calculated using equations (2.15) and (2.16).

Referring to Figure 2.7, the pressure at the parent artery is expressed by:

$$P = P_I + P_R = P_I\left(t - \frac{x}{c_0}\right) + RP_I\left(t + \frac{x}{c_0}\right) \tag{2.17}$$

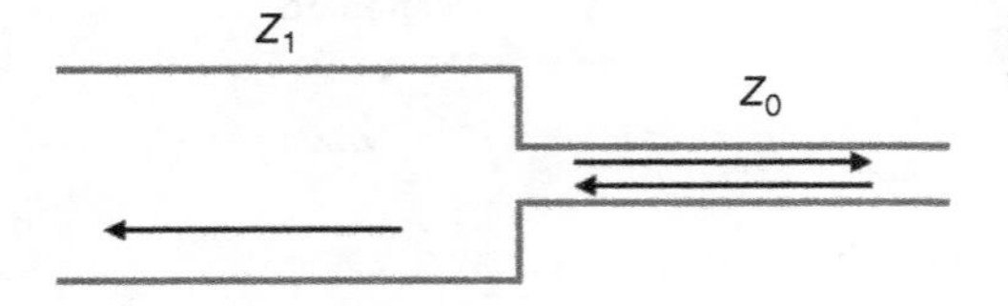

Figure 2.7 A schematic showing reflection from a small side branch.

The pressure at the daughter artery is defined as:

$$P_T = TP_I\left(t - \frac{x}{c_{1,2}}\right) \tag{2.18}$$

where $c_{1,2}$ is the wave velocity in either one of the daughter tubes. The earlier mentioned analysis is valid for any part of the arterial system; however, the site of measurement, namely the brachial artery, requires further details.

2.5 Lumped Modeling of the Aorta and Brachial Arteries

In general, the arteries do not take uniform circular shapes along their length. They are normally tapered. In this chapter, the aorta and the brachial arteries are considered. However, the wave equation is valid for a uniform circular element. Thus, lumped models are used for the aorta and the brachial artery. Each lump is considered of a uniform shape so that the linear wave equation can be used (Figure 2.8).

To determine the optimum functions for the radius and thickness of the aorta and brachial arteries, Westerhof et al. (1969) conducted experiments on human arteries and listed the thickness, radius, and the modulus of elasticity at 10 positions along the aorta and 7 positions along the subclavian and brachial arteries as given in Figure 2.9. These data varied smoothly and were fitted with appropriate functions. After several attempts, a suitable power function was found to describe the variation in thickness and radius of the aorta, while a linear function was used to describe the thickness and radius variation of the subclavian–brachial arteries.

To determine the appropriate number of lumps that would accurately describe wave propagation in the arteries, the model was simulated for different numbers of lumps using equation (2.10). Figure 2.10 shows the relationship between the time it takes for the wave to travel through the aorta and brachial artery versus the number of lumps used. It is clearly indicated that the time converged after 24 and 16 lumps for the aorta and the brachial artery, respectively. Hence, the aorta was divided into 24 lumps of 0.018 m length each, while the subclavian and brachial arteries were divided into 16 lumps of 0.026 m length each.

2.5.1 Input Signal

The left ventricle is the most muscular chamber of the heart. In normal conditions, when it contracts, it generates a pressure of approximately 120 mmHg (Dobrin 1978). When the electric excitation wave reaches the left ventricle, the ventricle myocardium starts to contract, and hence the pressure inside the ventricle increases. When the left ventricle pressure exceeds the left atrium pressure, blood starts to flow back into the left atrium and creates a backflow that closes the mitral valve. The aortic valve remains closed as long as the left

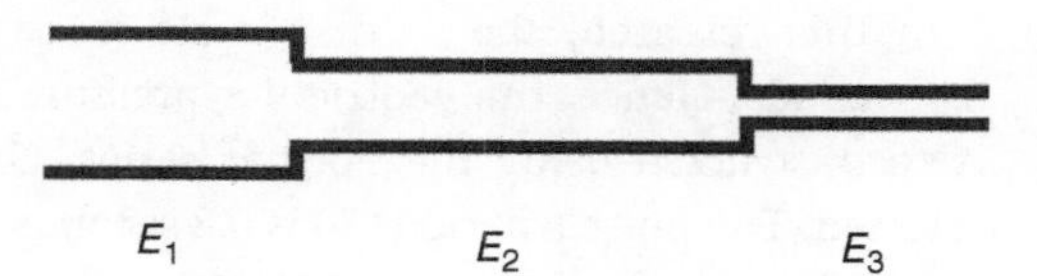

Figure 2.8 Lumped model of an artery. *Source:* Based on El-Aklouk et al. (2008).

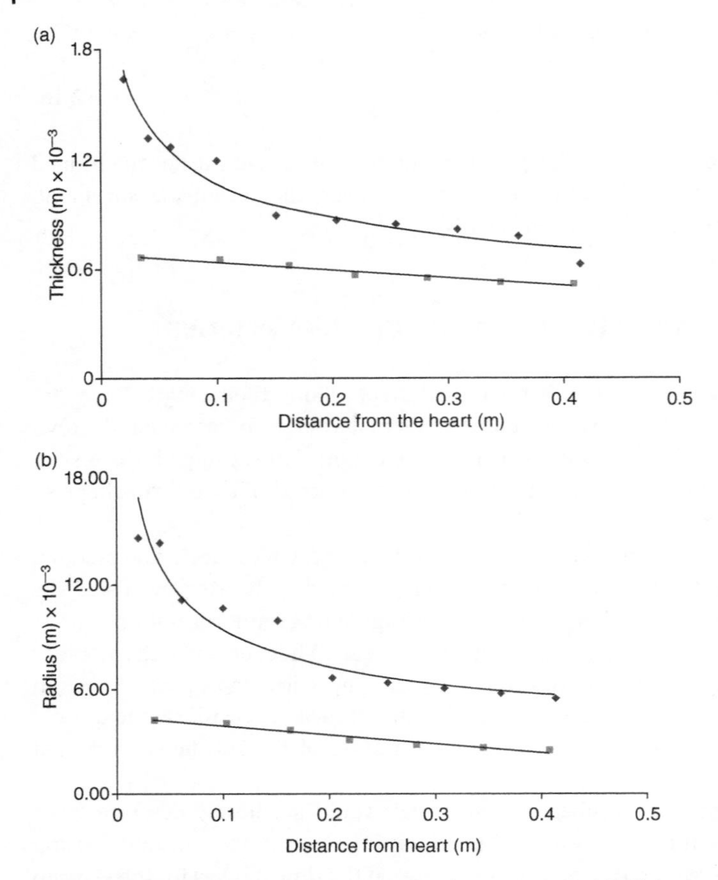

Figure 2.9 Power function describing the human artery: (a) thickness variation and (b) radius variation.

ventricle pressure remains less than the pressure in the aorta. During this period, both the mitral and aortic valves are closed, and thus the volume of the left ventricle stays constant. This period is called the isovolumic contraction period. However, when the left ventricular pressure becomes larger than the aortic pressure, the aortic valve opens and blood flows from the left ventricle into the aorta. As contraction continues, the left ventricle pressure rises, and blood accelerates into the aorta. Eventually, the contraction ceases, and the left ventricle pressure starts to decrease. Finally, the velocity of the blood drops to zero and begins to reverse. The backflow closes the aortic valve, which marks the end of the ventricular systole. The previous sequence of events can be seen in Figure 2.11 (Dobrin 1978). Hence, the pressure transmitted from the left ventricle to the ascending aorta can be assumed to be the supra-diastolic left ventricular pressure, as shown in Figure 2.12a.

In this research, the cardiac cycle is assumed to be periodic and repeated every 0.8 seconds. Hence, the ventricular pressure is considered to be a regular function and is reconstructed using the Fourier series. One supra-diastolic left ventricular pulse is selected. The pulse's period (T) is 0.8 seconds and is divided into 38 equally spaced intervals (N), as shown in Figure 2.12b.

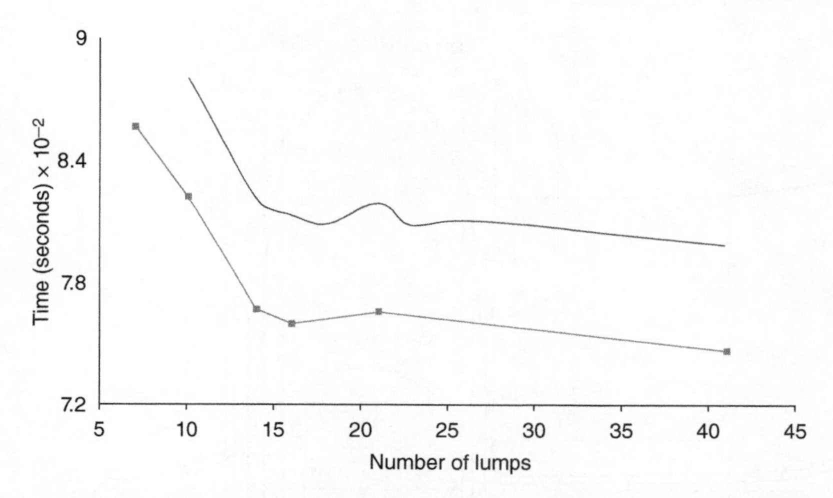

Figure 2.10 Pressure wave traveling time versus number of lumps: —— aorta; + brachial artery.

Each of the 38 ordinates is designated by the letter s. The Fourier coefficients A_0, A_n, and B_n are then calculated using:

$$A_n = \frac{2}{N}\sum_{s=0}^{N} y_s \cos n\omega t_s \quad \left(\text{for } n = 0, 1, 2, \ldots, \frac{N}{2}\right) \tag{2.19}$$

$$B_n = \frac{2}{N}\sum_{s=0}^{N} y_s \sin n\omega t_s \quad \left(\text{for } n = 0, 1, 2, \ldots, \frac{N}{2}\right) \tag{2.20}$$

$$A_0 = \frac{\sum y_s}{N} \tag{2.21}$$

where

$$\omega = \frac{2\pi}{T} \tag{2.22}$$

The function is then defined by a finite series where the function value (y_s) can be found for any time (t_s):

$$y_s = \frac{A_0}{2} + \left[\sum_{n=1}^{N/2} A_n \cos n\omega t_s + \sum_{n=1}^{N/2} B_n \sin n\omega t_s\right] \tag{2.23}$$

2.5.2 Wave Reflection Locations

The precise locations of the major reflection sites in the human aorta have been studied in several publications. It has been reported that the major reflection site is in the abdominal aorta and is located at the iliac bifurcation (Krafka 1939; Remington 1957; Ayer et al. 1958). Hence, in this work, the impedance mismatch between the abdominal aorta and the iliac arteries was assumed to be the only significant source of reflection for the forward traveling

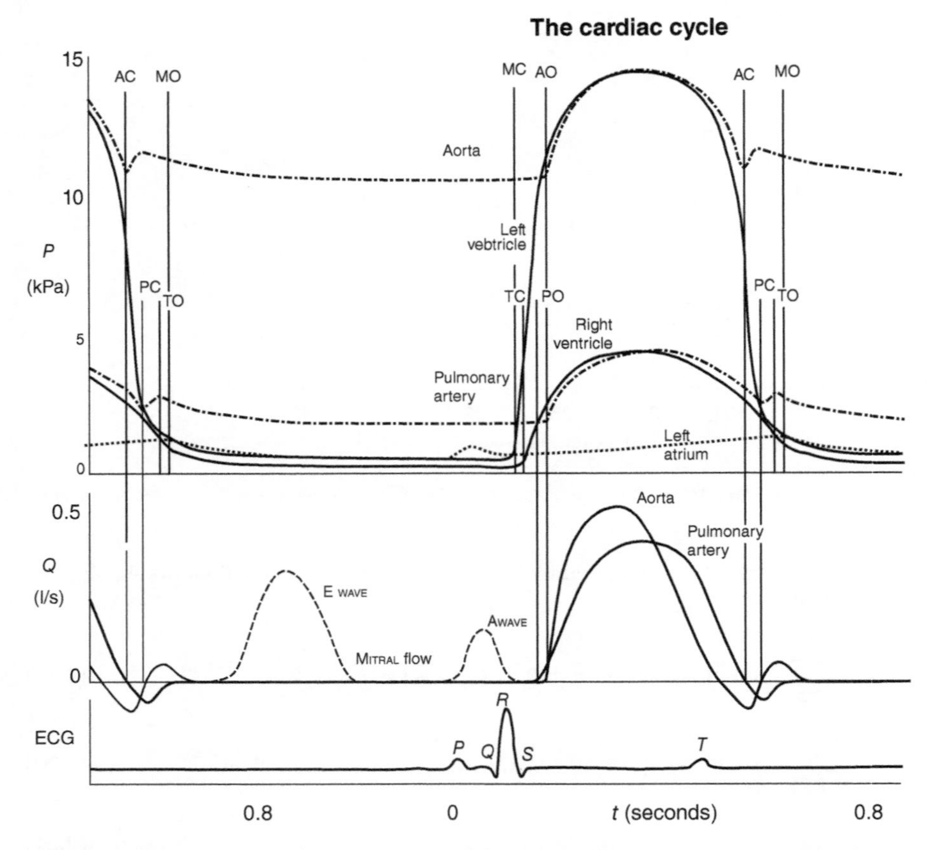

Figure 2.11 Tracings of pressure, flow, and ECG during the cardiac cycle–normal adult. *Source:* Based on Dobrin (1978).

waves in the lower body. The geometry and material properties for the aorta and the iliac arteries were obtained from experimental data.

The compression of the upper arm by the pneumatic cuff causes a sudden decrease in the brachial artery area, which results in a large impedance mismatch that leads to a reflection at this site. Furthermore, from the data reported by Westerhof et al. (1969), the reflected brachial wave will reflect negatively when it reaches the mouth of the subclavian artery because of the sudden increase in the area (Tortora 2018). Hence, in this work, the following wave reflection model is proposed.

Any wave generated by the heart travels via the aorta into the brachial artery and iliac bifurcation. The aortic wave is reflected at the iliac bifurcation and then travels back to the brachial artery. The original wave which traveled to the brachial artery reflects back due to the cuff and travels to the mouth of the subclavian artery where it reflects back negatively to the brachial artery. Secondary reflections are not considered in this work because their magnitude is too minute for the sensors on the cuff. Hence, the earlier mentioned three waves reach the brachial artery at different times, as shown in Figure 2.13.

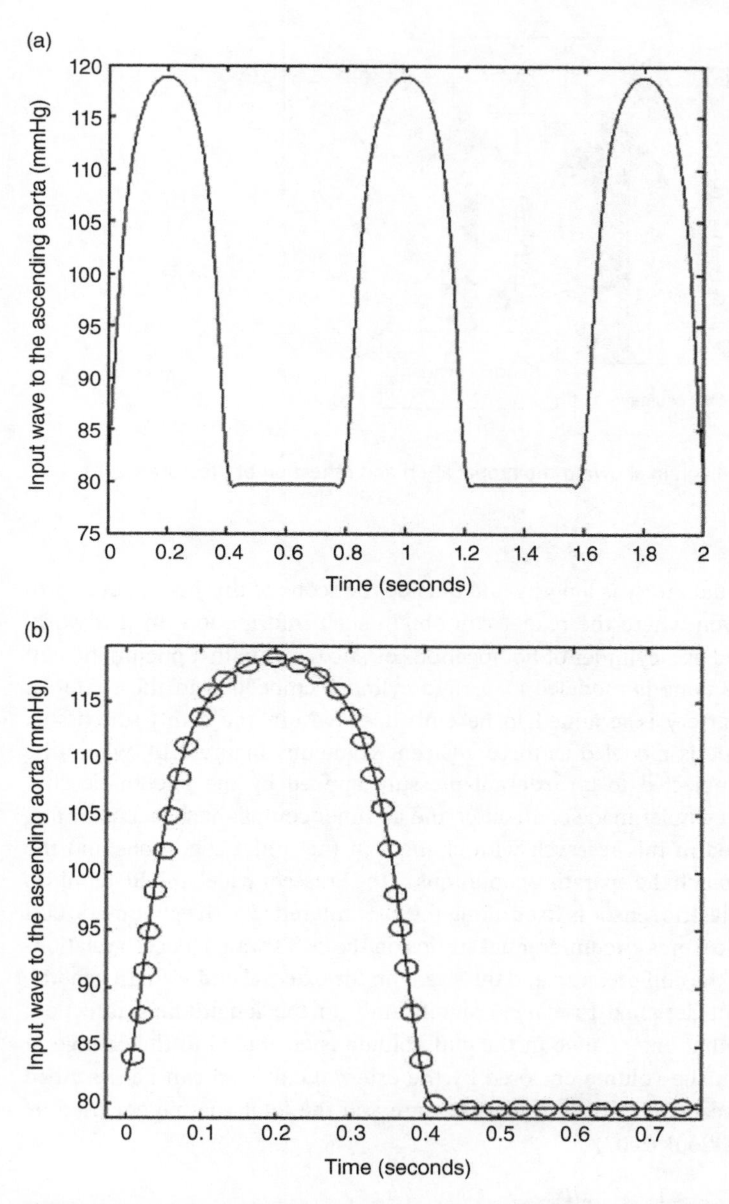

Figure 2.12 (a) The periodic pressure wave transmitted from the left ventricular to the ascending aorta. (b) One supra-diastolic pressure pulse divided into 38 equally spaced intervals (N).

2.5.3 Cuff–Soft Tissue–Brachial Artery Model

This section presents a numerical upper arm model to study the interactions between the pneumatic cuff, upper arm soft tissue, and brachial artery hemodynamic (Ursino and Cristalli 1995a, 1995b, 1996). Figure 2.14 shows the schematic of the upper arm with a pneumatic cuff wrapped around it. Deriving the separate governing equations for the cuff,

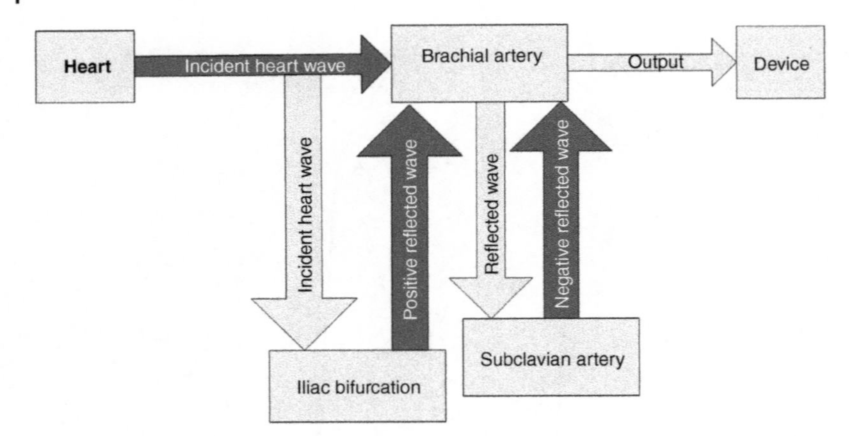

Figure 2.13 A schematic diagram showing the propagation and reflection of pressure waves in the system.

soft tissue, and the brachial artery is lengthy and outside the scope of this book, but appropriate references are given where the reader can obtain such information. In this work, the upper arm is modeled as a cylinder of homogenous elastic tissue with a pneumatic cuff wrapped around it. The bone is modeled as a rigid cylinder embedded in the center of the arm. The brachial artery is assumed to be embedded within the arm's soft tissue. The upper arm soft tissue is modeled as three adjacent segments, Figure 2.14, where the middle segment L_2 is subjected to an external pressure applied by the pneumatic cuff. The cuff is modeled as a cylinder made of an outer and an inner compliant sheet containing air. This model is utilized in this research with changes in the initial conditions and the simulation approach to match the operating conditions of the Pulsecor wideband *BP* monitor. In this monitor, a piezoelectric sensor is fixed along the circumference of the pneumatic cuff outer wall. The sensor measures circumferential strains on the cuff's wall. Hence, a relationship is required between the cuff pressure and the strain on the external cuff wall. In general, commercial cuffs are not designed to expand significantly in the longitudinal direction; hence, one can assume that any change in the cuff volume is attributed to the change in its radius. The change in the volume enclosed by the external cuff wall can be described by the relationship between the cuff's internal pressure and the total volume enclosed by its external wall as (El-Aklouk 2007):

$$\frac{dV_e}{dt} = C_e \frac{dP_c}{dt} = \frac{1}{\beta_e(P_c + P_{e0})} \frac{dP_c}{dt} \tag{2.24}$$

where V_e is the volume enclosed by the cuff external wall, C_e is the compliance of the cuff's external wall, P_c is the internal cuff pressure, and β_e and P_{e0} are constants that govern the cuff's external wall mechanics.

From the previous equation, the changes in the radius can be written as:

$$\frac{dR}{dt} = \frac{1}{2\pi R l_2 \beta_e(p_c + p_{e0})} \frac{dp_c}{dt} \tag{2.25}$$

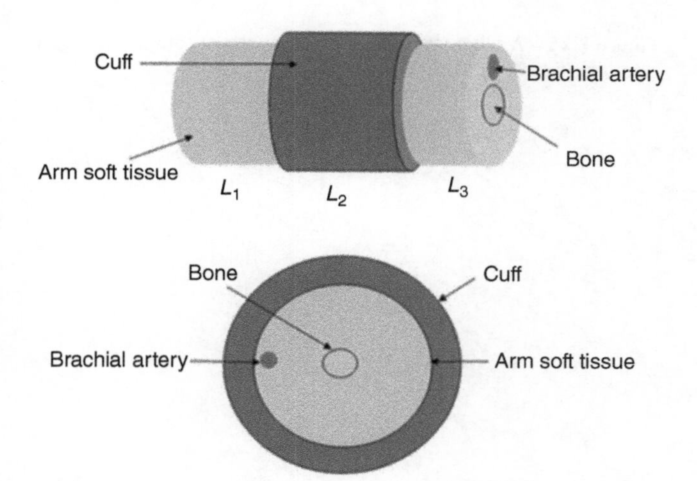

Figure 2.14 The upper arm with a pneumatic cuff wrapped around; L_2 is the middle segment which is subjected to an external pressure applied by the pneumatic cuff. *Source:* Based on Ursino and Cristalli (1995a, 1995b, 1996).

where R is the radius of the cuff outer wall and l_2 is the length of the middle segment. The circumferential strain (ε) can be defined as the ratio of change in radius divided by the original radius, i.e.:

$$\varepsilon = \frac{dR}{R_0} \tag{2.26}$$

where R_0 is the original radius.

A numerical model was developed in a MATLAB environment based on the preceding differential equations. The change in the volume of blood under the cuff (dV_b/dt) is unknown, which will be detailed in the following section.

2.5.4 Brachial Artery Model

Figure 2.15 displays a schematic diagram of the brachial artery compressed by a pneumatic cuff. The relationship between the transmural pressure ($P_a - P_e$) and the area of the brachial artery under the pneumatic cuff (A) is

$$(P_a - P_e)\sqrt{\frac{A}{\pi}} = \sigma h = (\sigma_e + \sigma_v)h \tag{2.27}$$

where P_a, P_e, and A are the brachial artery pressure and the artery area at different sections of the artery, and σ, σ_e, and σ_v are the circumferential stress, elastic stress, and viscous stress, respectively.

The relationship between the brachial artery elastic circumferential stress (σ_e) and circumferential strain (ε_a) is (Mohiaddin et al. 1993):

$$\sigma_e(A) = \frac{E_0}{\beta'_a}\left(e^{\beta'_a \varepsilon_a} + e^{\beta''_a \varepsilon_a^2} - 2\right) \tag{2.28}$$

where β'_a, β''_a, and E_0 are constants, and the circumferential strain $\varepsilon_a = \left(\sqrt{A} - \sqrt{A_0}\right)/\sqrt{A_0}$.

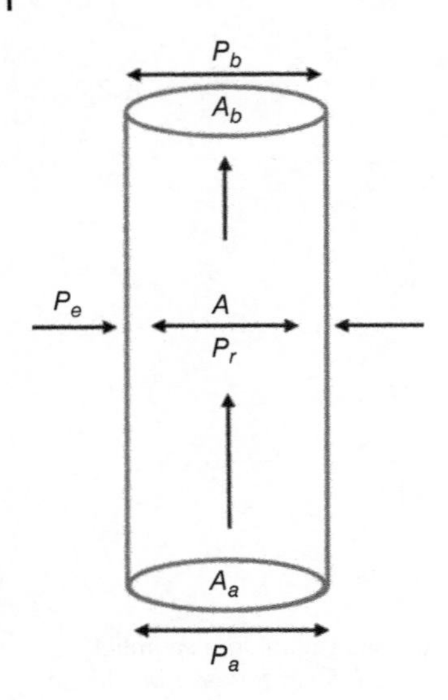

Figure 2.15 A schematic of the brachial artery compressed by a pneumatic cuff.

The wall thickness at positive transmural pressure is

$$h = -\sqrt{\frac{A}{\pi}} + \sqrt{\frac{A}{\pi} + 2h_0\sqrt{\frac{A_0}{\pi}} + h_0^2} \tag{2.29}$$

where h_0 and A_0 are the wall thickness and cross-sectional area at zero transmural pressure, respectively.

The viscous stress (σ_v) can be obtained:

$$\sigma_v = \eta\frac{d\varepsilon_a}{dt} = \frac{\eta}{2\pi\sqrt{\frac{A_0}{\pi}}\sqrt{\frac{A}{\pi}}}\frac{dA}{dt} \tag{2.30}$$

The relation for the rate of change of the brachial artery area at positive transmural pressure can be expressed as:

$$\frac{dV_b}{dt} = \frac{dA}{dt} = \frac{2(P_a - P_e)A_0\sqrt{\frac{A_0}{\pi}}}{\eta h_0} - \frac{2k_pA_0\sqrt{\frac{A_0}{\pi}}}{\eta h_0}\left(1 - \left(\frac{A}{A_0}\right)^{-3/2}\right) \tag{2.31}$$

where k_p is a constant that depends on the geometry and material properties of the brachial artery $k_p = h_0E_0/3r_{a0}$.

2.5.5 Combined Model

A combined, physiologically based numerical model was developed by combining the acoustic model and the cuff–soft tissue–brachial artery model. This model demonstrates the pressure contours in the brachial artery and the noninvasive estimation of these pressure contours by detecting the strain on the outside wall of the pneumatic cuff. It provides a quick and easy-to-use tool to investigate the effect of variations in the central arteries geometry and material properties on the brachial artery pressure contours and the strain on the cuff outside wall. The effect of variations in the pneumatic cuff and soft tissue properties on the acquired pressure and strain contours can also be investigated with this tool. The material properties and geometry for the cuff, arm soft tissue, and the brachial artery used in the earlier mentioned models were adopted from the values reported by Ursino and Cristalli (1995a, 1995b, 1996).

2.6 Arterial Blood Pressure

Arterial blood pressure, also called *BP*, is one of the principal vital signs and normally refers to the systemic arterial blood pressure when used without further specification (van Berge-Landry et al. 2008). Human's *BP* is expressed in terms of the *SP* over *DP*. An average normal resting *BP* of 120/80 millimeters of mercury (mmHg) was observed after studying 100 human subjects (Pesola et al. 2001). According to the American Heart Association (AAHA), *BP* can be classified into six groups for 18 years and older adults as described in Table 2.1 (Stage 2013). Figure 2.16 shows the *BP* changes in the different parts of the circulatory system (Tortora 2018). *BP* normally can be measured in the peripheral artery that is located in the upper arm; peripheral **blood pressure** (*PBP*) is usually higher than *CBP* because the peripheral site is closer to locations from which echoes reverberate. *BP* depends on a situation, activity, and disease status and is regulated by the nervous and endocrine systems. Clinical studies showed that people who maintain arterial pressure at the low end of those pressure ranges have much better long-term cardiovascular health.

The normal *BP* range in children is lower than that in adults and depends on height (Rosner et al. 1993). The reference blood values for children are based on the distribution of *BP* in children (Table 2.2) (Chiolero 2014). Different countries have different reference blood values for children.

CBP normally refers to the pressure in the aorta near the heart and is the pressure that the heart has to pump against to get blood to flow to the rest of the body. Key components of *BP* are displayed in Figure 2.17 (data were reproduced from Salvi 2012). Too high *CBP* will

Table 2.1 Classification of blood pressure for adults.

Category	Systolic (mm Hg)	Diastolic (mm Hg)
Hypotension	<90	<60
Desired	90–119	60–79
Prehypertension	120–139	80–89
Stage 1 hypertension	140–159	90–99
Stage 2 hypertension	160–179	100–109
Hypertensive emergency	≥180	≥110

Source: Based on Stage (2013).

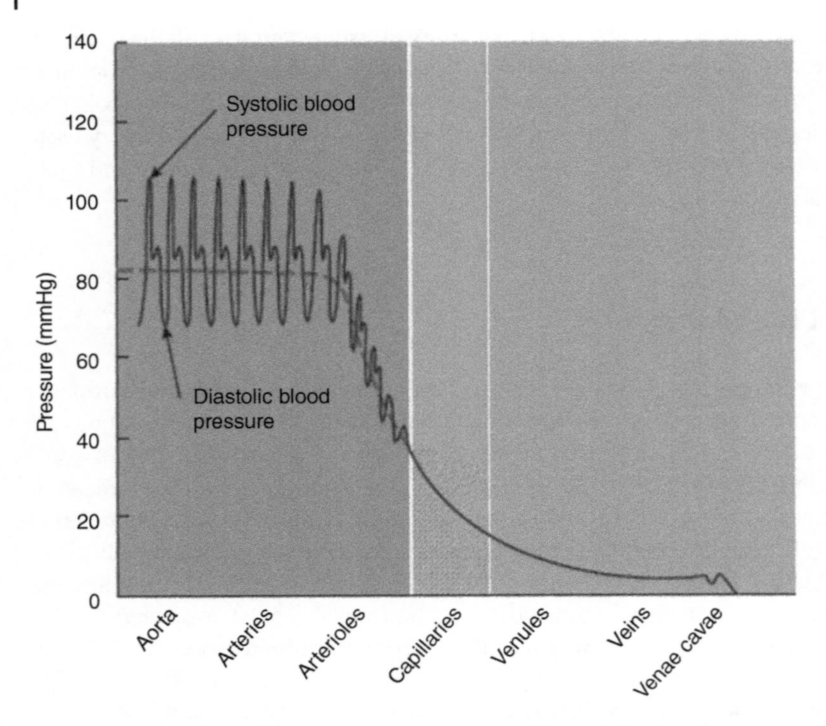

Figure 2.16 Normal blood pressure in different portions of the circulatory system; dashed line is the mean blood pressure in the aerorta, arteries, and arteriolies. *Source:* Tortora (2018)/with permission of John Wiley & Sons.

cause heart failure, aneurysms, and strokes. *CBP* also determines the pressure in the blood vessels feeding the brain. Clinical studies (Roman et al. 2007, 2009) suggested that *CBP* is more strongly related to vascular disease than the traditional upper arm *BP*. *CBP* can distinguish between the effects of different hypertension medications where upper arm *BP* and pulse wave analysis (*PWA*) failed (Avolio 2008).

2.6.1 Pulse Pressure

Figure 2.18 shows a curve of the arterial pressure during one cardiac cycle, and the up and down fluctuations of the arterial pressure result from the pulsatile nature of the cardiac

Table 2.2 Reference ranges for blood pressure in children.

Stage	Age (year)	*SP* (mm Hg)	*DP* (mm Hg)
Infants	≤1	75–100	50–70
Toddlers and preschoolers	1–5	80–110	50–80
School age	6–12	85–120	50–80
Adolescents	13–18	95–140	60–90

Source: Based on Rosner et al. (1993).

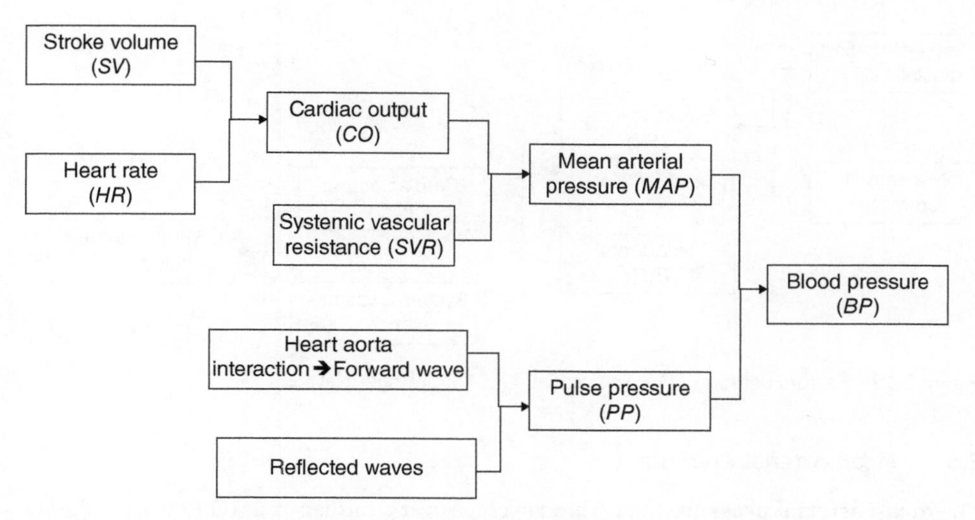

Figure 2.17 Key components of *BP*. *Source:* Data from Salvi (2012).

output, such as the heartbeat. Systolic pressure (*SP*) appears at the highest *BP* value when the heart pumps the blood, which is the heart's contraction. Diastolic pressure (*DP*) occurs when the *BP* falls to its minimum between each systole. The **pulse pressure** (*PP*) measures the difference between the measured *SP* and *DP* (O'Rourke 1982b):

$$PP \cong SP - DP \tag{2.32}$$

The *PP* is determined by the interaction of the stroke volume of the heart, compliance (ability to expand) of the aorta, and the resistance to flow in the arterial tree. By expanding under pressure, the aorta absorbs some of the force of the blood surge from the heart during a heartbeat. In this way, the pulse pressure is reduced from what it would be if the aorta were not compliant (O'Rourke and Avolio 1980; Nichols and O'Rourke 1991). The loss of arterial compliance that occurs with aging explains the elevated pulse pressures found in older patients. Previous studies found that the *SP* and *DP* are increasing with age.

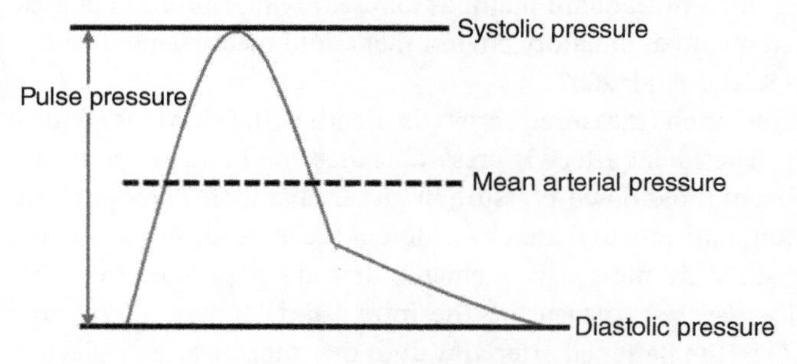

Figure 2.18 A curve of the arterial pressure during one cardiac cycle.

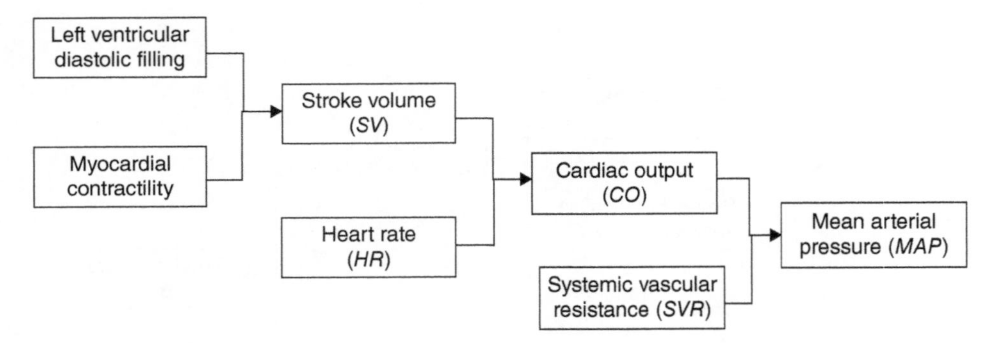

Figure 2.19 Factors defining *MAP*. *Source:* Salvi (2016)/Springer Nature.

2.6.2 Mean Arterial Pressure

The **mean arterial pressure** (*MAP*) is an average over a cardiac cycle (O'Rourke and Avolio 1980; Nichols and O'Rourke 1991). Figure 2.19 shows factors defining *MAP* (Salvi 2016):

$$MAP = (CO \cdot SVR) + CVP \tag{2.33}$$

where *CO* is the cardiac output, *SVR* is the systemic vascular resistance, and *CVP* is the central venous pressure. From *SP* and *DP*, we can also obtain *MAP*:

$$MAP \cong DP + \frac{1}{3}(SP - DP) \tag{2.34}$$

2.6.3 Noninvasive Blood Pressure Measurement Methods

There are several noninvasive *BP* measurement methods including auscultation (AUS), arterial tonometry (AT), electronic palpation (EP), unloading plethysmography (UP), volume oscillometry (VO), volume compensation (VC), and oscillometry (OSC).

AUS, also called Riva-Rocci/Korotkoff, uses an occluding cuff and a stethoscope to monitor acoustic pulses generated by an artery (Watson et al. 1998; Sebald et al. 2002). The pressures at which blood initially begins to flow through the brachial artery (*SP*) and at which normal blood flow returns (*DP*) (Sebald et al. 2002) are identified by the presence or absence of sound. Some limitations of this method include inconstant deflating rate, inappropriate cuff size for the arm diameter, different standard methods for identifying Korotkoff phases and observers, noises produced due to ambulatory environment, and measurement errors caused of patient movement (Sebald et al. 2002).

AT uses a transcutaneous simulation (measured across the depth of the skin) for continuously monitoring arterial *BP*. The radial artery is pressed against the radius bone by an array of pressure transducers, with hold-down pressure in an air chamber. The optimum hold-down pressure is given automatically to flatten a portion of the arterial wall and maximize the pulse pressure measured by the sensor elements that are positioned over the artery. The pressure transducer is used to measure the intra-arterial pressure (*P*), and the circumferential tension (*T*) in the flattened arterial wall to the transducer is neglected (Sato et al. 1993).

EP was first introduced in 1998 (Holejsovska et al. 2003). It uses an arm cuff around the brachial artery and monitors the pressure pulses from the radial artery on the wrist. A four-channel pulse sensor built in a wrist wrap is used as the radial artery sensor to monitor the *BP*. A simple but high accuracy algorithm was developed by Nissila et al. (1998) to identify the *SP* and *DP* points. The point where the *BP* pulse amplitude starts to drop determines the *DP* and the last pulse detected determines the *SP*. Measurements can be taken during inflating and deflating pressure model, but the inflating model provides more stable results.

UP measures *BP* based on the relationships between the cuff pressure, arterial pressure, and the arterial walls. Holejšovská et al. (2003) used the photoelectricity to measure blood flow in the finger and identified the character of the signal corresponding to the measured blood flow values by analyzing the plethysmography. However, there were measurement errors caused by improper setting of the reference value, changes in the tension of the arterial wall, and movement of the subject.

Yamakoshi et al. (1988) proposed **VO** and **VC** *BP* measurement methods based on the vascular unloading principle and the characteristics of the pressure–volume relationship in the artery. The photoelectric plethysmography was used in both VO and VC approaches to monitor the arteries' volume changes, which are different from OSC. VO method can be used for long-term ambulatory monitoring. The *SP* and *DP* values can be continuously measured, and the pressure waveform can be recorded using the VC method; however, the criterion of determining *DP* is not well established in this approach.

Based on the VC method, Tanaka et al. (2003, 2005) developed another technique to measure instantaneous *BP* of the radial artery. The radial artery was selected as a measuring site to avoid venous congestion during long-term monitoring, a disk-type cuff for local pressurization was selected, and a nozzle-flapper-type electro-pneumatic converter for the cuff pressure control was designed. The results showed that the radial artery could be completely compressed, and the nozzle-flapper-type electro-pneumatic converter has sufficient frequency response for *BP* measurement in human subjects. The instantaneous *BP* in rest and stressful conditions can be measured noninvasively using the developed prototype.

OSC is the most widely used noninvasive method for measuring *BP*. It is developed based on the pulsatile blood flowing through an artery that creates oscillations of the arterial wall. OSC measurement devices use an electronic pressure sensor with a numerical readout of *BP*. Once the blood flow is present but restricted, the cuff pressure will vary periodically in synchrony with the cyclic expansion and contraction of the brachial artery. The actual *SP*, *DP*, and *MAP* values can be calculated from the raw data using an algorithm (Wang et al. 2002). Height-based and slope-based algorithms are two commonly used methods to identify the *SP* and *DP* values. In the height-based approach, the ratio of the amplitude and maximum value of the amplitude is computed, and the *SP* and *DP* are identified by comparing the ratios before and after the maximum amplitude with a certain ratio (Lee et al. 2002; Lin et al. 2003). The slope-based approach applies the derivation of the oscillation amplitude curve with respect to the cuff pressure. There are several ratio selection criteria for the height-based approach. Many investigators take a fraction of 40 and 60% out of the maximum amplitude (Ball-Llovera et al. 2003). Nippon Colin Ltd (Sapinski 1994) selects a fraction of 55% for both *SP* and *DP* values. Geddes (1991) observed that fractions of 50% for the *SP* and 80% for the *DP* are the best correlated values with the AUS method. Cuff Link uses

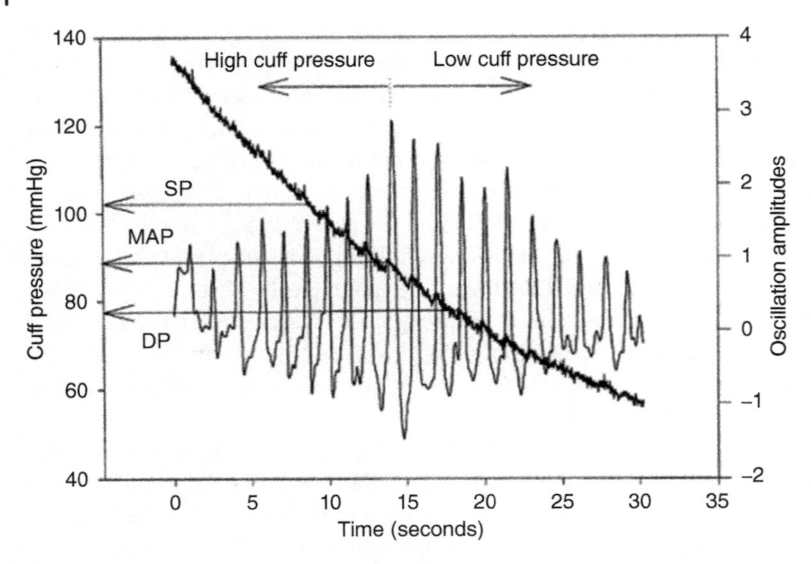

Figure 2.20 Cuff pressure signal (dense decreasing line) and oscillation waveform (thin line).

fractions of 50 and 67% for the *SP* and *DP*, respectively. *BP* pump takes 54 and 59% for the *SP* and *DP*, respectively (Kim-Gau 1997).

Figure 2.20 shows an example of cuff pressure measured using the OSC method. The air pressure of the arm cuff is recorded and plotted as an oscillation waveform. The *SP* and *DP* are obtained at the maximum and minimum slope of the curve (Sapinski 1997). Limitations of OSC include the motion artifacts (caused by respiration, specking, involuntary, and voluntary movement) and a large number of CVDs. For example, arrhythmia will lead to irregular oscillation amplitude.

AUS and OSC are the two most commonly used methods for measuring *BP* in commercial *BP* monitoring systems (Nissila et al. 1998; Ball Llovera et al. 2003; Lin et al. 2003). Various research groups have developed their new *BP* measurement algorithms to improve the measurement accuracy and stability based on AUS and OSC approaches. A group of researchers at the Institute of Biomedical Technologies (IBTec), Auckland University of Technology, New Zealand, has developed various new noninvasive *BP* measurement algorithms that are based on height-based, short-time Fourier transform (STFT) and ANN classification algorithms. After comparing and validating the newly developed methods with the Association for the Advancement of Medical Instrumentation (AAMI) and British Society of Hypertension (BSH) standard protocols, they selected the ANN classification algorithm using two ANNs in series and principle component analysis (PCA) techniques for further development. This newly developed ANN model has passed both AAMI (2003) and the BSH standard protocols, and the findings have been applied to a developed *BP* measurement device that is owned by Pulsecor Ltd.

2.6.4 Proposed Blood Pressure Measurement Method

Figure 2.21 shows an experimental system implemented by the AUT Institute of Biomedical Technologies (Mookerjee et al. 2007; El-Aklouk et al. 2008; Mookerjee et al. 2010). During data collection processing, raw data from the human subject's brachial artery were

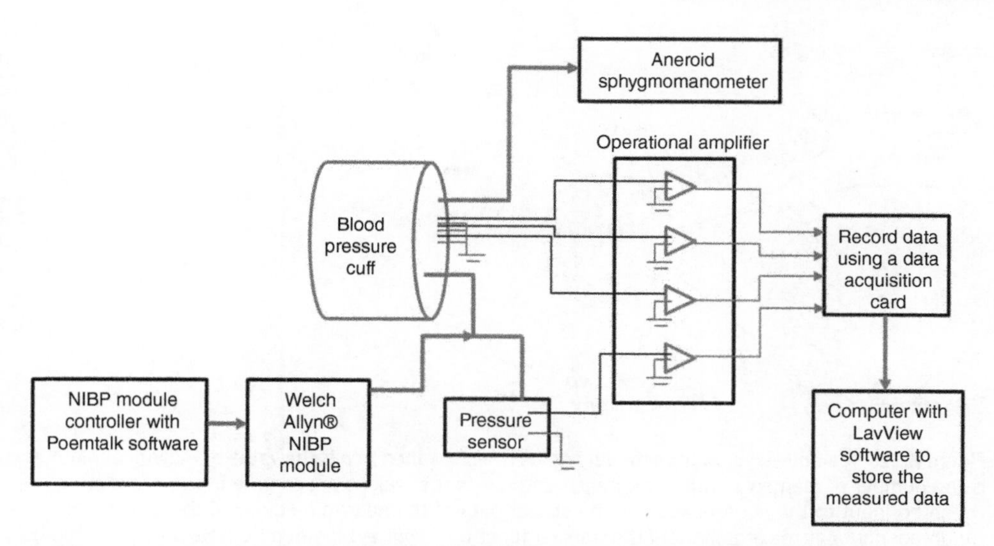

Figure 2.21 Flow chart of experimental system setup at IBTec for *BP* data collection, and the figure is reproduced based on the information described in Lin (2007). *Source:* Based on Mookerjee et al. 2007, 2010; El-Aklouk et al. 2008.

measured and enlarged through an amplifier, and a data acquisition card was then used to convert the analog signal to digital data. A computer with installed LabView software was used to store all measured data from the data acquisition card, and the MATLAB program was chosen to analyze the measured data (Lin 2007). In their experimental setups, three different sizes of *BP* cuffs were selected. Piezoelectric film sensors (DT1, DT2, and DT4) from Measurement Specialties (VA, USA) were attached to the surface of the cuff to collect signals from the brachial artery. DT1 and DT2 were placed on the inside wall of each cuff in the axial direction. The DT4 piezoelectric sensor was placed on the outside wall of each cuff in the circumferential direction. The inside and outside sensors were attached to the cuff using cyanoacrylate. Signals were measured using different voltages generated across the positive and negative terminals of each sensor.

2.7 Artificial Neural Network Classification

The mathematical model of ANN contains a number of several neuron processors, which is similar to the biological neurons in the brain. The ANN gains knowledge through a training process and a weighting process that store the obtained knowledge in the long-term memory. The neuron is the fundamental element of an ANN, which receives input data, calculates the weighted sum of the input data, computes the transfer function, and sends the output data (Negnevitsky 2002). Figure 2.22 displays a diagram of a single-layer neuron mode. An ANN is a multilayer structure that consists of an input layer of source neurons, at least one or more hidden layers of computational neurons, and an output layer of computational neurons (Negnevitsky 2002). An ANN model requires a training process, and weights are adjusted before it matches the target.

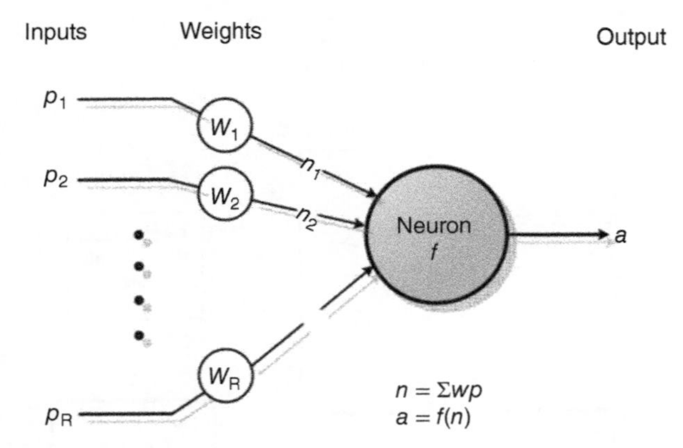

Figure 2.22 A single-layer neuron model and its transfer function p is the value of individual input, R is the number of elements within the input vector, w is the weight of separate input, n is the net weighted input to the neuron, and a is the output data of the neuron model, and the sum of the weighted input n is the argument of the transfer function f. *Source:* Modified from Negnevitsky (2002).

Figure 2.23 shows the new *BP* measurement model that the IBTec researchers developed. It has two ANN models. Figure 2.24 describes the flow chart of features extraction (Hu and Hwang 2002).

Three signals (inside, outside, and pressure sensors) were selected as the input data for the first ANN, and they were tested separately at each measurement. The second-order Butterworth low-pass filter (LPF) with corner frequencies 0.5–2 Hz was applied to the input signals to determine the heart rate (*HR*) as:

$$HR = \frac{1}{T_2 - T_1} \tag{2.35}$$

where T_1 and T_2 are the two peak values during the heartbeats.

The mean heart rate (*MHR*) is defined as:

$$MHR = \frac{\sum HR}{N} \tag{2.36}$$

where N is the total number of heartbeats.

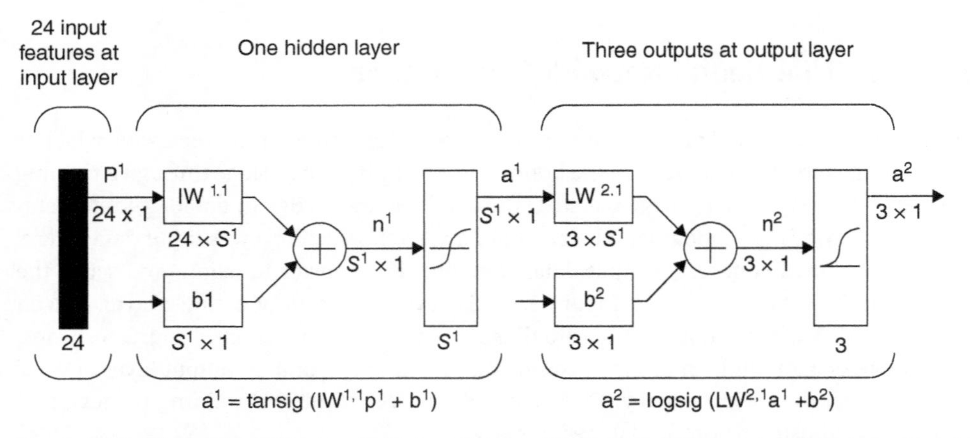

Figure 2.23 Network architecture built for ANN classification algorithms, where tansig and logsig are the Tan-Sigmoid and Log-Sigmoid transfer functions.

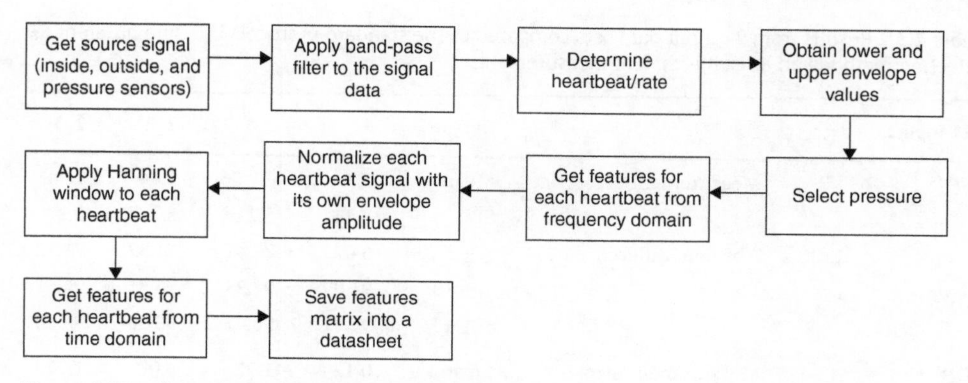

Figure 2.24 Flow chart of features extraction.

At each heartbeat, the upper and lower envelopes were generated, and the difference between the upper and the lower envelopes was calculated as the envelope amplitude. A MATLAB program was developed to select the pressure (Lin 2007), and features were obtained for each heartbeat from the frequency domain. The maximum envelope amplitude at each heartbeat was chosen as one of the input features. Each heartbeat signal was normalized using the equation:

$$y = \left(\frac{x - x_{min}}{x_{max} - x_{min}}\right)(y_{max} - y_{min}) + y_{min} \tag{2.37}$$

where x_{min} and x_{max} are the minimum and maximum measured input variable values, and y_{min} and y_{max} are the minimum and maximum scaled variable values.

The Hanning window was applied to the heartbeat signal before features were extracted from the time domain and brought down any signal outside the chosen range to zero.

The rate of change (*RoC*) in Figure 2.25 can be obtained by:

$$RoC = \frac{\Delta Amp}{\Delta t} = \frac{Change\ in\ amplitude}{Change\ in\ time} \tag{2.38}$$

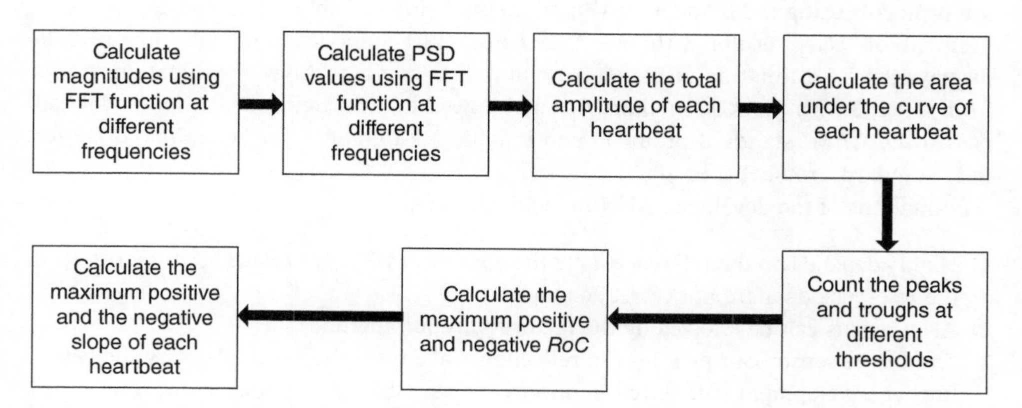

Figure 2.25 Feature extraction functions.

Table 2.3 Results from 21 input data sets compared to the standard protocols by using different *BP* selection methods on 86 subjects, 258 measurements.

21 Inputs		**Net**	**1**	**2_1**	**2_2**	**2_3**
SP	Measurement error	Mean	−3.17	−2.06	1.44	1.44
		SD	8.33	5.21	5.27	5.27
	Absolute difference (%)	≤±5	64.73	72.48	71.32	71.32
		≤±10	91.09	92.25	96.51	96.51
		≤±15	96.51	98.45	98.06	98.06
DP	Measurement error	Mean	0.12	1.77	5.02	1.77
		SD	7.30	6.17	6.33	6.17
	Absolute difference (%)	≤±5	66.67	63.95	45.35	63.95
		≤±10	89.15	89.53	81.40	89.53
		≤±15	96.90	96.12	94.96	96.12
Standard (*SP*/*DP*)	AAMI	Pass/fail	F/P	P/P	P/F	P/P
	BSH	Grade	A/A	A/A	A/C	A/A

Source: Based on Bland et al. (1999).

Various methods were applied for the training process to improve the generalization of the ANNs, including Levenberg–Marquardt backpropagation, quasi-Newton backpropagation, and Bayesian regularization (Negnevitsky 2002). PCA (Diamantaras and Kung 1996) was applied to reduce the number of input data, but it failed to provide good results. The output data of the first ANN were used as the input data for the second ANN, and the outside sensor signal was selected to analyze the measured *BP* value, which is the second value of the output data from the trained ANN. An adaptive linear neuron network and feedforward backpropagation network were applied to construct the second ANN. A total of 268 measurements from 86 human subjects were collected to validate the ANN model, and results were compared with the AAMI and BSH standard protocols using the Bland and Altman method (1999). Table 2.3 compares the measurement of ANN results with the AAMI and BSH standard protocols. Figure 2.26 displays the *BP* estimation of Bland–Altman plot to compare ANN classification results from Net 2_3 with the AUS method. Results showed that the developed ANN model had passed the AAMI standard protocol, and it achieved a grade A for both *SP* and *DP* for BSH standard protocol (Lin 2007).

Limitations of the developed ANN model including:

1) Highly depends on the reference data; the algorithm will not provide accurate results if the reference data are not correct enough.
2) ANN results can be affected by the feature selection methods.
3) The inside sensor can pick up the reflection wave from the brachial artery; however, further development is required to improve the sensitivity and noise reduction.

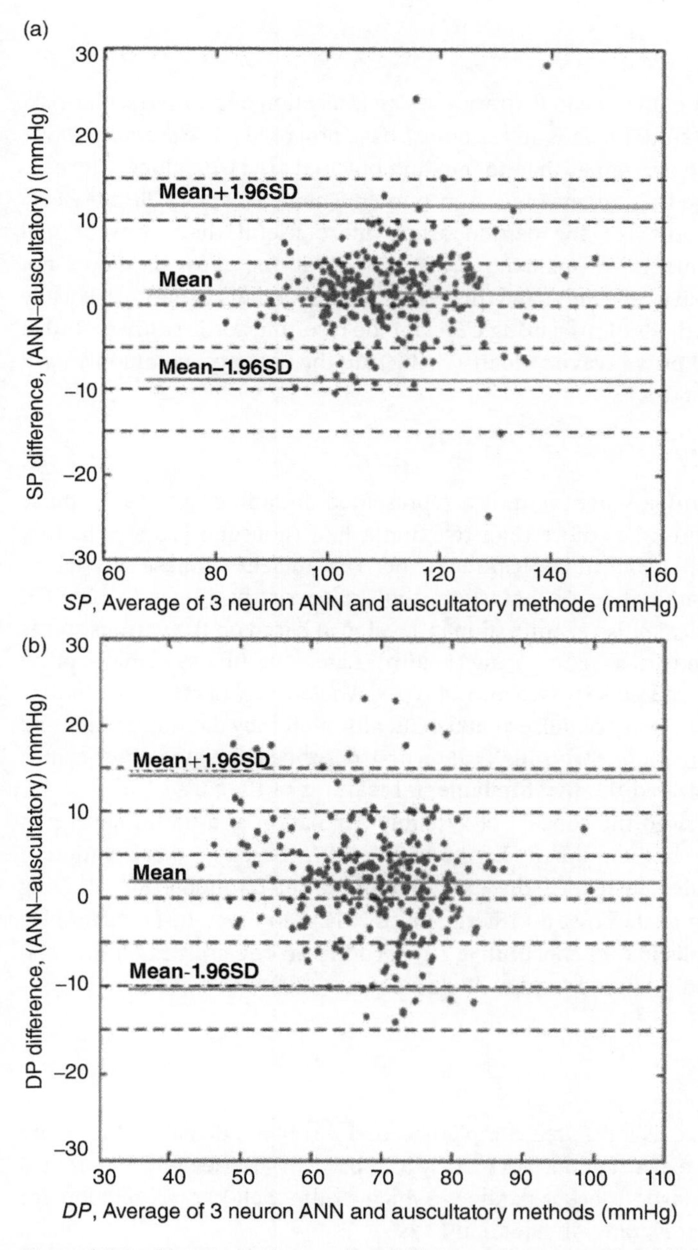

Figure 2.26 Bland and Altman plot of 21 input data sets with 3 hidden layer neurons in the ANN and AUS result comparison from 86 subjects, 258 measurements. (a) Systolic blood pressure (b) diastolic blood pressure.

2.8 Pulse Wave

Although *BP* measurement in the brachial artery gives an indication of cardiovascular risk, it has several limitations. Many clinicians and scientists have proposed pulse pressure waves for assessing cardiovascular risk rather than to measure brachial artery *BP* alone. The central pressure is the pressure that targets organs encounter. This makes the *CBP* measurements in the ascending aorta or the carotid artery more useful than conventional brachial pressure measurements for predicting CVDs. The *BP* in the upper limb does not represent the *CBP* due to wave reflection, and this fact made researchers to enthusiastically investigate noninvasive methods of measuring *CBP* and the resulting aortic stiffness. **Pulse wave analysis** (PWA) and **pulse wave velocity** (*PWV*) are the two most commonly used methods to monitor *CBP* and AT.

2.8.1 Pulse Wave History

Pulse wave, also called **pulse waveform,** is a represented digraph of a series of pulse data series. The arterial pulse has more than two and a half thousand years of history and is the most fundamental sign in clinical medicine. The ancient Chinese physicians described the physical examination of the arterial pulse at wrist by finger at 500 BCE (Fan et al. 2011). The physical pulse examination is developed based on the various stages of interaction between Yin (disease) and Yang (health). Later, the first systematic pulse literature (300 CE) categorized pulse waves into 24 types. Ancient Greeks started to notice the rhythm, strength, and velocity of pulse at 400 BCE, although they did not understand the notion of wave propagation. Struthius (1555) first described the pulse wave in a graphic form and demonstrated the five fundamental features of the pulse.

Harvey et al. first introduced the concept of systemic circulation and the properties of blood being pumped to the brain and body by the heart (1957). Based on conservation of mass and momentum, Euler developed the one-dimensional (1-D) model for studying arteries wave propagation (van de Vosse and Stergiopulos 2011). However, he failed to solve the equations due to unrealistic wall constitutive laws. Young first demonstrated the relation of wave speed to blood properties and wall elasticity as (1809):

$$c = \sqrt{\frac{A}{\rho C}} \tag{2.39}$$

where c is the wave speed, C is local area compliance, and ρ is blood density. Moens and Korteweg published their experimental work on wave speed in arteries and theoretical study on wave speed in elastic tubes separately, which is the well-known relation for thin-walled arteries as demonstrated in equation (2.39).

Bright conducted the first investigation of high arterial tension and its relationship with human health care (Bright 1827). A scientific basis did not arise until Marey (1860) and then Mahomed (Mackenzie et al. 2002) investigated a level-based sphygmograph to measure the arterial pulse. This was the first device that records the pulse wave. Riva-Rocci proposed the cuff sphygmomanometer for measuring the systolic and diastolic pressures to replace the graphic methods (1896). Fisher further confirmed that the *SP* has the ability to predict CVDs (1961).

The difference between pressure waves in central and peripheral arteries was recognized in Mahomed's studies. McDonald further studied introducing transfer functions to characterize properties of vascular beds in the frequency domain (1960). Womersley established the validity of assuming linearity in the arterial tree (1957). Their contributions have led to the techniques for PWA.

In the early twentieth century, *DP* was considered a better guide to elevated peripheral resistance and hypertensive disease due to its value closer to the mean pressure than *SP*, which Mackenzie endorsed (1902). The Framingham study group (1981) questioned this view (Franklin et al. 1999). They suggested that *DP* was the best predictor of CVD in subjects under 40 years of age; however, for people over 60 years old, *PP* was the best predictor (Siebenhofer et al. 1999). The Systolic Hypertension in the Elderly Program (SHEP) group (1991) showed that cardiovascular events are directly related to *SP* but not *DP* in older subjects, and for any given *SP*, events are inversely related to *DP* (Lasance et al. 1976). The results indicate that the increased *PP* plays a significant independent risk factor for cardiovascular events, which has been confirmed by many researchers worldwide.

Mackenzie (1902) used Mahomed's popular Dudgeon sphygmogram to reduce artifacts. Compared with this sphygmogram, the modern tonometer systems use piezoelectricity and are far more accurate, reliable, and easier to use. Drzewiecki (1983), Nichols and O'Rourke (1991) and O'Rourke et al. (1992) have adopted these systems for use in the vascular field. Since the beginning of the twentieth century, sphygmogram has been used to describe heart block and effects of some medication and hypertension (Mahomed 1874, 1877). It is also used for detecting arterial senility and increased risk of premature death (Postel-Vinay 1996).

2.8.2 Pulse Wave Types

The pulse wave is formed by combining the incident wave and the reflected wave from the periphery. Different vessels produce different pulse waves dependent on the pattern of left ventricular ejection, wave reflection, wave dispersion, viscoelastic properties of the artery, and the blood's viscosity. Dawber divided notch of pulse wave that is considered the indicator of arterial stiffness into four categories, as shown in Figure 2.27 (Dawber et al. 1973). Bates studied different types of pulse wave and discussed the cause of each pulse wave (see Table 2.4) (Bates and Hoekelman 1987).

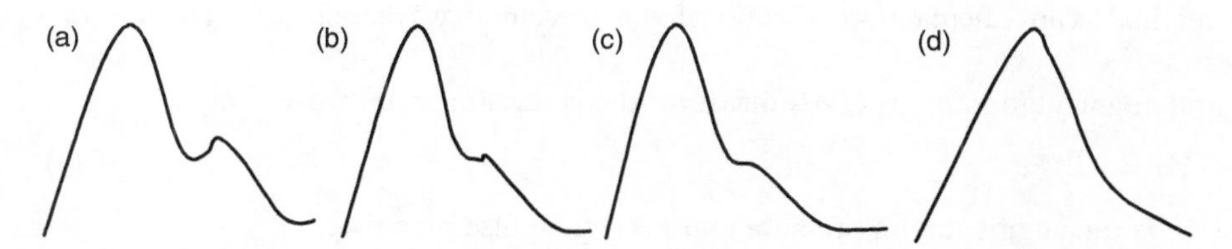

Figure 2.27 Waveforms based on dicrotic notch: (a) a distinct incisura is inscribed on the downward slope, (b) no incisura develops but the line of descent becomes horizontal, (c) no notch is present but a well-defined change in the angle of descent is observed, and (d) no evidence of a notch is seen. *Source:* Modified from Dawber et al. (1973).

Table 2.4 Pulse types.

Pulse type and shape	Physiological cause	Possible diseases
Normal		
Small, well possible diseases	Decreased stroke volume, increased peripheral resistance	Heart failure, hypovolemia, and severe aortic stenosis
Large and bounding	Decreased peripheral resistance and compliance	Fever, anaemia, hyperthyroidism, aortic regulation, bradycardia, heart block, and atherosclerosis
Bisferiens	Increased arterial pulse with double systolic peak	Aortic regulation, aortic stenosis, and regurgitation hypertropic
Pulsus alternans	Pulse amplitude varies from peak to peak, rhythm basically regular	Left ventricular failure

The pulse wave can be detected using tonometry by compressing the artery between a micro-manometer tapped probe and the underlying structures. Modern medical devices allow the computers to calculate the central aortic pressure and perform pulse pressure waveforms from the radial or carotid, and this process requires a "generalized transfer function." Many transfer functions were derived from recording both the peripheral and central ascending aortic waveforms.

2.8.3 Augmentation Index

AI measures the enhancement (augmentation) of the central aortic pressure by a reflected pulse wave (the area in the center of Figure 2.28). Both invasive and noninvasive measurement methods can calculate *AI*, a ratio of the measured wave reflection and arterial stiffness.

Central augmentation index (*CAI*) measures the central pressure waveform:

$$CAI = AP/PP \tag{2.40}$$

where *AP* is the augmentation pressure and *PP* is the pulse pressure.

Peripheral augmentation index (*PAI*) can be expressed as:

$$PAI = P2/P1 \tag{2.41}$$

where *P*1 and *P*2 are the early and the late *SP*, respectively, as shown in Figure 2.29.

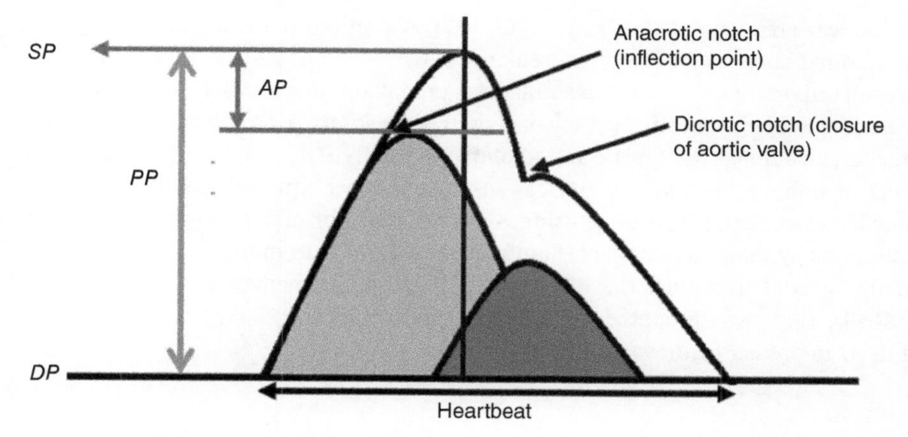

Figure 2.28 Reflected pulse wave.

AI normally depends on: (i) length of the cardiac cycle, which depends on the heartbeat rate; (ii) duration of systole and diastole; (iii) *PWV*, (iv) incident waveform; and (v) amplitude of the reflected pulse wave. *AI* is also influenced by gender, age, height, heartbeat rate, body fat content, and many other factors. On average, *AI* value is higher in women than that in men. Both *AI* and *CBP* increase with age and determine the arterial stiffness (Munir et al. 2008). The reflected wave returns to the heart earlier in stiffer arteries, and the augmentation of the forward wave occurs at lower pressure, which causes the increased *AI* (Fantin et al. 2006). Clinical trials suggest that the timing of the reflected wave also relates to the patient's height. Shorter subjects require a faster return time for reflected waves, leading to an increase in central pressure augmentation (Fantin et al. 2006). *CAI* increases with a slower heartbeat rate (Mcgrath et al. 2001) and positively correlates with cholesterol (Wilkinson et al. 2002; Wilkinson et al. 2004). Although *AI* is not affected by body mass index (BMI), increased body fat content is strongly associated with increased large artery

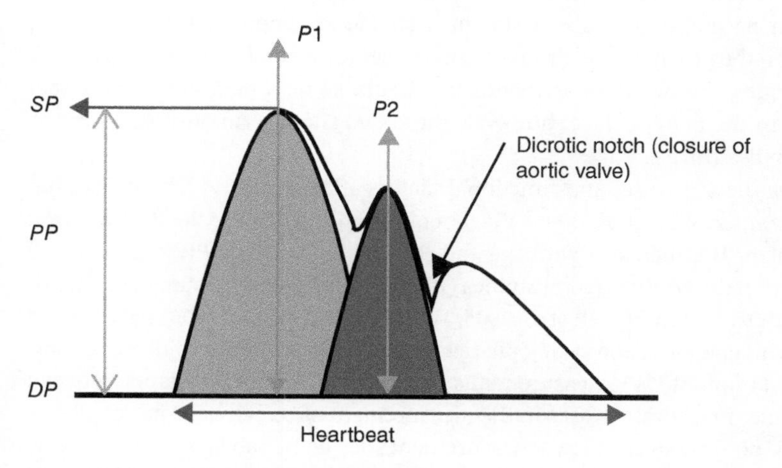

Figure 2.29 An example of pulse wave.

stiffness and wave reflection (Nürnberger et al. 2002). Variations in the brachial artery also produce minimal changes for the measurement of *AI*.

AI plays an important role in cardiovascular risk predictions due to: (i) *AI* is a predictor of adverse cardiovascular events in a variety of patient populations and higher *AI* is associated with target organ damage (O'Rourke and Avolio 1980); and (ii) *AI* can distinguish between the effects of different vasoactive medications when upper arm *BP* and *PWV* do not (Wykretowicz et al. 2007). However, using *AI* as an indicator of vascular properties has raised questions by many researchers (Fantin et al. 2007) due to more than one reflected wave in the arterial tree, then the definition of a single augmentation is not accurate (Brown 1999). *PWV* was proposed by O'Rourke and Franklin (2006) as an alternative method to *AI* to predict cardiovascular risk.

2.8.4 Pulse Wave Velocity

Pulse Wave Velocity (*PWV*) can be calculated using the Moens–Korteweg equation, which is widely used in clinical studies (Nichols and O'Rourke 1991):

$$PWV = \sqrt{Eh/2r\rho} \tag{2.42}$$

where E is Young's modulus of elasticity, h is the thickness of the vessel wall, r is the internal radius of the vessel, and ρ denotes the density of blood, approximately 1.05.

Based on the Moens–Korteweg approach, Bramwell and Hill (1922) proposed a new method to measure the velocity of the forward propagated arterial pressure wave, which is well known as the Bramwell–Hill model:

$$PWV = \sqrt{1/\rho D} \tag{2.43}$$

where ρ is the blood density (1059 kg/m^3) and D is dispensability:

$$D = \Delta V/(\Delta PV) \tag{2.44}$$

Practically, *PWV* can be calculated as the ratio of the distance and traveling time of the wave between two measuring sites of the pulse. Using invasive measurement methods, *PWV* can be measured between the proximal and distal sites in the same line of the same artery. Using noninvasive techniques, the two measurement sites should lie on a peripheral artery that can be palpated from the body surface; however, these two sites do not always lie in the same line that the pulse travels.

Figure 2.30 shows the structural and functional factors affecting *PWV*. Clinical studies (Kelly et al. 1989; Vaitkevicius et al. 1993; Vlachopoulos and O'Rourke 2000; Wilkinson et al. 2001) found that *PWV* increases with age and *BP*, and both *AI* and *PWV* are reduced with the treatment of reduced *BP*. Several studies (Blacher et al. 1999; Meaume et al. 2001; Cruickshank et al. 2002; Sutton-Tyrrell et al. 2005; Willum-Hansen et al. 2006) have proved that *PWV* can be an independent and strong predictor of cardiovascular morbidity and mortality. However, limitations of *PWV* measurements including: (i) the short distance between the two points provides inaccurate results in the measurement of velocity; (ii) it is challenging to measure *PWV* noninvasively as two measurement sites do not always lie in the same line that the pulse travels; and (iii) *PWV* is also affected by the age and *BP*.

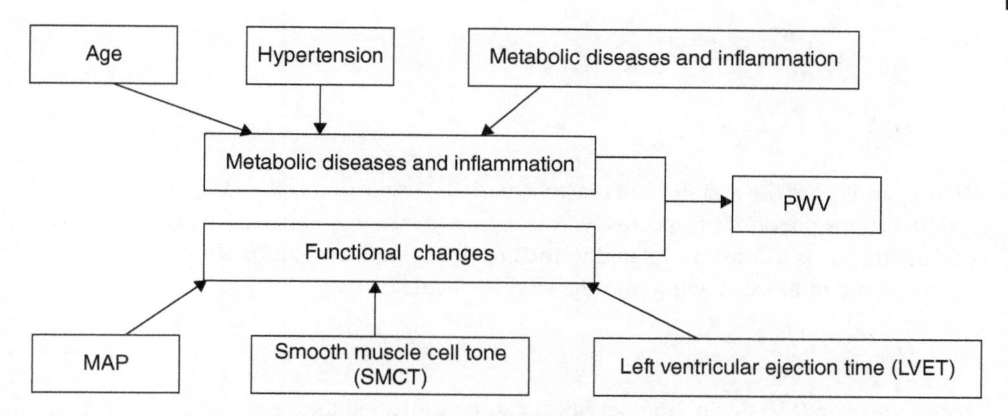

Figure 2.30 Structural and functional factors affecting *PWV*.

IBTec researchers have developed a computer model (Mookerjee et al. 2010) that can simulate the pressure wave propagation in the human central arterial tree. With applying the clinical data, the model calculates and compares the carotid–femoral *PWV* using three techniques: phase slope method, foot-to-foot technique, and the theoretical principle of *PWV*. Simulation results suggested that the phase-slope method is more suitable for implementation in *PWV* monitors. Details of the related work are reported (Mookerjee et al. 2007; El-Aklouk et al. 2008).

2.8.5 Arterial Stiffness Index

Arterial stiffness index (*ASI*) was recently proposed as an evolution for atrial stiffness in older patients with essential hypertension (Altunkan et al. 2005). It is an estimate of the *PWV* about artery stiffness, which is highly related to the time delay between the systolic and diastolic components of the waveform and the subject's height:

$$ASI = h/\Delta T_{DVP} \tag{2.45}$$

where h is the subject's height, and ΔT_{DVP} is the time interval between the systolic and diastolic peaks of the pulse wave contour.

Previous research findings (Altunkan et al. 2005) showed that *ASI* is correlated with age, weight, and *SP*, and higher *ASI* appears typically in people who have CVD compared to healthy people. In general, illness and risk factors have more impact on *ASI*, which makes *ASI* perceptible in diagnosing arterial stiffness. It is shown that *ASI* is a significant factor in *PWA* to detect the degree of arterial stiffness. Mars Medical (Taiwan) developed a medical device "VitalVision" to measure the *ASI* in the upper arm using computerized oscillometry.

2.8.6 Cardiac Output

Cardiac output (*CO*) is the amount of blood ejected by the left ventricle in one minute and is highly related to the heart rate, contractility, preload, and afterload. It can be calculated using the pulse contour analysis method (Vincent 2008), and the volume of blood flowing into the artery Q_{in} is defined as:

$$Q_{in} = AC\frac{dP}{dt_1} + \frac{P - P_v}{R} \tag{2.46}$$

$$AC\frac{dp}{dt_2} + \frac{P - P_v}{R} = 0 \tag{2.47}$$

where t_1 and t_2 are the systolic and diastolic periods, respectively, P and P_v are the arterial and the venous pressures, respectively; R is the peripheral resistance of the cardiovascular system, and AC is a constant related to the arterial compliance, $AC = dV/dP$.

The volume of blood flowing into the vein is calculated as:

$$Q_{out} = \frac{P - P_v}{R} \tag{2.48}$$

Stroke volume (*SV*) during the heartbeat can be expressed as:

$$SV = AC(SP^* - DP) + \frac{A_s}{R} \tag{2.49}$$

$$AC \cdot (DP - SP^*) + \frac{A_d}{R} = 0 \tag{2.50}$$

where *DP* and *SP* are diastoles and systolic *BP*s.

Researchers at Pathophysiological Laboratory Netherlands investigated continuous *CO* monitoring with pulse contour during cardiac surgery (Jansen et al. 1990). They used pulse contour and thermodilution to calculate *CO* 8–12 times during the process. Results suggested a linear regression between pulse contour and thermodilution methods. The *CO* is calculated by pulse wave factors and is accurate even when the heart rate, *BP*, and total peripheral resistance change. Rödig et al. (1999) applied pulse wave factors and thermodilution techniques to calculate the *CO* from two groups of patients with different ejection fractions (ejection fraction in Group 1 is greater than 45% and it is less than 45% in Group 2). Results showed no meaningful difference for *CO*, but the significant difference was achieved when systemic vascular resistance increased by 60% in the short period after an operation. These findings indicated that *CO* is a good complement to SI for analyzing cardiovascular events. *CO* is highly related to age, weight, and *SP*.

2.8.7 Pulse Wave Analysis Methods

PWA can accurately measure *CBP* and identify the difference between each measured pulse wave component. Many researchers have used *PWA* because it is easier and safer to obtain than other signals and can achieve higher accuracy results considering the related conditions. Referring to Figure 2.31, there are several parameters used to analyze central *PWA* (Dulbecco 1997):

1) Augmentation index (*AI*): the percentage increase in *BP* because of the earliness in the reflected wave about the pulse pressure, $AI = AP \times 100\,\% / CPP$.
2) Augmentation rate (*AR*): the percentage increase in *BP* because of the earliness in the reflected wave with relation to the forward pressure, $AR = AP \times 100\,\% / P_i \cdot DBP$.
3) Augmentation pressure (*AP*): the increase in *BP* to the earliness of the reflected wave, $AP = CSBP - P_i$.

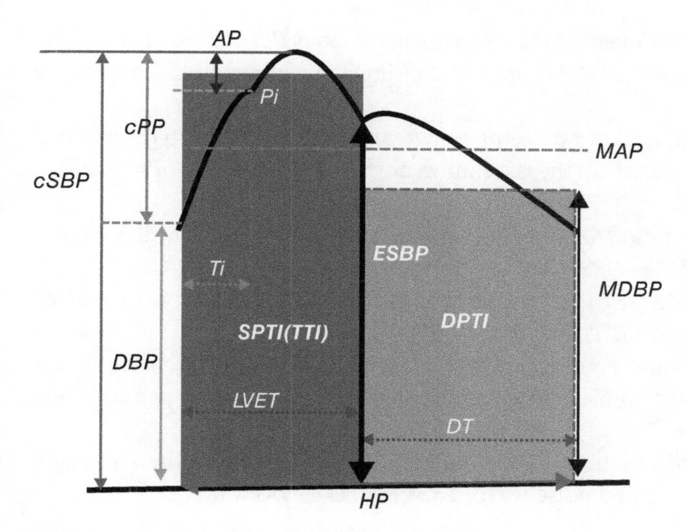

Figure 2.31 Parameters defined in central *PWA*.

4) Blood pressure at the inflection point (P_i): the *BP* corresponding to the point where the backward wave starts superimposing onto the forward wave.
5) Central pulse pressure (*CPP*): the pulse pressure such as the system-diastolic change in arterial pressure, $CPP = CSBP - DBP$.
6) Central systolic blood pressure (*CSBP*), which is the maximum *BP* in systole.
7) Diastolic blood pressure (*DBP*): the *BP* in end diastole.
8) End systolic blood pressure (*ESBP*): the *BP* at the end of systole.
9) Travel time of the reflected wave (T_i): the time delay of the reflected waveform corresponding to P_i.

Several critical parameters in *PWA* including:

1) Amplification phenomenon (*Ampl*): the difference between *SBP* measured in the brachial artery (*pSBP*) to *SBP* in the ascending aorta (*cSBP*), $Ampl = pSBP - cSBP$.
2) Diastolic pressure time index (*DPTI*): area included between pulse wave in ascending aorta and pressure wave in the left ventricle in diastole (*LVDP*), which represents the myocardial oxygen supply, $DPTI = (MDBP - LVDP) \cdot DT$.
3) Diastolic time (*DT*): duration of the diastolic phase.
4) Diastolic time fraction (*DTF*): diastolic time as a fraction of the heart period.
5) Form factor (*FF*): the ratio between *MPP* and *PP* is an attempt to "quantify" pulse waveform, $FF = MPP/cPP$.
6) Heart period (*HP*): duration of the cardiac cycle, corresponding to the R'– R′ interval of electrocardiography (ECG).
7) Left ventricular ejection time (*LVET*): duration of the systolic phase.
8) Mean arterial pressure (*MAP*): mean of the single instantaneous *BP*.
9) Mean diastolic blood pressure (*MDBP*): mean of the single instantaneous *BP* during the diastolic phase.

10) Mean pulse pressure (*MPP*): mean of the single instantaneous *PP*, $MPP = MAP - DBP$.
11) Mean systolic blood pressure (*MSBP*): mean of the single instantaneous *BP* during the systolic phase.
12) Pulse pressure amplification (*PPA*): percentage increase in *PP* measured in the brachial artery (*pPP*) for *PP* measured in the ascending aorta (*cPP*), $PPA = (pPP - cPP) \times 100\,\% / cPP$.
13) Subendocardial viability ratio (*SEVR*): the balance between subendocardial oxygen supply and demand, $SEVR = DPTI/SPTI$.
14) Systolic pressure time index (*SPTI*): area subtending the systolic phase; it means the myocardial oxygen demand, $SPTI = MSBP \cdot LVET$.
15) Tonometric diastolic pressure time index (*TDPTI*): area subtending the diastolic pulse wave; it is basically a *DPTI* which does not take ventricular diastolic pressure into account, $TDPTI = MDBP \cdot DT$.
16) Tonometric subendocardial viability ratio (*TSEVR*): the ratio between the areas subtending by diastolic and systolic pulse wave, $TSEVR = TDPTI/SPTI$.

PWA derives an aortic pulse pressure wave from the radial artery wave via a mathematical transfer function, and most *PWA* studies focused on the time domain (Baker et al. 1997; Dulbecco 1997; Lee et al. 2002), the frequency domain (Sapinski 1994, 1997), and harmonic analysis of pulse waves (Farrow and Stacy 1961; Wang et al. 2010). Fourier transform and wavelet transform are two widely used transforms in the analysis of pulse waves. Using Fourier transform, amplification of the central pressure wave between two sites can be plotted in terms of amplitude and phase. Significant features can be detected using Fourier transform from similar series data with big differences in the frequency domain. Inverse Fourier transform is also used to recover the series signal after analysis of the frequency domain data. Due to the relatively small diastolic size, which is easily affected by systolic, it is challenging to detect features of the diastolic pulse wave at the frequency domain. Previous studies (Silva et al. 2009) suggested that fast Fourier transform (FFT) is useful in detecting cardiac diseases, but high accuracy results would not be obtained with the frequency domain only method.

Wavelet transform uses the time and frequency domains together to describe the variability. It has been widely applied in many fields, including coherent structures in turbulent flows (Farge 1992), atmospheric cold fronts (Gamage and Blumen 1993), and dispersion of ocean waves (Meyers et al. 1993), to analyze tropical convection (Weng and Lau 1994), wave growth and breaking (Liu 1994), El Niño–Southern Oscillation (Gu and Philander 1995), and central England temperature (Baliunas et al. 1997). Wavelet provides multi-resolution analysis to the source data that makes the result adequate for feature detection. Clinical trial results observed that wavelet transform could easily detect diastolic components, and it has a significant impact on slope changes of continuous values. In the nineteenth century, clinicians and scientists obtained *PWA* information based on basic mathematic algorithms, which divide the wave into two parts (increasing and decreasing) and calculate the wave's height and area. Other techniques such as calculus, hemodynamic, biomathematics, and pattern recognition are also applied in *PWA* (Fan et al. 2011).

With the application of model techniques, *PWA* information can be obtained using two main approaches: point-based and area-based approaches. The point-based method picks up top and bottom points from different components of the waveform or derivative curve and calculates the medical significance of those points. Area-based analysis specialized in blood volume monitoring such as *CO*. The attempt for getting *CO* from pulse waves was first studied around 100 years ago (Erlanger and Hooker 1904). The pulse wave is the result of the interaction between stroke volume and arteries resistance. Arterial tree models help to obtain *CO* information from pulse waves. The simplest model used in clinics contains single resistance. Other elements should be involved in the calculation, including the capacitance element and resistance element (Cholley et al. 1995).

2.9 Medical Applications of Pulse Wave Analysis

This section briefly presents various medical applications of *PWA* with particular focus on diagnosis and detection of diseases such as CVD and lung disease, and applications in traditional Chinese medicine (TCM) and prediction of preeclampsia were also reviewed.

2.9.1 Pulse Wave Analysis for the Early Detection of Cardiovascular Disease

Published clinical studies on hypertension, postmenopausal women with symptomatic coronary artery disease (CAD), and appropriate control subjects confirmed that *PWA* is useful in screening subjects for early evidence of CVD and monitoring the response to therapy (O'Rourke et al. 2001; Ghasemzadeh and Zafari 2011). Increased AT is associated with CAD and predicts cardiovascular outcomes in various populations, including healthy subjects (11–13). AT can be assessed noninvasively by measures including aortic (carotid–femoral) *PWV*, *AI*, and peripheral (brachial) *PP*. Aortic *PWV* is considered the most clinically relevant measure of AT and independently predicts cardiovascular risk.

Type 1 Diabetes (T1D) is associated with a high mortality risk from premature CAD, including coronary artery calcification (CAC) and lower extremity arterial disease (LEAD) (Laing et al. 2003). This increased risk is especially apparent in women with T1D, virtually eliminating the traditional female advantage seen in the absence of diabetes (Brooks et al. 1999). *AI*, one parameter applied in *PWA*, is elevated in T1D (Nichols and Singh 2002), which has also been found to correlate with traditional CAD risk factors, cardiovascular outcomes, CAC, and LEAD (Nurnberger et al. 2002; Weber et al. 2004; Altunkan et al. 2005) in the general population (Covic et al. 2005; Khaleghi and Kullo 2007). Prince et al. (2010) first examined the association between pulse wave reflection measurements and prevalent CVD in childhood-onset T1D and assessed the potential impact of medications that affects *PWA* measurements (Protogerou et al. 2009). Their findings suggested that greater *AP* is independently associated with prevalent CAD and estimated myocardial perfusion with low ankle-to-brachial index in T1D. The achieved measurement results may help better characterize the risk of CVD in T1D, and further examination is required.

Clinical studies (Marchais et al. 1993; Saba et al. 1993; Vlachopoulos et al. 2001; Pauca et al. 2004; Hashimoto et al. 2006) demonstrated that pulse wave contour analysis is helpful in determining the left ventricular (LV) load. Evidence (Kelly et al. 1990; Asmar et al. 2001; Hirata et al. 2005; William et al. 2006) is accumulating that antihypertensive treatment can reduce LV load through decreasing peripheral wave reflection. However, large numbers of patients are generally required to associate regression of LV mass with a decrease in brachial cuff *BP* in the treatment of hypertension. Hashimoto et al. (2006) examined the potential superiority of *PWA* over conventional *BP* measurement in predicting treatment LV mass reduction. Their research findings suggested that reduction in wave reflection is an important therapeutic strategy for reducing LV mass, which can be predicted with modest subject numbers. The major limitation of this study is that only a small number of subjects were involved.

2.9.2 Pulse Wave Analysis in Chronic Obstructive Pulmonary Disease

Chronic obstructive pulmonary disease (COPD) has been predicted to be the third most common cause of death worldwide by 2020 (Murray and Lopez 1996). It has been studied by a number of research groups. Sabit et al. (2007) recently reviewed the relationship between *PWA* and COPD. They examined the hypothesis that patients with COPD would have increased arterial stiffness, which would be associated with osteoporosis and systemic inflammation. Among these patients, 75 patients with a range of severity of airway obstruction and 42 healthy smoker or ex-smoker subjects free of CVD were involved in their investigation. All subjects underwent spirometry, measurement of aortic *PWV* and *AI*, dual-energy X-ray absorptiometry, and blood sampling for inflammatory mediators. They observed that increased arterial stiffness is related to the severity of airflow obstruction, and the patients with osteoporosis have increased aortic *PWV*. These findings may be a factor in the excess risk for CAD in COPD and increased aortic *PWV* association with systemic inflammation. Increased aortic *PWV* association with systemic inflammation indicates that age-related bone and vascular changes occur prematurely in COPD.

Limitations of this research are as follows: the study did not allow causal relationships between COPD, systemic inflammation, arterial stiffness, and CVD to be inferred, but has shown strong relationships that could be explored further, and this study did not test the relationship between aortic *PWV* and a large number of independent variables.

2.9.3 Pulse Wave in Traditional Chinese Medicine (TCM)

The pulse examination is one of the most important symbols and diagnosis diagrams in TCM, and it became a specialized technique around 250 BC (TAO 1952). Chinese medicine practitioners (Yang 1997; Li 1998) diagnosed the patient by feeling their pulse beating at the measuring point of the radial artery; however, the reliability and repeatability of pulse diagnosis highly depend on the clinician's knowledge and skills. *PWA* emerged as a quantitative evaluation of the pulse waves' morphology for clinical pulse examination in the last decades.

Several researchers have made great efforts in *PWA* and developed a new quantitative system for pulse diagnosis to build the mapping from pulse parameters to pulse types (Chiu et al. 1996; Yoon et al. 2000; Wang and Cheng 2006). Clinical studies showed that the recently developed pulse diagnosis system (Wang and Cheng 2006) (Figure 2.32) could keep enhancing the predictive accuracy of pulse types by learning the experiences of experts of TCM. The probabilistic reasoning module contains discovering dependency relationship model and parameter learning and reasoning module that is implemented via Clique Tree Propagation (CTP) algorithm (Lauritzen and Spiegelhalter 1988). This allows computation sharing among multiple queries and can satisfy the requirement of pulse diagnosis.

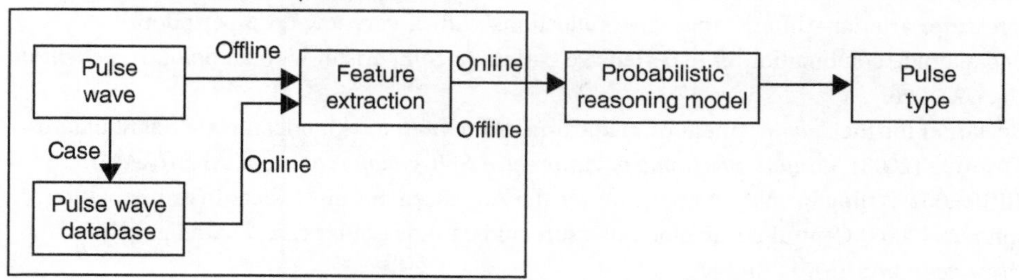

Figure 2.32 The pulse diagnosis system. *Source:* Modified from Wang and Cheng (2006.

2.9.4 Pulse Wave Analysis for the Prediction of Preeclampsia

Normal pregnancy is associated with several important changes to the maternal cardiovascular system, including a rise in *HR*, intravascular volume, and *CO*, and a corresponding reduction in vascular resistance. Few research groups (Smith et al. 2004; Macedo et al. 2008; Kaihura et al. 2009) used *PWA* to examine maternal wave reflections and arterial stiffness, and they observed that lower central *SP*, *DP*, and *AI* occur in normal pregnancy and not in a nonpregnant state. Hausvater et al. (2012) and North et al. (2011) reported that preeclampsia is associated with increased arterial stiffness indices, both during and after pregnancy, potentially contributing to the increased long-term cardiovascular risk seen in affected women. However, a recent study conducted by Carty et al. (2014) presents different results.

Carty et al. (2014) examined 180 women with ≥2 risk factors for preeclampsia at gestational weeks 16 and 28, and 17 (9.4%) developed preeclampsia. Researchers also examined an additional 30 healthy nonpregnant women to study the effects of pregnancy itself. *PWA* was performed during and after pregnancy using SphygmoCor, and *AI* was measured using EndoPAT-2000 to verify the findings. The research findings suggest that: (i) *PWA* does not provide additional information beyond brachial *BP* and maternal risk factor profile about the risk of future development of preeclampsia, when examined at gestational week 16 or 28; and (ii) *PWA* does not demonstrate that women who have suffered from preeclampsia have ongoing vascular dysfunction when examined at six to nine months postnatally.

Bibliography

Altunkan, S., Oztas, K., and Seref, B. (2005) Arterial stiffness index as a screening test for cardiovascular risk: a comparative study between coronary artery calcification determined by electron beam tomography and arterial stiffness index determined by a VitalVision device in asymptomatic subjects. *European Journal of Internal Medicine* 16, 580–584.

van Andel, C.J., Pistecky, P.V., and Borst, C. (2003) Mechanical properties of porcine and human arteries: implications for coronary anastomotic connectors. *The Annals of Thoracic Surgery* 76(1): 58–64.

Asmar, R.G., London, G.M., O'Rourke, M.F., and Safar, M.E. (2001) Improvement in blood pressure, arterial stiffness and wave reflections with a very-low-dose perindopril/ indapamide combination in hypertensive patient: a comparison with atenolol. *Hypertension* 38, 922–926.

Association for the Advancement of Medical Instrumentation. American National Standard Institute (2003) *Manual, Electronic or Automated Sphygmomanometers*, ANSI/AAMI SP10:2002. Arlington, VA: Association for the Advancement of Medical Instrumentation.

Avolio, A. (2008) Central aortic blood pressure and cardiovascular risk a paradigm shift? *Hypertension* 51(6): 1470–1471.

Avolio, A.P., Chen, S.-G., Wang, R.-P. et al. (1983) Effects of aging on changing arterial compliance and left ventricular load in a northern Chinese urban community. *Circulation* 68(1): 50–58.

Ayer, J.P., Hass, G.M., and Philpott, D.E. (1958) Aortic elastic tissue; isolation with use of formic acid and discussion of some of its properties. *AMA Archives of Pathology* 65(5): 519.

Baker, P.D., Westenskow, D.R., and Kück, K. (1997) Theoretical analysis of non-invasive oscillometric maximum amplitude algorithm for estimating mean blood pressure. *Medical and Biological Engineering and Computing* 35(3): 271–278.

Baliunas, S., Frick, P., Sokoloff, D., and Soon, W. (1997) Time scales and trends in the central England temperature data (1659–1990): a wavelet analysis. *Geophysical Research Letters* 24 (11): 1351–1354.

Ball Llovera, A., Del Rey, R., Ruso, R. et al. (2003) An experience in implementing the oscillometric algorithm for the non-invasive determination of human blood pressure. *Proceedings of the Annual International Conference of the IEEE Engineering in Medicine and Biology Society*, Cancun, Mexico (17–21 September 2003), pp. 3173–3175.

Bates, B., Hoekelman, R.A., Thompson, J.B., and Bickley, L.S. (1987) *A Guide to Physical Examination and History Taking*. Philadelphia, PA: Lippincott.

Baulmann, J., Schillings, U., Rickert, S. et al. (2008) A new oscillometric method for assessment of arterial stiffness: comparison with tonometric and piezo-electronic methods. *Journal of Hypertension* 26(3): 523–528.

Benetos, A. (1999) Pulse pressure and cardiovascular risk. *Journal of Hypertension. Supplement* 17, S21–S24.

Benetos, A., Adamopoulos, C., Bureau, J.-M. et al. (2002) Determinants of accelerated progression of arterial stiffness in normotensive subjects and in treated hypertensive subjects over a 6-year period. *Circulation* 105(10): 1202–1207.

Bergel, D.H. (1961) The static elastic properties of the arterial wall. *The Journal of Physiology* 156 (3): 445.

van Berge-Landry, H.M., Bovbjerg, D.H., and James, G.D. (2008) Relationship between waking–sleep blood pressure and catecholamine changes in African–American and European–American women. *Blood Pressure Monitoring* 13(5): 257.

Bessette, P.L. and Glagov, S. (1970) Relation of medial structure to tension in mammalian coronary and renal arteries. *Federation Proceedings* 29(2): A284.

Blacher, J., Asmar, R., Djane, S. et al. (1999) Aortic pulse wave velocity as a marker of cardiovascular risk in hypertensive patients. *Hypertension* 33, 1111–1117.

Blacher, J., Safar, M.E., Guerin, A.P. et al. (2003) Aortic pulse wave velocity index and mortality in end-stage renal disease. *Kidney International* 63, 1852–1860.

Bland, J.M. and Altman, D.G. (1999) Measuring agreement in method comparison studies. *Statistical Methods in Medical Research* 8(2): 135–160.

Bortel, V., Luc, M., Duprez, D. et al. (2002) Clinical applications of arterial stiffness, Task Force III: recommendations for user procedures. *American Journal of Hypertension* 15(5): 445–452.

Boutouyrie, P., Tropeano, A.I., Asmar, R. et al. (2002) Aortic stiffness is an independent predictor of primary coronary events in hypertensive patients: a longitudinal study. *Hypertension* 39, 10–15.

Boutouyrie, P., Achouba, A., Trunet, P., and Laurent, S. (2010) Amlodipine-valsartan combination decreases central systolic blood pressure more effectively than the amlodipine-atenolol combination: the EXPLOR study. *Hypertension* 55(6): 1314–1322.

Bramwell, J.C. and Hill, A.V. (1922) The velocity of the pulse wave in man. *Proceedings of the Royal Society of London. Series B, Containing Papers of a Biological Character* 93, 298–306.

Bright, R. (1827) *Reports of Medical Cases, Selected with a View to Illustrate the Symptoms and Cure of Diseases by a Reference to Morbid Anatomy*. London: Longman, Rees, Orme, Brown, and Green.

Brooks, B., Molyneaux, L., and Yue, D.K. (1999) Augmentation of central arterial pressure in type 1 diabetes. *Diabetes Care* 22, 1722–1727.

Brown, M.J. (1999) Similarities and differences between augmentation index and pulse wave velocity in the assessment of arterial stiffness. *QJM: An International Journal of Medicine* 92(10): 595–600.

Bund, S.J., Oldham, A.A., and Heagerty, A.M. (1996) Mechanical properties of porcine small coronary arteries in one-kidney, one-clip hypertension. *Journal of Vascular Research* 33(2): 175–180.

Carmines, D.V., McElhaney, J.H., and Stack, R. (1991) A piece-wise non-linear elastic stress expression of human and pig coronary arteries tested in vitro. *Journal of Biomechanics* 24(10): 899–906.

Carty, D.M., Neisius, U., Rooney, L.K. et al. (2014) Pulse wave analysis for the prediction of preeclampsia. *Journal of Human Hypertension* 28, 98–104.

Cecelja, M. and Chowienczyk, P. (2012) Role of arterial stiffness in cardiovascular disease. *JRSM Cardiovascular Disease* 1(4): 11.

Chang, C.C. and Bülent Atabek, H. (1961) The inlet length for oscillatory flow and its effects on the determination of the rate of flow in arteries. *Physics in Medicine and Biology* 6(2): 303.

Chen, C.-H., Nevo, E., Fetics, B. et al. (1997) Estimation of central aortic pressure waveform by mathematical transformation of radial tonometry pressure Validation of generalized transfer function. *Circulation* 95(7): 1827–1836.

Chiolero, A. (2014) The quest for blood pressure reference values in children. *Journal of Hypertension* 32(3): 477–479.

Chiu, C.-C., Yeh, S.-J., and Yu, Y.-C. (1996) Classification of the pulse signals based on self-organizing neural network for the analysis of the autonomic nervous system. *Chinese Journal of Medical and Biological Engineering* 16, 461–476.

Cholley, B.P., Shroff, S.G., Sandelski, J. et al. (1995) Differential effects of chronic oral antihypertensive therapies on systemic arterial circulation and ventricular energetics in African-American patients. *Circulation* 91(4): 1052–1062.

Chowienczyk, P.J., Kelly, R.P., MacCallum, H. et al. (1999) Photoplethysmographic assessment of pulse wave reflection: blunted response to endothelium-dependent beta2-adrenergic vasodilation in type II diabetes mellitus. *Journal of the American College of Cardiology* 34(7): 2007–2014.

Covic, A., Haydar, A.A., Bhamra-Ariza, P. et al. (2005) Aortic pulse wave velocity and arterial wave reflections predict the extent and severity of coronary artery disease in chronic kidney disease patients. *Journal of Nephrology* 18, 388–396.

Cruickshank, K., Riste, L., Anderson, S.G. et al. (2002) Aortic pulse-wave velocity and its relationship to mortality in diabetes and glucose intolerance: an integrated index of vascular function? *Circulation* 106, 2085–2090.

Dawber, T.R., Emerson Thomas, H., and McNamara, P.M. (1973) Characteristics of the dicrotic notch of the arterial pulse wave in coronary heart disease. *Angiology* 24(4): 244–255.

De Mey, J.G. and Brutsaert, D.L. (1984) Mechanical properties of resting and active isolated coronary arteries. *Circulation Research* 55(1): 1–9.

Diamantaras, K.I. and Kung, S.Y. (1996) *Principal Component Neural Networks: Theory and Applications*. New York: Wiley.

Dobrin, P.B. (1978) Mechanical properties of arteries. *Physiological Reviews* 58(2): 397–460.

Drzewiecki, G.M., Melbin, J., and Noordergraaf, A. (1983) Arterial tonometry: review and analysis. *Journal of Biomechanics* 16, 141–153.

Dulbecco, R. (1997) *Encyclopedia of Human Biology*, 2nde. San Diego, CA: Academic Press.

El-Aklouk, E. (2007) An investigation into the acoustic response of the arteries. Master thesis, Auckland University of Technology.

El-Aklouk, E., Al-Jumaily, A.M., and Lowe, A. (2008) Pressure waves as a noninvasive tool for artery stiffness detection. *Journal of Medical Devices* 2, 1–8.

Erlanger, J. and Hooker, D.R. (1904) An experimental study of blood pressure and of pulse pressure in man. *Johns Hopkins Hospital Reports* 12, 357.

Fan, Z., Zhang, G., and Liao, S. (2011) *Pulse Wave Analysis*, Advanced Biomedical Engineering, 21–40. Rijeka: Intech.

Fantin, V.R., St-Pierre, J., and Leder, P. (2006) Attenuation of LDH-A expression uncovers a link between glycolysis, mitochondrial physiology, and tumor maintenance. *Cancer Cell* 9(6): 425–434.

Fantin, F., Mattocks, A., Bulpitt, C.J. et al. (2007) Is augmentation index a good measure of vascular stiffness in the elderly? *Age and Ageing* 36(1): 43–48.

Farge, M. (1992) Wavelet transforms and their applications to turbulence. *Annual Review of Fluid Mechanics* 24(1): 395–458.

Farrow, R.L. and Stacy, R.W. (1961) Harmonic analysis of aortic pressure pulses in the dog. *Circulation Research* 9(2): 395–401.

Fisher, C.M. (1961) Anticoagulant therapy in cerebral thrombosis and cerebral embolism: a national cooperative study, interim report. *Neurology* 4(part 2): 119–131.

Franklin, S.S., Khan, S.A., Wong, N.D. et al. (1999) Is pulse pressure useful in predicting risk for coronary heart disease? The Framingham Heart Study. *Circulation* 100, 354–360.

Gamage, N. and Blumen, W. (1993) Comparative analysis of low-level cold fronts: wavelet, Fourier, and empirical orthogonal function decompositions. *Monthly Weather Review* 121(10): 2867–2878.

Gaszner, B., Lenkey, Z., Illyés, M. et al. (2012) Comparison of aortic and carotid arterial stiffness parameters in patients with verified coronary artery disease. *Clinical Cardiology* 35(1): 26–31.

Gavallér, H., Sepp, R., Csanády, M. et al. (2011) Hypertrophic cardiomyopathy is associated with abnormal echocardiographic aortic elastic properties and arteriograph-derived pulse-wave velocity. *Echocardiography* 28(8): 848–852.

Geddes, L.A. (1991) *Handbook of Blood Pressure Measurement*. Totowa, NJ: Humana Press.

Ghasemzadeh, N. and Zafari, A.M. (2011) A brief journey into the history of the arterial pulse. *Cardiology Research and Practice* 2011, 164832.

Gow, B.S. and Hadfield, C.D. (1979) The elasticity of canine and human coronary arteries with reference to postmortem changes. *Circulation Research* 45(5): 588–594.

Gow, B.S., Schonfeld, D., and Patel, D.J. (1974) The dynamic elastic properties of the canine left circumflex coronary artery. *Journal of Biomechanics* 7(5): 389–395.

Groenink, M., de Roos, A., Mulder, B.J. et al. (1998) Changes in aortic distensibility and pulse wave velocity assessed with magnetic resonance imaging following beta-blocker therapy in the Marfan syndrome. *The American Journal of Cardiology* 82, 203–208.

Gu, D. and Philander, S.G.H. (1995) Secular changes of annual and interannual variability in the tropics during the past century. *Journal of Climate* 8(4): 864–876.

Guerin, A.P., Blacher, J., Pannier, B. et al. (2001) Impact of aortic stiffness attenuation on survival of patients in end-stage renal failure. *Circulation* 103(7): 987–992.

Harvey, W. (1957) *Movement of the Heart and Blood in Animals* (trans. K.J. Franklin). Oxford: Blackwell (1628).

Hashimoto, J., Watabe, D., Hatanaka, R. et al. (2006) Enhanced late-systolic blood pressure augmentation in hypertensive patients with left ventricular hypertrophy. *American Journal of Hypertension* 19, 27–32.

Hausvater, A., Giannone, T., Sandoval, Y.-H.G. et al. (2012) The association between preeclampsia and arterial stiffness. *Journal of Hypertension* 30(1): 17–33.

Hirata, K., Vlachopoulos, C., Adji, A., and O'Rourke, M.F. (2005) Benefits from angiotensin-converting enzyme inhibitor "beyond blood pressure lowering": beyond blood pressure or beyond the brachial artery? *Journal of Hypertension* 23, 551–556.

Holejsovska, P., Peroutka, Z., and Cengery, J. (2003) Non-invasive monitoring of the human blood pressure. *Proceedings of the 16th IEEE Symposium Computer-Based Medical Systems*, New York (26–27 June 2003), pp. 301–306. IEEE.

Holzapfel, G.A., Sommer, G., Gasser, C.T., and Regitnig, P. (2005) Determination of layer-specific mechanical properties of human coronary arteries with nonatherosclerotic intimal thickening and related constitutive modeling. *American Journal of Physiology-Heart and Circulatory Physiology* 289(5): H2048–H2058.

Hope, S.A., Meredith, I.T., and Cameron, J.D. (2008) Arterial transfer functions and the reconstruction of central aortic waveforms: myths, controversies and misconceptions. *Journal of Hypertension* 26(1): 4–7.

Horvath, I.G., Nemeth, A., Lenkey, Z. et al. (2010) Invasive validation of a new oscillometric device (Arteriograph) for measuring augmentation index, central blood pressure and aortic pulse wave velocity. *Journal of Hypertension* 28(10): 2068–2075.

Hu, Y.H. and Hwang, J.-N. (2002) *Handbook of Neural Network Signal Processing*. Boca Raton, FL: CRC Press.

Jansen, J.R.C., Wesseling, K.H., Settels, J.J., and Schreuder, J.J. (1990) Continuous cardiac output monitoring by pulse contour during cardiac surgery. *European Heart Journal* 11(Suppl I): 26–32.

Jatoi, N.A., Mahmud, A., Bennett, K., and Feely, J. (2009) Assessment of arterial stiffness in hypertension: comparison of oscillometric (Arteriograph), piezoelectronic (Complior) and tonometric (SphygmoCor) techniques*. *Journal of Hypertension* 27(11): 2186–2191.

Kaihura, C., Savvidou, M.D., Anderson, J.M. et al. (2009) Maternal arterial stiffness in pregnancies affected by preeclampsia. *American Journal of Physiology-Heart and Circulatory Physiology* 297(2): H759–H764.

Kasjanovs, V., Ozolanta, I., and Purina, B. (1999) Features of biomechanical properties of human coronary arteries. *Mechanics of Composite Materials* 35(2): 155–168.

Kelly, R., Hayward, C., Avolio, A., and O'Rourke, M. (1989) Noninvasive determination of age-related changes in the human arterial pulse. *Circulation* 80, 1652–1659.

Kelly, R.P., Gibbs, H.H., O'Rourke, M.F. et al. (1990) Nitroglycerin has more favourable effects on left ventricular afterload than apparent from measurement of pressure in a peripheral artery. *European Heart Journal* 11, 138–144.

Khaleghi, M. and Kullo, I.J. (2007) Aortic augmentation index is associated with the ankle–brachial index: a community-based study. *Atherosclerosis* 195, 248–253.

Kim-Gau, N. (1997) Blood pressure measurement. *Medical Electronics* February.

Kracht, D., Shroff, R., Baig, S. et al. (2011) Validating a new oscillometric device for aortic pulse wave velocity measurements in children and adolescents. *American Journal of Hypertension* 24(12): 1294–1299.

Krafka, J. Jr. (1939) Comparative study of the histo-physics of the aorta. *American Journal of Physiology-Legacy Content* 125, 1–14.

Laffon, E., Marthan, R., Montaudon, M. et al. (2005) Feasibility of aortic pulse pressure and pressure wave velocity MRI measurement in young adults. *Journal of Magnetic Resonance Imaging* 21, 53–58.

Laing, S.P., Swerdlow, A.J., Slater, S.D. et al. (2003) Mortality from heart disease in a cohort of 23,000 patients with insulin-treated diabetes. *Diabetologia* 46, 760–765.

Lally, C., Reid, A.J., and Prendergast, P.J. (2004) Elastic behavior of porcine coronary artery tissue under uniaxial and equibiaxial tension. *Annals of Biomedical Engineering* 32(10): 1355–1364.

Lasance, H.A.F., Wesseling, K.H., and Ascoop, C.A. (1976) Peripheral pulse contour analysis in determining stroke volume. Progress Report 5, Inst. Med Phys. Da Costakade 45, Utrecht, Netherlands, pp 59–62.

Laurent, S., Caviezel, B., Beck, L. et al. (1994) Carotid artery distensibility and distending pressure in hypertensive humans. *Hypertension* 23(6 Pt 2): 878–883.

Laurent, S., Boutouyrie, P., Asmar, R. et al. (2001) Aortic stiffness is an independent predictor of all-cause and cardiovascular mortality in hypertensive patients. *Hypertension* 37, 1236–1241.

Lauritzen, S.L. and Spiegelhalter, D.J. (1988) Local computations with probabilities on graphical structures and their application to expert systems. *Journal of the Royal Statistical Society. Series B (Methodological)* 50, 157–224.

Lee, J.Y., Kim, J.K., and Yoon, G. (2002) Digital envelope detector for blood pressure measurement using an oscillometric method. *Journal of Medical Engineering and Technology* 26(3): 117–122.

Li, S. (1998) *The Lakeside Master's Study of the Pulse*. Boulder, CO: Blue Poppy Press.

Lin, H.C.V. (2007) Specialised non-invasive blood pressure measurement algorithm. Master thesis, Auckland University of Technology.

Lin, C.T., Liu, S.-H., Wang, J.-J., and Wen, Z.-C. (2003) Reduction of interference in oscillometric arterial blood pressure measurement using fuzzy logic. *IEEE Transactions on Biomedical Engineering* 50(4): 432–441.

Liu, P.C. (1994) Wavelet spectrum analysis and ocean wind waves. *Wavelets in Geophysics* 4, 151–166.

Lowe, A. (2010) Method and apparatus for producing a central pressure waveform in an oscillometry blood pressure system. US Patent US20100152593 A1.

Lustig, J.V. and Glagov, S. (1969) Relation of medial architecture to tangential tension in mammalian pulmonary trunks. *Federation Proceedings* 28(2): 685.

Macedo, M.L., Luminoso, D., Savvidou, M.D. et al. (2008) Maternal wave reflections and arterial stiffness in normal pregnancy as assessed by applanation tonometry. *Hypertension* 51(4): 1047–1051.

Mackenzie, J. (1902) *The Study of the Pulse: Arterial, Venous and Hepatic, and of the Movements of the Heart*. Edinburgh: Young J. Pentland.

Mackenzie, I.S., Wilkinson, I.B., and Cockcroft, J.R. (2002) Assessment of arterial stiffness in clinical practice. *QJM: An International Journal of Medicine* 95(2): 67–74.

Mahomed, F. (1874) The aetiology of Bright's disease and the prealbuminuric stage. *Medico-Chirurgical Transactions* 57, 197–228.

Mahomed, F.A. (1877) On the sphygmographic evidence of arterio-capillary fibrosis. *Transactions of the Pathological Society of London* 28, 394–397.

Marchais, S.J., Guerin, A.P., Pannier, B.M. et al. (1993) Wave reflections and cardiac hypertrophy in chronic uremia. Influence of body size. *Hypertension* 22, 876–883.

Marey, E.-J. (1860) *Recherches sur le pouls au moyen d'un nouvel appareil enregistreur le sphygmographie....* E. Thunot et cie.

McDonald, D.A. (1960) *Blood Flow in Arteries*. London: Edward Arnold.

McDonald, D.A. (1968) Regional pulse-wave velocity in the arterial tree. *Journal of Applied Physiology* 24, 73–78.

McEniery, C.M., Hall, I.R., Qasem, A. et al. (2005) Normal vascular aging: differential effects on wave reflection and aortic pulse wave velocity The Anglo-Cardiff Collaborative Trial (ACCT). *Journal of the American College of Cardiology* 46(9): 1753–1760.

Mcgrath, B.P., Liang, Y.L., Kotsopoulos, D., and Cameron, J.D. (2001) Impact of physical and physiological factors on arterial function. *Clinical and Experimental Pharmacology and Physiology* 2001(28): 1104–1107.

Meaume, S., Benetos, A., Henry, O.F. et al. (2001) Aortic pulse wave velocity predicts cardiovascular mortality in subjects >70 years of age. *Arteriosclerosis, Thrombosis, and Vascular Biology* 21, 2046–2050.

Meyers, S.D., Kelly, B.G., and O'Brien, J.J. (1993) An introduction to wavelet analysis in oceanography and meteorology: with application to the dispersion of Yanai waves. *Monthly Weather Review* 121(10): 2858–2866.

Millasseau, S.C., Kelly, R.P., Ritter, J.M., and Chowienczyk, P.J. (2002) Determination of age-related increases in large artery stiffness by digital pulse contour analysis. *Clinical Science* 103(4): 371–378.

Mitchell, G.F., Parise, H., Benjamin, E.J. et al. (2004) Changes in arterial stiffness and wave reflection with advancing age in healthy men and women the Framingham Heart Study. *Hypertension* 43(6): 1239–1245.

Mohiaddin, R.H., Firmin, D.N., and Longmore, D.B. (1993) Age-related changes of human aortic flow wave velocity measured noninvasively by magnetic resonance imaging. *Journal of Applied Physiology* 74, 492–497.

Mookerjee, A., Al-Jumaily, A., and Lowe, A. (2007) The effect of various physical phenomena on wave propagations in the human aorta. *ASME 2007 International Design Engineering Technical Conferences and Computers and Information in Engineering Conference*, Las Vegas, NV (4–7 September 2007), pp. 353–356. American Society of Mechanical Engineers.

Mookerjee, A., Al-Jumaily, A.M., and Lowe, A. (2010) Arterial pulse wave velocity measurement: different techniques, similar results–implications for medical devices. *Biomechanics and Modeling in Mechanobiology* 9(6): 773–781.

Munir, S., Guilcher, A., Kamalesh, T. et al. (2008) Peripheral augmentation index defines the relationship between central and peripheral pulse pressure. *Hypertension* 51(1): 112–118.

Murray, C.J. and Lopez, A.D. (1996) *The Global Burden of Disease*. Cambridge, MA: Harvard School of Public Health, Harvard University Press.

Negnevitsky, M. (2002) *Artificial Intelligence: A Guide to Intelligent Systems*, 1e. New York: Addison Wesley.

Nemes, A., Takacs, R., Gavaller, H. et al. (2010) Correlations between aortic stiffness and parasympathetic autonomic function in healthy volunteers. *Canadian Journal of Physiology and Pharmacology* 88(12): 1166–1171.

Nemes, A., Takács, R., Gavallér, H. et al. (2011) Correlations between Arteriograph-derived pulse wave velocity and aortic elastic properties by echocardiography. *Clinical Physiology and Functional Imaging* 31(1): 61–65.

Nichols, W.W. (2005) Clinical measurement of arterial stiffness obtained from noninvasive pressure waveforms. *American Journal of Hypertension* 18(S1): 3S–10S.

Nichols, W.W. and Edwards, D.G. (2001) Arterial elastance and wave reflection augmentation of systolic blood pressure: deleterious effects and implications for therapy. *Journal of Cardiovascular Pharmacology and Therapeutics* 6, 5–21.

Nichols, W.W. and O'Rourke, M.F. (1991) *McDonald's Blood Flow in Arteries: Theoretic, Experimental and Clinical Principles*, 3e, 283–329. London: Edward Arnold Publisher.

Nichols, W.W. and Singh, B.M. (2002) Augmentation index as a measure of peripheral vascular disease state. *Current Opinion in Cardiology* 17, 543–551.

Nissila, S., Sorvisto, M., Sorvoja, H., Vieri-Gashi, E., and Myllyla, R. (1998) Non-invasive blood pressure measurement based on the electronic palpation method. *Proceedings of the 20th Annual International Conference of the IEEE Engineering in Medicine and Biology Society*, Hong Kong, China (1 November 1998), 4, pp. 1723–1726.

North, R.A., McCowan, L.M.E., Dekker, G.A. et al. (2011) Clinical risk prediction for pre-eclampsia in nulliparous women: development of model in international prospective cohort. *British Medical Journal* 342, d1875.

Nürnberger, J., Keflioglu-Scheiber, A., Opazo Saez, A.M. et al. (2002) Augmentation index is associated with cardiovascular risk. *Journal of Hypertension* 20(12): 2407–2414.

Nussbacher, A., Gerstenblith, G., O'Connor, F.C. et al. (1999) Hemodynamic effects of unloading the old heart. *American Journal of Physiology-Heart and Circulatory Physiology* 277, H1863–H1871.

Ohtsuka, S., Kakihana, M., Watanabe, H., and Sugishita, Y. (1994) Chronically decreased aortic distensibility causes deterioration of coronary perfusion during increased left ventricular contraction. *Journal of the American College of Cardiology* 24, 1406–1414.

Opgaard, O.S. and Edvinsson, L. (1997) Mechanical properties and effects of sympathetic co-transmitters on human coronary arteries and veins. *Basic Research in Cardiology* 92(3): 168–180.

O'Rourke, M.F. (1982a) *Arterial Function in Health and Disease*, 94–132. Edinburgh: Churchill Livingstone, 185–243.

O'Rourke, M.F. (1982b) Vascular impedance in studies of arterial and cardiac function. *Physiological Reviews* 62(2): 570–623.

O'Rourke, M.F. (1982) *Arterial Function in Health and Disease*, 159. Edinburgh; New York: Churchill Livingstone.

O'Rourke, M.F. and Avolio, A.P. (1980) Pulsatile flow and pressure in human systemic arteries: studies in man and in multibranched model of the human systemic arterial tree. *Circulation Research* 46, 363–372.

O'Rourke, M.F. and Franklin, S.S. (2006) Arterial stiffness: reflections on the arterial pulse, *European Heart Journal* 27, 2497–2498, doi: https://doi.org/10.1093/eurheartj/ehl312

O'Rourke, M.F., Avolio, A.P., and Kelly, R.P. (1992) *The Arterial Pulse*. Baltimore, MD: Lea & Febiger.

O'Rourke, N., Pauca, A., and Jiang, X.J. (2001) Pulse wave analysis. *British Journal of Clinical Pharmacology* 51, 507–522.

Ozolanta, I., Tetere, G., Purinya, B., and Kasyanov, V. (1998) Changes in the mechanical properties, biochemical contents and wall structure of the human coronary arteries with age and sex. *Medical Engineering and Physics* 20(7): 523–533.

Palombo, C., Kozakova, M., Morizzo, C. et al. (2011) Circulating endothelial progenitor cells and large artery structure and function in young subjects with uncomplicated type 1 diabetes. *Cardiovascular Diabetology* 10(88): 1475–2840.

Pannier, B.M., Avolio, A.P., Hoeks, A. et al. (2002) Methods and devices for measuring arterial compliance in humans. *American Journal of Hypertension* 15(8): 743–753.

Patel, D.J. and Janicki, J.S. (1970) Static elastic properties of the left coronary circumflex artery and the common carotid artery in dogs. *Circulation Research* 27(2): 149–158.

Pauca, A.L., O'Rourke, M.F., and Kon, N.D. (2001) Prospective evaluation of a method for estimating ascending aortic pressure from the radial artery pressure waveform. *Hypertension* 38(4): 932–937.

Pauca, A.L., Kon, N.D., and O'Rourke, M.F. (2004) The second peak of the radial artery pressure wave represents aortic systolic pressure in hypertensive and elderly patients. *British Journal of Anaesthesia* 92, 651–657.

Pesola, G.R., Pesola, H.R., Nelson, M.J., and Westfal, R.E. (2001) The normal difference in bilateral indirect blood pressure recordings in normotensive individuals. *The American Journal of Emergency Medicine* 19(1): 43–45.

Postel-Vinay, N. (1996) The essential contribution of life insurance companies to the discovery of risk. In: *A Century of Arterial Hypertension 1896–1996*, 31–48. New York: Wiley.

Prince, C.T., Secrest, A.M., Mackey, R.H. et al. (2010) Pulse wave analysis and prevalent cardiovascular disease in type 1 diabetes. *Atherosclerosis* 213(2): 469–474.

Protogerou, A.D., Stergiou, G.S., Vlachopoulos, C. et al. (2009) The effect of antihypertensive drugs on central blood pressure beyond peripheral blood pressure. Part II: evidence for specific class-effects of antihypertensive drugs on pressure amplification. *Current Pharmaceutical Design* 15, 272–289.

Remington, J.W. (ed.) (1957) Extensibility behavior and hysteresis phenomena in smooth muscle tissues. In: *Tissue Elasticity*, 138–153. Washington, DC: American Physiological Society.

Rezai, M.-R., Goudot, G., Finn Conchubhair Winters, J.D. et al. (2011) Calibration mode influences central blood pressure differences between SphygmoCor and two newer devices, the Arteriograph and Omron HEM-9000. *Hypertension Research* 34(9): 1046–1051.

Riva-Rocci, S. (1896) Un nuovo sfigmomanometro. *Gazzet Med Torino* 47, 981–1001.

Roach, M.R. and Burton, A.C. (1957) The reason for the shape of the distensibility curves of arteries. *Canadian Journal of Biochemistry and Physiology* 35(8): 681–690.

Rödig, G., Prasser, C., Keyl, C. et al. (1999) Continuous cardiac output measurement: pulse contour analysis vs thermodilution technique in cardiac surgical patients. *British Journal of Anaesthesia* 82(4): 525–530.

Roman, M.J., Devereux, R.B., Kizer, J.R. et al. (2007) Central pressure more strongly relates to vascular disease and outcome than does brachial pressure: the Strong Heart Study. *Hypertension* 50(1): 197–203.

Roman, M.J., Devereux, R.B., Kizer, J.R. et al. (2009) High central pulse pressure is independently associated with adverse cardiovascular outcome the Strong Heart Study. *Journal of the American College of Cardiology* 54(18): 1730–1734.

Rosner, B., Prineas, R.J., Loggie, J.M.H., and Daniels, S.R. (1993) Blood pressure nomograms for children and adolescents, by height, sex, and age, in the United States. *The Journal of Pediatrics* 123(6): 871–886.

Saba, P.S., Roman, M.J., Pini, R. et al. (1993) Relation of arterial pressure waveform to left ventricular and carotid anatomy in normotensive subjects. *Journal of the American College of Cardiology* 22, 1873–1880.

Sabit, R., Bolton, C.E., Edwards, P.H. et al. (2007) Arterial stiffness and osteoporosis in chronic obstructive pulmonary disease. *American Journal of Respiratory and Critical Care Medicine* 175(12): 1259–1265.

Safar, M.E., Henry, O., and Meaume, S. (2002) Aortic pulse wave velocity: an independent marker of cardiovascular risk. *The American Journal of Geriatric Cardiology* 11, 295–298.

Salvi, P. (2012) *Pulse Waves: How Vascular Hemodynamics Affects Blood Pressure*. Milan; New York: Springer.

Salvi, P. (2016) *Pulse Waves: How Vascular Hemodynamics Affect Blood Pressure*. Cham: Springer.

Salvi, P., Safar, M.E., Labat, C. et al. (2010) Heart disease and changes in pulse wave velocity and pulse pressure amplification in the elderly over 80 years: the PARTAGE Study. *Journal of Hypertension* 28(10): 2127–2133.

Sapinski, A. (1994) Standard algorithm of blood-pressure measurement by the oscillometric method. *Medical and Biological Engineering and Computing* 32(5): 599–600.

Sapinski, A. (1997) Theoretical basis for proposed standard algorithm of blood pressure measurement by the sphygmooscillographic method. *Journal of Clinical Engineering* 22(3): 171–174.

Sato, T., Nishinaga, M., Kawamoto, A. et al. (1993) Accuracy of a continuous blood pressure monitor based on arterial tonometry. *Hypertension* 21(6 Pt 1): 866–874.

Sebald, D.J., Bahr, D.E., and Kahn, A.R. (2002) Narrowband auscultatory blood pressure measurement. *IEEE Transactions on Biomedical Engineering* 49(9): 1038–1044.

SHEP Cooperative Research Group (1991) Prevention of stroke by antihypertensive drug treatment in older persons with isolated systolic hypertension: final results of the Systolic

Hypertension in the Elderly Program (SHEP). *JAMA: The Journal of the American Medical Association* 265, 3255–3264.

Siebenhofer, A., Kemp, C.R.W., Sutton, A.J., and Williams, B. (1999) The reproducibility of central aortic blood pressure measurements in healthy subjects using applanation tonometry and sphygmocardiography. *Journal of Human Hypertension* 13, 625–629.

Silva, G.J.J., Ushizima, M.R., Lessa, P.S. et al. (2009) Critical analysis of autoregressive and fast Fourier transform markers of cardiovascular variability in rats and humans. *Brazilian Journal of Medical and Biological Research* 42(4): 386–396.

Smith, S.A., Morris, J.M., and Gallery, E.D.M. (2004) Methods of assessment of the arterial pulse wave in normal human pregnancy. *American Journal of Obstetrics and Gynecology* 190(2): 472–476.

Stage, Hypertension (2013) Understanding blood pressure readings. Heart (2013): 17.

Stewart, A.D., Millasseau, S.C., Kearney, M.T. et al. (2003) Effects of inhibition of basal nitric oxide synthesis on carotid-femoral pulse wave velocity and augmentation index in humans. *Hypertension* 42(5): 915–918.

Stewart, A.D., Jiang, B., Millasseau, S.C. et al. (2006) Acute reduction of blood pressure by nitroglycerin does not normalize large artery stiffness in essential hypertension. *Hypertension* 48(3): 404–410.

Struthius, J. (1555) Sphygmicae artis iam mille ducentos annos perditae et desideratae Libri V [Lessons about the pulse in five books written by Josephus Struthius]. Oporinus, Basileae.

Sutton-Tyrrell, K., Najjar, S.S., Boudreau, R.M. et al. (2005) Health ABC Study: elevated aortic pulse wave velocity, a marker of arterial stiffness, predicts cardiovascular events in well-functioning older adults. *Circulation* 111, 3384–3390.

Szekeres, M.A.R., Nadasy, G.L., Dezsi, L. et al. (1998) Segmental differences in geometric, elastic and contractile characteristics of small intramural coronary arteries of the rat. *Journal of Vascular Research* 35(5): 332–344.

Tajaddini, A., Kilpatrick, D.L., Paul Schoenhagen, E. et al. (2005) Impact of age and hyperglycemia on the mechanical behavior of intact human coronary arteries: an ex vivo intravascular ultrasound study. *American Journal of Physiology-Heart and Circulatory Physiology* 288(1): H250–H255.

Takazawa, K., Tanaka, N., Fujita, M. et al. (1998) Assessment of vasoactive agents and vascular aging by the second derivative of photoplethysmogram waveform. *Hypertension* 32(2): 365–370.

Tanaka, S., Gao, S., Nogawa, M., and Yamakoshi, K. (2003) A new non-invasive device for measuring instantaneous blood pressure in radial artery based on the volume-compensation method. *Proceedings of the Annual International Conference of the IEEE Engineering in Medicine and Biology Society* 2003, 3149–3152.

Tanaka, S., Gao, S., Nogawa, M., and Yamakoshi, K.-I. (2005) Noninvasive measurement of instantaneous radial artery blood pressure. *IEEE Engineering in Medicine and Biology Magazine* 24(4): 32–37.

TAO, L. (1952) Achievements of Chinese medicine in the Ch'in (221–207 BC) and Han (206 BC–219 AD) dynasties. *Chinese Medical Journal* 71(5): 380–396.

Tortora, G.J. (2018) *Principles of Anatomy and Physiology, 2nd Asia-Pacific Edition.* Melbourne: Wiley. ProQuest Ebook Central. https://ebookcentral.proquest.com/lib/AUT/detail.action?docID=5561269.

Ursino, M. and Cristalli, C. (1995a) Mathematical modeling of noninvasive blood pressure estimation techniques—part I: pressure transmission across the arm tissue. *Journal of Biomechanical Engineering* 117(1): 107–116.

Ursino, M. and Cristalli, C. (1995b) Mathematical modeling of noninvasive blood pressure estimation techniques—part II: brachial hemodynamics. *Journal of Biomechanical Engineering* 117(1): 117–126.

Ursino, M. and Cristalli, C. (1996) A mathematical study of some biomechanical factors affecting the oscillometric blood pressure measurement. *IEEE Transactions on Biomedical Engineering* 43(8): 761–778.

Vaitkevicius, P.V., Fleg, J.L., Engel, J.H. et al. (1993) Effects of age and aerobic capacity on arterial stiffness in healthy adults. *Circulation* 88, 1456–1462.

Van Andel, C.J., Pistecky, P.V., and Borst, C. (2001) Mechanical properties of coronary arteries and internal mammary arteries beyond physiological deformations. *Proceedings of the 23rd Annual International Conference of the IEEE Engineering in Medicine and Biology Society*, Argentina Buenos Aires (31 August to 4 September 2010), Vol. 1, pp. 113–115. IEEE.

Verbeke, F., Segers, P., Heireman, S. et al. (2005) Noninvasive assessment of local pulse pressure importance of brachial-to-radial pressure amplification. *Hypertension* 46(1): 244–248.

Veress, A.I., Vince, D.G., Anderson, P.M. et al. (2000) Vascular mechanics of the coronary artery. *Zeitschrift für Kardiologie* 89(2): S092–S100.

Vincent, J.-L. (2008) Understanding cardiac output. *Critical Care* 12(4): 174.

Vlachopoulos, C. and O'Rourke, M. (2000) Genesis of the normal and abnormal pulse. *Current Problems in Cardiology* 25, 297–368.

Vlachopoulos, C., Hirata, K., and O'Rourke, M.F. (2001) Pressure-altering agents affect central aortic pressures more than is apparent from upper limb measurements in hypertensive patients. The role of arterial wave reflections. *Hypertension* 38, 1456–1460.

van de Vosse, F.N. and Stergiopulos, N. (2011) Pulse wave propagation in the arterial tree. *Annual Review of Fluid Mechanics* 43, 467–499.

Vulliemoz, S., Stergiopulos, N., and Meuli, R. (2002) Estimation of local aortic elastic properties with MRI. *Magnetic Resonance in Medicine* 47, 649–654.

Wang, W.K. (1986) Modern view of Chinese Medicine. *Annual Report of the Institute of Physics* 16, 269–275.

Wang, H. and Cheng, Y. (2006) A quantitative system for pulse diagnosis in traditional Chinese medicine. *Proceedings of the 27th Annual International Conference of the IEEE Engineering in Medicine and Biology Society, 2005. IEEE-EMBS 2005*, Shanghai, China (17–18 January 2006), pp. 5676–5679. IEEE.

Wang, J.-J., Lin, C.-T., Liu, S.-H., and Wen, Z.-C. (2002) Model-based synthetic fuzzy logic controller for indirect blood pressure measurement. *IEEE Transactions on Systems, Man and Cybernetics* 32(3): 306–315.

Wang, C., Zhang, W., and Kassab, G.S. (2008) The validation of a generalized Hooke's law for coronary arteries. *American Journal of Physiology-Heart and Circulatory Physiology* 294(1): H66–H73.

Wang, Y.-Y.L., Hsu, T.-L., Jan, M.-Y., and Wang, W.-K. (2010) Review: theory and applications of the harmonic analysis of arterial pressure pulse waves. *Journal of Medical and Biological Engineering* 30(3): 125–131.

Wassertheurer, S., Kropf, J., Weber, T. et al. (2010) A new oscillometric method for pulse wave analysis: comparison with a common tonometric method. *Journal of Human Hypertension* 24(8): 498–504.

Watson, S., Wenzel, R.R., Di Matteo, C. et al. (1998) Accuracy of a new wrist cuff oscillometric blood pressure device comparisons with intra-arterial and mercury manometer measurements. *American Journal of Hypertension* 11(12): 1469–1474.

Weber, T., Auer, J., O'Rourke, M.F. et al. (2004) Arterial stiffness, wave reflections, and the risk of coronary artery disease. *Circulation* 109, 184–189.

Weng, H. and Lau, K.M. (1994) Wavelets, period doubling, and time-frequency localization with application to organization of convection over the tropical western Pacific. *Journal of the Atmospheric Sciences* 51(17): 2523–2541.

Westerhof, N., Bosman, F., De Vries, C.J. et al. (1969) Analog studies of human systemic arterial tree. *Journal of Biomechanics* 2, 121–143.

Wilkinson, I.B., Franklin, S.S., Hall, I.R. et al. (2001) Pressure amplification explains why pulse pressure is unrelated to risk in young subjects. *Hypertension* 38, 1461–1466.

Wilkinson, I.B., Prasad, K., Hall, I.R. et al. (2002) Increased central pulse pressure and augmentation index in subjects with hypercholesterolemia. *Journal of the American College of Cardiology* 39(6): 1005–1011.

Wilkinson, I.B., Maccallum, H., Flint, L. et al. (2004) The influence of heart rate on augmentation index and central arterial pressure in humans. *The Journal of Physiology* 525(1): 263–270.

William, B., Lacy, P.S., Thom, S.M. et al. (2006) Differential impact of blood pressure-lowering drugs on central aortic pressure and clinical outcomes: principal results of the Conduit Artery Function Evaluation (CAFE) study. *Circulation* 113, 1213–1225.

Williams, M.J.A., Stewart, R.A.H., Low, C.J.S., and Wilkins, G.T. (1999) Assessment of the mechanical properties of coronary arteries using intravascular ultrasound: an in vivo study. *The International Journal of Cardiac Imaging* 15(4): 287–294.

Willum-Hansen, T., Staessen, J.A., Torp-Pedersen, C. et al. (2006) Prognostic value of aortic pulsewave velocity as index of arterial stiffness in the general population. *Circulation* 113, 664–670.

Wolinsky, H. and Glagov, S. (1967) A lamellar unit of aortic medial structure and function in mammals. *Circulation Research* 20(1): 99–111.

Wolinsky, H. and Glagov, S. (1969) Comparison of abdominal and thoracic aortic medial structure in mammals. *Circulation Research* 25(6): 677–686.

Womersley, J.R. (1957) The mathematical analysis of the arterial circulation in a state of oscillatory motion. Wright Air Development Center, Technical Report Wade-TR. 56-614, Dayton, OH.

Wood, G.C. (1954) Some tensile properties of elastic tissue. *Biochimica et Biophysica Acta* 15(3): 311–324.

Wykretowicz, A., Adamska, K., Guzik, P. et al. (2007) Indices of vascular stiffness and wave reflection in relation to body mass index or body fat in healthy subjects. *Clinical and Experimental Pharmacology and Physiology* 34, 1005–1009.

Yamakoshi, K., Rolfe, P., and Murphy, C. (1988) Current developments in non-invasive measurement of arterial blood pressure. *Journal of Biomedical Engineering* 10(2): 130–137.

Yang, S. (1997) *The Pulse Classic: A Translation of the Mai Jing*. Boulder, CO: Blue Poppy Enterprise.

Yoon, Y.-Z., Lee, M.-H., and Soh, K.-S. (2000) Pulse type classification by varying contact pressure. *IEEE Engineering in Medicine and Biology Magazine* 19(6): 106–110.

Young, T. (1809) The Croonian lecture. On the function of the heart and arteries. *Philosophical Transactions of the Royal Society of London* 99, 1–31.

Young, T. (1809) The Croonian lecture. On the functions of the heart and arteries. *Philosophical Transactions of the Royal Society of London* 99, 1–31.

3

Radiation Force

3.1 Introduction

All the chapters in this book except this one focus on either the cardiovascular or the respiratory system. For the sake of generality, this chapter is introduced to summarize some other biomedical applications of the pressure wave propagation. These applications are based on the radiation force (RF) as a phenomenon used for the purpose of diagnostics.

Over several centuries, clinicians have used manual palpation of tissue as an essential method to examine patients. It is simple and used for estimating mechanical properties of the tissue. The technique employs a force exerted on the body surface that spreads over the tissues. If the response to palpation is sufficiently different from the surrounding tissues, the clinician may identify abnormalities. For example, clinicians may use the palpation method to detect some breast tumors through the overlying tissue. However, this method fails if the abnormality lies deep in the body or it is too small to be resolved by touch. To address this problem, various research groups worldwide have developed several free acoustic radiation force (ARF) techniques, and elasticity imaging (EI) approaches and applied for medical imaging and diagnostics, including breast tumor detection.

The RF is another application of pressure waves that has been extensively investigated for a couple of centuries, and it has been applied to a many fields, including medical applications. Several new radiations force-based EI methods have been proposed in the past four decades. These techniques are based on applying dynamic RFs to an object and measuring the object's dynamic displacement response to estimate its mechanical properties. An example of these methods is vibro-acoustography (VA). In this technique, an ARF is applied to an elastic body, and the reflection is visualized to determine healthy from unhealthy conditions. In this method, an ultrasound-based wave strikes the tissue, and the reflected waves are mapped to determine any inhomogeneity in the tissue. For a homogeneous healthy tissue, the reflected map will be homogeneous too. However, if any abnormality exists in the tissue, it will record different mechanical characteristics. This technique has been widely studied and applied in industry and biomedical areas, including medical imaging and diagnostics.

This chapter briefly describes the history of ARF and its applications and provides an overview of ARF-based EI approaches. Examples of ARF-based EI–VA imaging and multi-frequency VA techniques and their applications in medical evaluation are discussed. Several advantages and disadvantages of VA, comparison between VA and pulse-echo systems, as well as future directions are also presented.

Pressure Oscillation in Biomedical Diagnostics and Therapy, First Edition. Ahmed Al-Jumaily and Lulu Wang.

3.2 Acoustic Radiation Force

ARF is another application of pressure waves for diagnostic purposes. In principle, an acoustic pressure wave is generated and applied to an object. As discussed in Chapter 1, the behavior of the wave depends on the speed of sound, which is a function of the geometry of the target object, mechanical properties, and the propagating medium. The force is a physical phenomenon caused by momentum transfer from the acoustic wave to the propagation medium. The biological processes leading to ARF are very complex. Generally, it is defined as a period of average force exerted on the medium by the sound wave. The magnitude of the force F (dyn/(1000 cm)^3 or (kg/(s cm)^2) exerted by an acoustic plane wave with an average intensity I (W/cm^2) at any given location is defined as (Nyborg 1965; Torr 1984):

$$|F| = \frac{2\alpha I}{c} \tag{3.1}$$

where α is the absorption coefficient in Np/cm and c is the speed of sound in the medium in cm/s.

Several objects, including spheres, drops, and bubbles, have been manipulated using ARF (Hasegawa and Yosioka 1969; Crum 1971; Marston 1980; Mitri 2005; Mitri and Chen 2005; Mitri and Fatemi 2005; Lee and Shung 2006). Further, the acoustic output of several ultrasound transduces have been calibrated using RF balances (Carson et al. 1978; Fick and Breckenridge 1996; Shou et al. 2006). In general, ARF can be either static or dynamic with respect to its time dependence. A static ARF is produced on an object by steady-state acoustic waves due to the nonlinear nature of wave propagation which brings up a DC component in the spectrum of the pressure wave associated with the static RF (Borgnis 1953). On the other hand, a dynamic RF is produced on an object by the momentum flux of any acoustic wave or pulse. This force is also nonlinear similar to the static force (Silva 2010), and it has been applied in measuring absorption in liquids (McNamara and Beyer 1953), the power of ultrasound transducers (Greenspan et al. 1978), and inducing oscillation in liquid drops (Marston 1980). It has also been studied for manipulation of objects including spheres, and spherical and cylindrical shells (Chen et al. 2005; Mitri 2005; Mitri and Fellah 2006).

3.2.1 Types of Radiation Force

RF can be produced by various physical effects:

a) **Change in the propagating wave density of energy due to absorption and scattering:** This type of RF was proposed by Carl Eckart (Carl Eckart 1948) who derived equations for the average force and motion in a homogeneous viscous fluid. This mechanism uses the RF of focused ultrasound to generate shear stress in tissues (Fatemi and Greenleaf 1998, 1999; Sarvazyan et al. 1998; Nightingale et al. 2002; Bercoff et al. 2004).
b) **Spatial variations of energy density in standing acoustic wave:** This type of RF was first experimentally demonstrated by Kundt (Kundt 1874). The RF was generated in the standing acoustic waves. Such RF is currently widely used for applications related to manipulating particles (Brandt 1989, 2001; Coakley et al. 2000; Hawkes et al. 2004; Haake et al. 2005; Hultström et al. 2007).

c) **Reflection from inclusions, walls, or other interfaces:** Wood and Loomis introduced ultrasound RF balance, the oldest in applications of RF (Wood and Loomis 1927). It is the basis of many applications employing VA principles developed by Greenleaf, Fatemi, and their coworkers in the last decade (Alizad et al. 2006; Urban et al. 2006; Pislaru et al. 2008; Mitri et al. 2009).
d) **Spatial variations in propagation velocity:** Ostrovsky et al. proposed "medium inhomogeneity" as a non-dissipative mechanism of RF generation, and the corresponding medium response was calculated for rubber-like medium (Ostrovsky et al. 2007).

3.2.2 Acoustic Radiation Force History

Sarvazyan et al. gives a comprehensive historical review of RF (Sarvazyan et al. 2010). The idea of RF was first introduced in 1619 by Kepler who suggested that the solar ray pressure is responsible for the deflection of the comet tails. Faraday proposed the streaming induced by sound waves (Faraday 1831) and observed that light powder moves above the vibrating plate because of local air circulation. Kundt & Leham experimentally described the ARF acting on particles in standing acoustic waves (Kundt and Lehman 1874).

Dvorak and Mayer independently observed that the pressure on the closed end of an acting resonator is greater than at the open end (Mayer and Woodword 1916). Rayleigh first described the pressure of vibration (Rayleigh 1902). Altberg used the ARF pressure to measure the intensity of sound radiating from the end of a Kundt's tube. He also measured the spectrum of acoustic waves emitted by the spark discharged from a condenser in 1907 (Altberg 1903). Bjerknes published an acoustic phenomenon related to RF acting on bubbles in the acoustic field (Bjerknes 1909). Wood and Loomis introduced the basic concept of the ARF balance (Wood and Loomis 1927), and its important practical application was published two years later. This system became the prototype for the most common instrument for calibration of the therapeutic transducers. Many researchers explored the forces acting on particles in an acoustic field later. King analyzed the incompressible particles in acoustic fields (King 1934), Yosioka and Kawasima calculated the forces on compressible particles in-plane acoustic waves (Yosioka and Kawasima 1955), and their contributions were summarized and published by Gorkov (Gorkov 1962).

Between the 1930s and 1940s, several groups have conducted studies related to medical applications of ultrasound. In the early 1950s, the development of ultrasonic methods and devices for medical diagnostics and therapy has created a major surge in RF application. Wild and Reid built the first real-time ultrasonic scanner (Wild and Reid 1952a, 1952b, 1954). Later on, several groups developed ultrasonic diagnostic devices based on pulse-echo and Doppler techniques (Hueter and Bolt 1951; Joyner and Reid 1963; Evans et al. 1966; Kossoff 1966; Kossoff et al. 1968; Robinson et al. 1968). Various scientists also explored ultrasonic techniques for surgical and therapeutic applications (Wall et al. 1951; Lehmann and Herrick 1953; Schwan et al. 1954; Fry and Fry 1960; Kossoff et al. 1967; Dyson et al. 1968). Many groups have investigated the ultrasonic properties of biological tissues and fluids (Fry 1952; Piersol et al. 1952; Carstensen and Schwan 1959; Dunn and Fry 1961; Dunn 1962; Dunn et al. 1969; Lizzi et al. 1970). Mechanism of ultrasonic bio-effects was one of the RF-related research areas. Dyson et al. discovered that red blood cells in the blood vessels in vivo could be collected in standing acoustic waves in bands half a wavelength apart. They demonstrated the biomedical significance of this effect in 1971 (Dyson et al. 1971). Since that time, the biological effects of ultrasound, including the effects of ARF, have been extensively studied. Nyborg reported the biological effects of ultrasound, including the effects of ARF (Nyborg 1965). Analyses of the physical basis of biomedical applications of ultrasound RF were detailed in Nyborg (1968, 1982).

3.2.3 Applications of Acoustic Radiation Force

Biomedical applications of ARF have become the major research area since early 1990s, which included EI (Fatemi and Greenleaf 1998, 1999; Sarvazyan et al. 1998; Nightingale et al. 2002; Bercoff et al. 2004), monitoring therapy (Lizzi et al. 2003; Bercoff et al. 2004; Maleke and Konofagou 2008), targeted drug and gene delivery (Dayton et al. 1999, 2006), molecular imaging (Zhao et al. 2004), acoustic tweezers (Wu 1991; Hu and Santoso 2004; Lee et al. 2005), and for increasing the sensitivity of biosensors and immunochemical tests (Wiklund and Hertz 2006; Kuznetsova and Coakley 2007). Medical diagnostics has the widest field of biomedical applications of RF, in particular for the EI. However, measuring the viscoelastic properties of tissues and fluids has been one of the main areas of application. Manipulating biological cells and particles in standing ultrasonic wave fields is another actively explored area of RF applications.

3.2.4 Acoustic Radiation Force-Based Elasticity Imaging Techniques

EI is a noninvasive quantitative method that measures the mechanical properties of an object. In medical applications, EI is the general field of quantitative approaches to measure the mechanical properties of tissue and map the elastic properties of tissues in an anatomically meaningful picture to provide useful diagnostic information. Although both dynamic (vibration) and static can be the excitation stress, the former excitation provides more comprehensive tissue properties information over a spectrum of frequencies. Some of the existing medical imaging modalities, such as ultrasound scanners and MRI, can be used as measurement devices.

ARF-based EI techniques that are currently being widely applied in medical imaging include but are not limited to: (i) acoustic streaming in diagnostic ultrasound (Dymling et al. 1991; Nightingale et al. 1995), (ii) sonorheometry (Viola et al. 2004), (iii) acoustic radiation force impulse (ARFI) imaging (Nightingale et al. 2000, 2001; Fahey et al. 2004, 2005a, 2005b, 2008a, 2008b; Trahey et al. 2004), (iv) shear wave elasticity imaging (SWEI) (Sarvazyan et al. 1998), (v) supersonic shear imaging (SSI) (Bercoff et al. 2004), (vi) shear wave spectroscopy (SWS) (Deffieux et al. 2009), (vii) spatially modulated ultrasound radiation force (SMURF) (McAleavey et al. 2007), (viii) VA (Fatemi et al. 2001, 2002), (ix) harmonic motion imaging (HMI) (Konofagou and Hynynen 2003; Konofagou 2004; Vappou et al. 2009), (x) shear wave ultrasound dispersion ultrasound vibrometry (SDUV) (Chen et al. 2004), and (xi) crawling wave spectroscopy (CWS) (Hah et al. 2010).

Table 3.1 summarizes the ARF-based EI methods in diagnostic ultrasound. Sarvazyan et al. first investigated SWEI, the quantification of tissue shear modulus using transient, impulse-like focused ARFs to generate shear waves within tissues (Sarvazyan et al. 1998). They used high frequency focused ultrasound (HIFU) pistons to generate RF and applied magnetic resonance imaging methods to monitor the shear wave propagation.

Table 3.1 Overview of ARF-based EI methods.

	Pushing beam	Tracking beam	Excitation	Detection
ASS	A single push beam is used to generate acoustic streaming in a fluid medium.	Multiple tracking beams at a single location are used to monitor the displacement response is monitored.	Impulse	
ARFI	This push and receive sequence is swept across the lateral field of view.	The deformation response is monitored on-axis at a single location, with a tracking beam aligned with the pushing beam axis.	Impulse	Modified linear array Transducer
SWEI	A single pushing beam is used to generate shear waves	Multiple off-axis tracking beams at lateral locations are used to monitor shear waves.	Oscillatory force (low kHz)	Ultrasound/ acoustic Detector
SSI & SWS	Create a near plane wave supersonic shear front by rapidly focusing the push at multiple axial depths.	Monitor shear wave propagation off-axis at multiple locations.	Impulse	Modified 1-D array Transducer
SMURF	Create a shear wave at each of two push locations separated from the single-track location by known distances.	Monitor the propagation of shear waves generated by each push at the track location.		
VA	Transmit two pushing beams at slightly offset frequencies of vibrating overlap region at the beat frequency Δf.	Phase and amplitude of vibration are recorded with a hydrophone.	Oscillatory force (low kHz)	Hydrophone
HMI	Harmonic vibrations are generated with amplitude modulation.	The amplitude of displacement through time is monitored with a separate transducer.	Oscillatory force (10–50 Hz)	Ultrasound/ phased-array transducer
SDUV	Use amplitude modulation to create monochromatic shear waves with known frequency f_S	Record phase at two different track locations separated by known distance Δx	Oscillatory force	Ultrasound/ Vibrometry
CWS	Transmit two pushing beams at slightly offset frequencies to create slow-moving crawling waves.	Monitor interference pattern with Doppler techniques.		

Using the same diagnostic ultrasound array, Nightingale et al. initially used inversion of the Helmholtz equation to quantify shear wave speed in humans in vivo (Nightingale et al. 2003).

VA is a speckle-free imaging technique that has a significant advantage over traditional ultrasound imaging. Fatemi and Greenleaffirst investigated VA by using two different ultrasound beams to produce low-frequency vibration of an object or tissue (Fatemi and Greenleaf 1998, 1999). The acoustic emission field (amplitude or phase) resulting from object vibration is recorded by a hydrophone and used to form an image by repeating the procedure at multiple points across the focus. High ultrasound frequency waves produce the acoustic emission (low acoustic frequency wave), often two orders of magnitude smaller than that in ultrasound frequency. VA has been applied in multiple scenarios, including vascular imaging (Pislaru et al. 2008), prostate imaging (Mitri et al. 2009), and breast imaging (Fatemi et al. 2002). The theory of VA and recent advances in various applications of this technique will be described in the following sections.

Nightingale et al. proposed the ARFI imaging method (Nightingale et al. 2000). They used short-duration ARF to generate localized displacements in tissues and used pulse-echo ultrasound to track the displacements during relaxation. Images of tissue were created when short impulses were applied at many locations in the tissue of interest. The tissue response to the transient excitation was complicated and depended on tissue geometry, RF field geometry, and tissue mechanical and acoustic properties. Three pulses reference, excitation (pushing), and tracking consisted of a typical ARFI at a lateral signal location were applied. Reference pulses established a baseline position of the tissue before the ARF that was generated by the excitation pulses to induce localized deformation, then the tracking pulses were immediately applied to monitor the deformation response and recovery of the tissue (see Figure 3.1). ARFI images have spatial resolution comparable to that of B-mode, often with greater contrast, providing matched, adjunctive information. Multiple clinical applications of ARFI imaging include detecting and characterizing a wide variety of soft tissue lesions and identifying and characterizing atherosclerosis, plaque, and thrombosis.

Chen et al. investigated the SDUV that quantifies both elasticity and viscosity from the frequency dispersion of shear wave propagation speed (Chen et al. 2004). SDUV uses

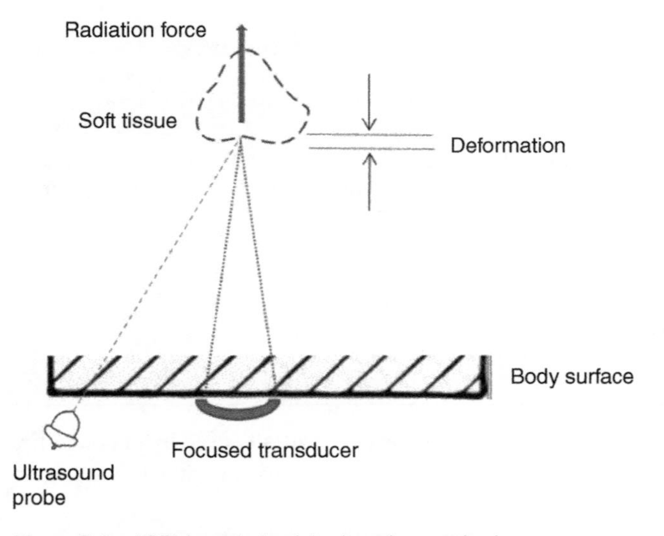

Figure 3.1 ARFI-based elasticity imaging method.

time-varying RF to generate harmonic shear wave and ultrasonic tracking to measure the tissue motion at known distances from the vibration source. Referring to the shear modulus formula $G = \rho(\lambda f)^2$, where ρ is the tissue density, it can be seen that dispersion is assessed by analyzing shear wave speed differences for different shear wave frequencies. SDUV has the advantage of narrowband excitation, and thus filtering methods can be employed to isolate the tissue response to the SDUV excitation, facilitating the use of lower acoustic output. SDUV has been applied in the study of tissues such as the liver (Chen et al. 2009) and blood vessels (Zhang et al. 2005).

Konofagou and Hynynen established HMI that generates harmonic oscillations by the beat frequency of confocal beams as in VA or by amplitude modulation of a single excitation beam at the desired frequency (Konofagou and Hynynen 2003; Konofagou 2004; Vappou et al. 2009). Conventional ultrasonic correlation-based techniques were used to monitor the displacement response. The peak-to-peak amplitude of the vibrating tissue was used to determine the relative differences in tissue stiffness. Previous studies used two separate elements focused on ultrasound transducers to demonstrate the HMI technique both theoretically and experimentally. An investigation based on the finite element method (FEM) and Monte Carlo simulation to perform an oscillatory displacement within an applied force frequency range of 200–800 Hz was conducted. The force was applied to the specific nodes within the focal region of the FEM model (Maleke et al. 2010). The results showed that HMI has a promise in imaging HIFU-generated thermal ablation lesions (Maleke et al. 2006). However, the acoustic pressure field had not been taken into account until recently; therefore, the applied RF amplitude was not realistic (Hou et al. 2011).

Bercoff et al. introduced SSI, as an ultrasound-based technique for real-time visualization of soft tissue viscoelastic properties by tracking the progress RF-induced shear waves with an ultrafast scanner (Bercoff et al. 2004). Referring to Figure 3.2, an ultrafast scanner is used to create an ultrasound-focused beam in elastic agar–gelatine phantom at a selected location. To create SSI image, an ultrasonic reference image of the medium is realized by a first plane wave in sonification, and this can be used to calculate the displacements induced by the shear wave propagation. Then the supersonic source is activated by successively focusing a “pushing” beam (typically 100 μs) at sequence of depths. To monitor the activation of the supersonic source, a set of ultrasonic images are acquired at a high-frame rate (3000 Hz). Variations in shear modulus distort the propagating shear waves. An inversion algorithm is then applied to the tracked motion data to recover the shear modulus. Medical applications of SSI include the noninvasive assessment of liver fibrosis. Previous studies (Bavu et al. 2011) and (Ferraioli et al. 2012) compared elastography (ET) with SSI in the evaluation of liver fibrosis. Results suggested that SSI performed significantly better in diagnosing significant fibrosis, whereas transient ET and SSI performed similarly in predicting liver cirrhosis. This has been further proved in Sirli et al. (2013), where three ET methods for liver fibrosis assessment were compared, and results suggest that SSI had better accuracy in the diagnosis of liver cirrhosis than that in the diagnosis of significant fibrosis, whereas for ARFI ET the accuracy rates were similar.

SMURF imaging introduced by McAleavey (McAleavey et al. 2007) is a method to use RF to generate a shear wave (unknown wavelength) in the material of unknown shear modulus and determine the material’s shear modulus by measuring the temporal frequency of the propagating shear wave. The frequency of the wave depends on its point of generation

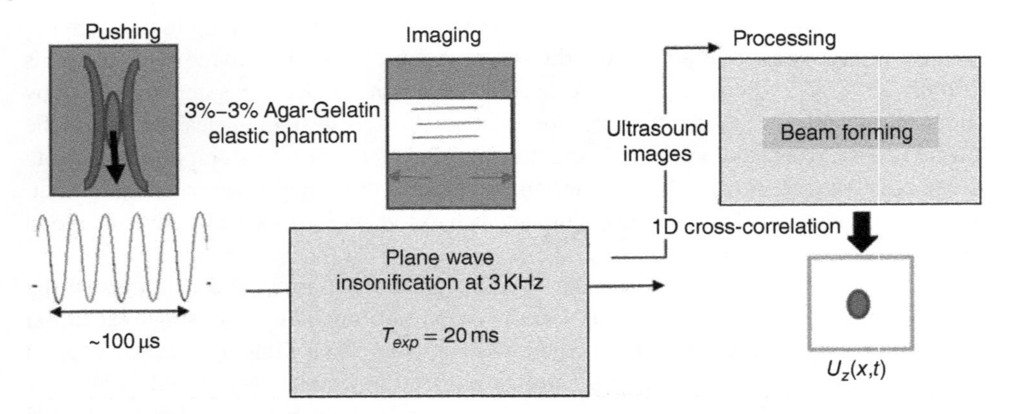

Figure 3.2 SSI working principle.

and is calculated to determine the shear modulus of the tissue. Different from SDUV and SSI, SMURF tracking takes place at a single location. The location of a speckle at a particular depth may bias the arrival time estimate of a shear wave from one pushing location. Studies have shown that SMURF imaging can provide rapid assessments of shear modulus in good agreement with values obtained through standard mechanical testing methods.

3.2.5 Commercial Implementations of Acoustic Radiation Force-Based Imaging

Commercial implementations of RF-based imaging methods employing transient RF excitations have been developed by Siemens Medical Solutions and Supersonic Imagine. These have been designed such that the pulse sequences are within the diagnostic limits for non-obstetric imaging. Siemens implemented a version of ARFI imaging on their ACUSON S2000 ultrasound scanner as the Virtual Touch Tissue Imaging tool that provides qualitative images of relative differences in tissue displacement in response to transient ARF excitation as for ARFI imaging. This imaging technique has been evaluated by clinical studies in Asia and Europe (Clevert et al. 2009; Cho et al. 2010; Gallotti et al. 2010; Stock et al. 2010; D'onofrio et al. 2011). Siemens also introduced a tool for RF-based shear wave speed estimations, the Virtual Touch Tissue Quantification tool, which is actively being studied in the context of abdominal and thyroid tissue stiffness quantification (Fierbinteanu-Braticevici et al. 2009; Friedrich-Rust et al. 2009; Haque et al. 2010; Goertz et al. 2011; Piscaglia et al. 2011).

An Aixplorer ultrasound scanner that generates quantitative elasticity images based on the SSI technology was developed by Siemens Medical Solutions and Supersonic Imagine (Bercoff et al. 2004). It employs shear wave speed reconstruction algorithms similar to those reported by McLaughlin and Renzi (McLaughlin and Renzi 2006a, 2006b). Initial clinical applications of the Aixplorer have been in characterizing breast lesions (Tanter et al. 2008), and the system is actively being translated to applications in the liver, thyroid, and prostate diagnostics. HemoSonics, LLC, in collaboration with StarFish Medical, has begun the development and clinical testing of a portable point of care analyzer based on sonorheometry to characterize blood hemostasis potential in a variety of clinical settings.

3.3 Vibro-Acoustography

VA is a noninvasive technique that uses the ultrasound RF (pressure wave) to harmonically vibrate soft tissue and measure the resulting acoustic emission field. Single-frequency and multifrequency VA methods have been proposed and applied in medical and material evaluation areas. Figure 3.3 describes a simplified VA imaging system. The confocal transducer introduces two ultrasound beams (with slightly different frequencies) to the same object. At the focal point, the interference of these two beams causes the object to vibrate at the beat frequency. A sensitive hydrophone detects the acoustic emission (acoustic field resulting from object vibration) which is used to form the image of the object. Finally, a 2D image of the object can be created by scanning the focal point of the transducer in a raster manner.

3.3.1 Soft Tissue Material Properties

The dynamic response of the soft tissue to a force is a valuable parameter in medical diagnosis. The stiffness of tissues can be described by their elastic moduli, in compression/tension the Young's modulus of elasticity, E, and in shear the shear modulus, μ,. These represent the material resistance to deformation. Higher elastic moduli tissues (muscle and fibrous tissue) are more resistant to deformation than more compliant tissues (fat). Tissue deformation occurs in response to stress (σ) being applied to the tissue, in the case of manual palpation, this stress is related to the force exerted by the clinician's fingers over the surface area of an organ or a mass. The deformation that occurs in response to applied stress is defined as the strain (ε) (Lai et al. 2009):

$$\varepsilon = \left((\nabla u)^T + \nabla u\right)/2 \tag{3.2}$$

where T is the transpose operation, and ∇u is the spatial displacement gradient.

Cross-correlation and Doppler-based autocorrelation are usually used to monitor the dynamic displacement response of soft tissues (Kasai et al. 1985). The resolution of ultrasonic displacement tracking methods is anisotropic. It is typically an order of magnitude better in the axial direction (i.e. fractions of a micrometer (Walker and Trahey 1995) than the lateral direction (tens of micrometers using two-dimensional cross-correlation methods). Conveniently, the micrometer scale displacements generated by RF-based methods occur along the direction of wave propagation (in the axial direction); thus, axial displacements are typically the only component of displacement that is monitored using these methods. While soft tissues are very complex heterogeneous materials, many assumptions are

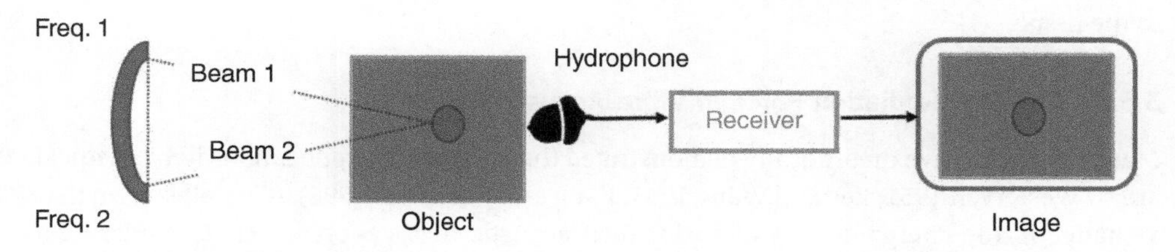

Figure 3.3 Diagram of a VA imaging system.

made in the field to simplify the analysis and interpreting the elasticity images. Common material assumptions include that the tissue is linear, elastic, and isotropic. Under these assumptions, stress and strain can be related to each other as (Lai et al. 2009):

$$\sigma = E\varepsilon \tag{3.3}$$

Monitoring the propagation of shear waves also determines the elastic properties of a material. In contrast to ultrasonic or compressive waves that propagate in the same direction as the tissue displacement, shear waves propagate orthogonally to the induced tissue displacement direction. Under these simplifying material assumptions, shear wave propagation is governed by (Lai et al. 2009):

$$\mu\nabla^2 u - \rho\frac{\partial^2 u}{\partial t^2} = 0 \tag{3.4}$$

where ρ denotes the material density, ∇^2 is the Laplacian operator, and t is the time.

The speed of propagating shear (or transverse) waves (C_T) can be obtained:

$$C_T = \sqrt{\mu/\rho} \tag{3.5}$$

where μ is the shear modulus.

The shear modulus is related to Young's modulus by (Lai et al. 2009):

$$\mu = \frac{E}{2(1+\nu)} \tag{3.6}$$

where ν is Poisson's ratio.

Equation (3.4) is another form of the 3D wave equation introduced in Chapter I. Soft tissues are commonly considered to be incompressible, with Poisson's ratio of 0.5 in materials with the assumptions stated earlier, leading to:

$$\mu = E/3 \tag{3.7}$$

Modeling the tissue as viscoelastic and/or nonlinear are the two standard deviations from the simplifying assumptions used to derive the earlier mentioned equations. The introduction of viscosity to the tissue description introduces a dependence of the tissue stiffness on the excitation frequency, where higher frequency excitations yield a stiffer tissue response compared with lower frequency excitations (i.e. the elastic moduli are a function of frequency ($E(f)$ and $\mu(f)$) (Nyborg 1965). Referring to equation (3.4), a frequency-dependent shear modulus would result in a frequency-dependent shear wave speed, called dispersion. These viscous mechanisms also result in the absorption of energy by the tissue. Tissue nonlinearities imply that the strain in response to applied stress is dependent on the absolute pressure that is applied to the tissue.

3.3.2 Dynamic Radiation Force in Vibro-Acoustography

Many scientists have theoretically demonstrated the ARF and its applications in biomedical areas (Westervelt 1951; Lee and Wang 1993; Jiang and Greenleaf 1996). It is well known that a change in the energy density of an incident acoustic field generates RF. Considering a plane traveling wave interacting with a 2D object of arbitrary shape and boundary impedance that scatters and absorbs, the force vector, F, is defined as:

$$F = d_r S\langle E\rangle \tag{3.8}$$

where d_r represents the vector drag coefficient with a component both parallel and perpendicular to the incident beam direction, and S is the projected area of the object.

The energy density, E, in equation (3.8) is given by:

$$E(t) = p^2(t)/\rho c^2 \tag{3.9}$$

where p is the unit vectors parallel to the beam direction, and c is the sound speed in the medium.

The drag coefficient can be expressed as (Westervelt 1951):

$$d_r = \frac{p}{S}\left(\Pi_a + \Pi_s - \int r\cos(\theta_s)dS\right) + \frac{q}{S}\int \gamma \sin(\theta_s)dS \tag{3.10}$$

where q is the unit vectors perpendicular to the beam direction, Π_a and Π_s are the total absorbed and scattered power, γ denotes the scattered intensity, θ_s is the angle between the incident and scattered intensity, and dS is the area element.

The drag coefficient can also be interpreted as the ratio of the RF magnitude on a given object to the corresponding value if the object were replaced by an absorbing object of similar size. This is because $|d_r| = 1$ for the totally absorbing object. This coefficient can be determined for objects of different shapes and sizes. If a planar object is normal to the beam axis, the transverse component vanishes; thus, the drag coefficient will have only a part normal to the object surface, which is denoted by scalar $d_r(F)$. Using the diameter and the wavelength to represent the values of d_r for spheres are conducted in Westervelt (1951).

An amplitude-modulated beam can be used to produce a dynamic RF. Assume $\Delta\omega \ll \omega_0$, the amplitude-modulated incident (ultrasonic) pressure field, $p(t)$, can be obtained:

$$p(t) = P_{\omega_0}\cos(\Delta\omega t/2)\cos\omega_0 t \tag{3.11}$$

where P_{ω_0} is the pressure amplitude, $\Delta\omega t/2$ represents the modulating frequency, and ω_0 is the center frequency.

The short-term time average operator is defined as:

$$\langle\xi(t)\rangle_T = 1/T\int_{t-T/2}^{t+T/2}\xi(\tau)d\tau \tag{3.12}$$

Setting $T \to \infty$, then the long-term time average can be obtained.

The short-term time average $p^2(t)$ is obtained under the condition of $2\pi/\omega_0 \ll T \ll 4\pi/\Delta\omega$:

$$\langle p^2(t)\rangle_T = \left(P_{\omega_0}^2/4\right)(1 + \cos\Delta\omega t) \tag{3.13}$$

The time-varying component of the short-term time average of the energy density is denoted as:

$$e\Delta\omega(t) = \left(P_{\omega_0}^2/4\rho c^2\right)\cos\Delta\omega t \tag{3.14}$$

This component of energy density produces a time-varying RF on the target at frequency $\Delta\omega$. The amplitude of this force is rewritten as:

$$F_{\Delta\omega} = P^2_{\omega_0} S d_r / 4\rho c^2 \tag{3.15}$$

Equation (3.15) states that the amplitude of time-varying force is proportional to the square of incident ultrasound pressure, equivalent to the incident power. If the object moves in response to this force, then the high-frequency ultrasound energy would convert to low-frequency mechanical energy.

3.3.3 Acoustic Emission

The acoustic emission is usually measured by a hydrophone and can be defined as:

$$P_{\Delta\omega} = 4\rho c^2 H_{\Delta\omega}(l) Q_{\Delta\omega} F_{\Delta\omega} \tag{3.16}$$

where $H_{\Delta\omega}(l)$ is propagation medium transfer function, $Q_{\Delta\omega}$ represents the mechanical response to the dynamic RF applied to the object at frequency $\Delta\omega$, and $F_{\Delta\omega}$ denotes the electromotive force.

The medium transfer function $H_{\Delta\omega}(l)$ in equation (3.16) is represented as:

$$H_{\Delta\omega}(l) = j\frac{\Delta\omega}{c^2} \times \frac{e^{\frac{j\Delta\omega l}{c}}}{4\pi l}\left[\frac{2J_1(b\Delta\omega/c)\sin\vartheta)}{(b\Delta\omega/c)\sin\vartheta} \times \frac{\cos\vartheta}{\cos\vartheta + \beta_B}\right] \tag{3.17}$$

where l is the distance from the observation point to the center of the piston, ϑ is the angle between this line and the piston axis, β_B is the specific acoustic admittance of the boundary surface, and b represents the object size.

The function $Q_{\Delta\omega}$ in equation (3.17) is given by:

$$Q_{\Delta\omega} = 2\pi b^2 Y_{\Delta\omega} = 2\pi b^2 / Z_{\Delta\omega} \tag{3.18}$$

where $Y_{\Delta\omega}$ is mechanical admittance at the acoustic frequency, $Z_{\Delta\omega} = Z'_m + Z_r$ is comprised of the mechanical impedance of the object in a vacuum Z'_m and the radiation impedance of the object Z_r, all defined at $\Delta\omega$. Modeling the object as a mass-spring system, Z'_m can be written in terms of $\Delta\omega$ as (Morse and Ingard 1968; Morse 1981):

$$Z'_m = R'_m - j(m\Delta\omega - K'/\Delta\omega) \tag{3.19}$$

where m represents the mass, R'_m denotes mechanical resistance, and K' is the spring constant of the object.

The radiation impedance of the piston can be written as (Morse 1981):

$$Z_r = \pi b^2 (R_r - jX_r) \tag{3.20}$$

where $R_r = \rho c[1 - (c/\Delta\omega b)J_1(c/2\Delta\omega b)$ and

$$X_r = (4\rho c/\pi)\int_0^{\pi/2} \sin\left[\left(\frac{2b\Delta\omega}{c}\right)\cos\alpha\right]\sin^2\alpha d\alpha,$$

Here, J_1 is the first-order Bessel function of the first kind.

The object vibration results in an acoustic field in the medium which is related to the object shape, size, and viscoelastic properties. The project area also determines the extent of the force applied to the object. The vibrating area, however, influences the total acoustic outflow in the medium caused by object vibration. The effects of the dynamic RF and parametric effects were studied to understand better the acoustic emission (Calle et al. 2002; Silva et al. 2005, 2006; Malcolm et al. 2008). Previous studies showed that parametric effects arise when two different frequency waves mix in a nonlinear acoustic medium (Calle et al. 2002; Silva et al. 2005, 2006; Malcolm et al. 2008). This mixture produces waves at the sum and difference frequencies of the original waves, respectively. Silva et al. explored parametric amplification of the RF produced in a fluid due to the interaction of sound beams of different frequencies (Silva et al. 2005).

Malcolm et al. conducted a numerical model to demonstrate the acoustic emission at much lower frequencies (Malcolm et al. 2008a, 2008b). The model addressed the computational needs of simulating ultrasound waves at high spatial and temporal resolution in a nonlinear medium to find the RF and the resulting acoustic emission information at much lower frequencies. A simulation image of an aluminum rod presented a good agreement with the theory.

3.3.4 Ultrasound Beamforming

The purpose of beamforming is to produce a resolution cell as small as possible. An amplitude-modulated single-focused beam can provide a resolution cell that is small in diameter but has a long in-depth direction. A superior strategy that can achieve a small resolution cell in all dimensions is to use two unmodulated focused beams at slightly different frequencies and allow them to cross each other at their focal regions. Figure 3.4 demonstrates the plane wave for VA beamforming. The small planar target is placed at R_1 tangent to the surface S. The resulting ultrasound beam is focused at R_0. The acoustic emission by the target is detected at R_2. This is accomplished by projecting two coaxial confocal continuous wave (CW) ultrasound beams on the object. Assuming the beams are propagating in a lossless medium, the total pressure field on the $z = 0$ plane is defined by:

$$p(t) = P_1(r)\cos(\omega_1 t + \psi_1(r)) + P_2(r)\cos(\omega_2 t + \psi_2(r)) \tag{3.21}$$

where the radial distance $r = \sqrt{x^2 + y^2}$, $\omega_1 = \omega_0 - \Delta\omega/2$ and $\omega_2 = \omega_0 + \Delta\omega/2$

The amplitude of the two beams are:

$$P_1(r) = \rho c U_{01}\left(\pi a_1^2/\lambda_1 z_0\right) jinc(r a_1/\lambda_1 z_0) \tag{3.22}$$

and

$$P_2(r) = \rho c U_{02}(\pi/\lambda_2 z_0)\left[a_2^2 jinc(r a_2/\lambda_2 z_0) - a_2'^2 jinc\left(r a_2'/\lambda_2 z_0\right)\right] \tag{3.23}$$

where $\lambda_i = 2\pi/\omega_i$, $i = 1, 2$ is the ultrasound wavelength, U_{0i} is the particle velocity amplitude at the ith transducer element surface, and $jinc(X) = J_1(2\pi X)/\pi X$. The phase function, $\psi_i(r) = -\pi r^2/\lambda_i z_0$, for $i = 1, 2$ is conveniently set to be 0 at the origin.

A unit point target at the position (x_0, y_0) on the focal plane with a drag coefficient distribution is obtained:

$$d_r(x, y) = \delta(x - x_0, y - y_0) \tag{3.24}$$

where $d_r(x, y)dxdy$ is unity at (x_0, y_0) and zero elsewhere.

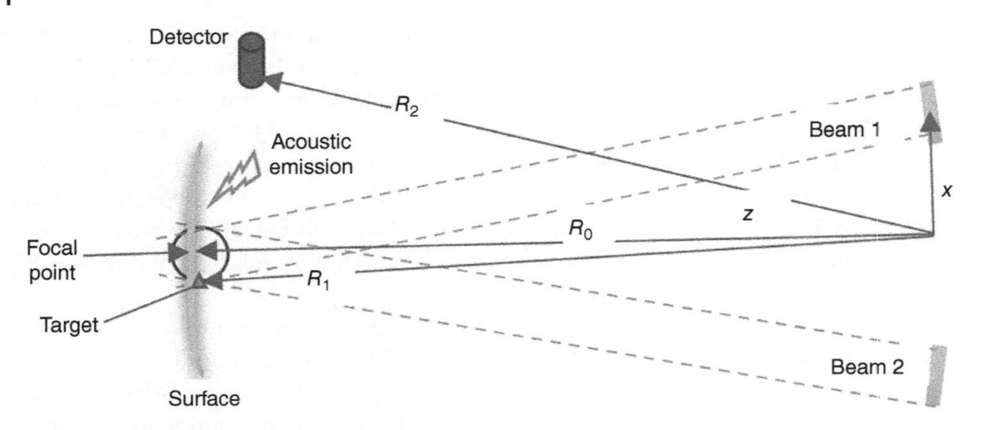

Figure 3.4 The plane wave model for VA beamforming.

Equation (3.24) is merely used as a mathematical model because d_r is physically finite. In this case, the projected area can be $S = dxdy$. Combing equations (3.8), (3.11), (3.15), and (3.21), the complex amplitude of the force on the unit point target can be produced:

$$F_{\Delta\omega}(x_0, y_0) = \rho U_{01} U_{02} \left(\pi a_1^2 / 4\lambda_1 z_0\right) jinc(r_0 a_1 / \lambda_1 z_0)$$
$$\left[\left(\pi a_2^2/\lambda_2 z_0\right) \cdot jinc(r_0 a_2/\lambda_2 z_0) - \left(\pi a'^2_2/\lambda_2 z_0\right) jinc\left(r_0 a'_2/\lambda_2 z_0\right)\right] \exp\left(j r_0^2 \Delta\omega / 2cz_0\right) \tag{3.25}$$

where the arguments x_0 and y_0 are added to denote the position of the target point and $r_0 = \sqrt{x_0^2 + y_0^2}$.

Equation (3.25) describes the spatial distribution of the force (the stress field). It shows that the stress field is confined to the regions near the beam axis ($r_0 = 0$) and decays as the radial distance r_0 increases. The lateral extent of the stress field, and hence the resolution cell diameter, would be smaller at higher ultrasound frequencies (smaller λ_1 and λ_2). The total force on an arbitrary object can be computed by integrating the force over the projected area. The axial extent of the resolution cell (depth resolution or the depth of field) can be determined by calculating the force $F_{\Delta\omega}$ as a function of the depth variable, as outlined in equations (3.21)–(3.25).

Different beamforming strategies were investigated to produce radiation stress (Chen et al. 2004; Silva et al. 2004, 2005). Three different beamforming methods applied in the study conducted by Chen et al. are amplitude modulation with a single-element transducer, confocal beams with a two-element transducer, and X-focal with two separate single-element transducers. Results show that all these methods produce beams with sub-millimeter lateral resolution. However, the X-focal arrangement could provide more improved axial solution than the other two approaches. Calle et al. (2002) examined the use of annular arrays with multiple rings. In this analysis, the authors investigated the field at Δf produced in the interaction region of the ultrasound beams and along the propagation path. Other types of transducers were explored for optimization of resolution and contrast as well as to move VA closer to clinical applications (Silva et al. 2004, 2005; Heikkila and Hynynen 2006). Urban et al. (Urban et al. 2006) used four simultaneously transmitted ultrasound frequencies to produce six unique frequencies. Their results showed that image visualization could be improved when combining various collected data in different ways.

3.3.5 Image Formation

An object (g) is scanned in a focal plane (x– y), and the complex amplitudes of the acoustic emission at different positions are recorded to produce a VA image. The acoustic emission data obtained by vibrating the object at point (x, y) are assigned to the corresponding point (x, y) in the image. The object function g(x, y) is defined (Fatemi and Greenleaf 1999):

$$g(x,y) = Q_{\Delta\omega}(x,y)d_r(x,y) \tag{3.26}$$

Variables x and y are added to denote the dependency of d_r and $Q_{\Delta\omega}$ in position. In particular, $Q_{\Delta\omega}(x, y)$ implies the total acoustic outflow by the object when unit force is applied at the point (x, y).

Point spread function (PSF) is a commonly used method to define the image of a point object in an imaging system. The normalized PSF of the coherent imaging system as the complex function can be defined as (Fatemi and Greenleaf 1999):

$$h(x,y) = P_{\Delta\omega}(x,y)/P_{\Delta\omega}(0,0) \tag{3.27}$$

Division by $P_{\Delta\omega}(0, 0)$ cancels the constant multipliers. PSF of the coherent imaging system can be rewritten referring to equations (3.16) and (3.25):

$$h(x,y) = \left(a_2^2 - a_2'^2\right)^{-1} jinc(ra_1/\lambda_1 z_0) \left[a_2^2 jinc(ra_2/\lambda_2 z_0) - a_2'^2 jinc(ra_2'/\lambda_2 z_0)\right] \exp\left(-jr^2\Delta\omega/2cz_0\right) \tag{3.28}$$

Equation (3.28) demonstrates that the system PSF is a circularly symmetric function with the peak at the origin, and decaying amplitude will increase the radial distance r. Amplitude decays faster for higher ultrasound frequency.

3.3.6 Experimental System

VA has been extensively investigated in laboratory environments. Figure 3.5 illustrates an example of the VA experimental setup that is described in Fatemi and Greenleaf (1999). A spherical piezoelectric cap is constructed as a confocal transducer. The two elements are constructed by dividing the back electrode of the piezoelectric wafer into a central disk and the outer ring. The elements have identical beam axes and focal lengths. Two stable radio frequency synthesizers drive transducer elements at two frequencies ($f_0 - \Delta f/2$) and ($f_0 + \Delta f/2$), where f_0 is the center frequency, and $\Delta f = \Delta\omega/2\pi$. The object is placed at the focal plane of the ultrasound beams in a water tank. The sound produced by the object vibration can be detected with a sensitive hydrophone (placed within the water tank). The received signal is filtered and amplified by a programmable filter to reject noise, then it is digitized by a digitizer at a rate sufficiently higher than the Nyquist rate for the particular Δf used. Data is recorded on a computer disk. For coherent imaging, which requires the phase information, the reference signal (i.e. cos$\Delta\omega t$) can be obtained by electronic downmixing of the two driving signals and is recorded along with the hydrophone signal. Discrete Hilbert transform is used to calculate the relative phase of the acoustic emission at each point (Fatemi and Greenleaf 1999). Experimental results are detailed in Fatemi and Greenleaf (1998).

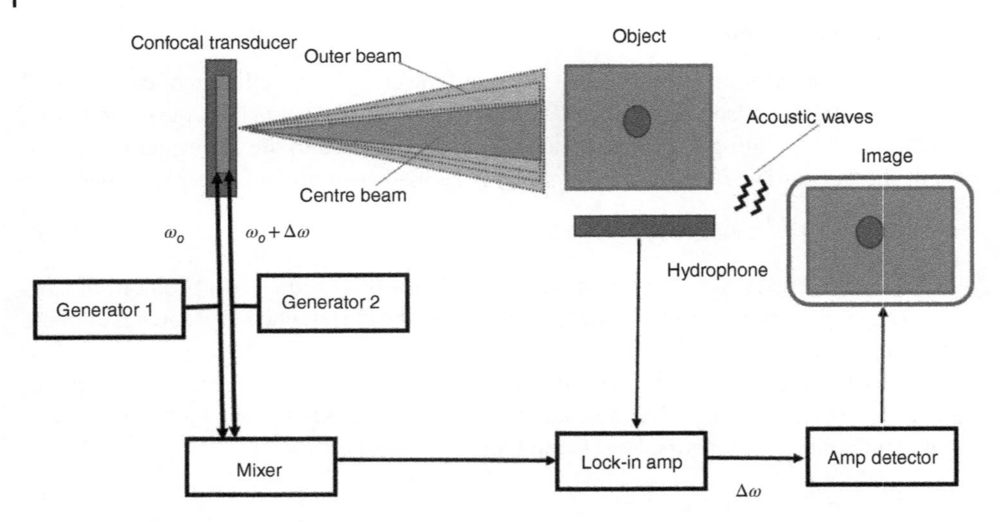

Figure 3.5 The plane wave model for VA beamforming. *Source:* Modified from Fatemi and Greenleaf (1999).

3.3.7 Multifrequency Vibro-Acoustography

It is well known that the RF can be created by two ultrasound beams at different frequencies, and the frequency difference (Δf) is typically in the kilohertz range. Multiple scans are required to obtain images of the object with different spectral content. Thus, creating images of the object to varying values of Δf may help to better understand spectral characteristics of the object. To solve this problem, Urban et al. proposed a multifrequency VA with different frequencies to produce a multifrequency stress and analyzed the created multifrequency stress field.

In the multifrequency VA, the ultrasound waves are described in terms of the velocity potential $\hat{\phi}(r,t)$, N intersecting ultrasound beams focused at the same spatial location are used to generate multifrequency dynamic radiation stress, each beam has a frequency ω_i, where $i = m, n$. The incident beams were assumed to resemble plane waves in the focal region. Based on previous findings (Silva et al. (2005) and Silva et al. (2007)), the dynamic radiation at a different frequency ω_{nm} on a small sphere of radius a is defined (Urban et al. 2006):

$$f(r,t) = \pi a^2 E_0 \sum_{(m \neq n) = 1}^{N} \hat{y}_{mn} \hat{\phi}_m(r) \hat{\phi}_n^*(r) e^{j\omega_{mn} t} \tag{3.29}$$

where $\omega_{nm} = \omega_m - \omega_n$, $m = 2, \ldots, N$, and $n = 1, 2, \ldots, N$, it is assumed $\omega_m > \omega_n$ when $m > n$. $E_0 = \varepsilon^2 \rho c^2/2$ is the energy density at the acoustic source, ε is the Mach number, $\hat{y}_{mn}$ is the RF function for the sphere, and the symbol denotes the complex conjugate.

The radiation stress is obtained with the RF divided by the projected area of the sphere:

$$\sigma(r,t) = \sum_{(m \neq n) = 1}^{N} \hat{\sigma}_{mn}(r) e^{j\omega_{mn} t} \tag{3.30}$$

where $\hat{\sigma}_{mn}(r) = E_0 \hat{y}_{mn} \hat{\phi}_m \hat{\phi}_n^*(r)$.

In a study conducted by Urban et al., force on a small sphere was used to describe the radiation stress created by the interaction of the acoustic waves. The spatial distribution of the force on a point target was represented by a sphere, which allowed to examine the stress exerted by the ultrasound in terms of the PSF. The PSF at each value of ω_{nm} is obtained (Urban et al. 2006):

$$\hat{h}_{mn} = A\hat{\phi}_m\hat{\phi}_n^*(r) \tag{3.31}$$

where A is a normalization constant for $a \rightarrow 0$, such as a point object.

Urban et al. conducted the multifrequency VA method by implementing vibrometry and VA (Urban et al. 2006). In vibrometry, images of the velocity response of an excited object were created by the multifrequency stress field (Fatemi et al. 2003). The magnitude M and phase P of the object under test can be given by the PSF and the frequency response of the object, (ω):

$$M(r, \omega_{mn}) = Re\left\{|V(\omega_{mn})|\hat{h}_{mn}(r) \otimes s(r)\right\} \tag{3.32}$$

$$P(r, \omega_{mn}) = Re\left\{\angle V(\omega_{mn})\hat{h}_{mn}(r) \otimes s(r)\right\} \tag{3.33}$$

where $|V(\omega_{mn})|$ is the magnitude of the velocity at ω_{mn}, $\angle V(\omega_{mn})$ is the phase of velocity at ω_{mn}, $\otimes$ is a spatial convolution, $Re\{\quad\}$ takes the real part of the argument, and $s(r)$ is the real function to represent the spatial sphere.

For VA method, the acoustic emission $\Phi(r, \omega_{mn})$ is obtained:

$$\Phi(r, \omega_{mn}) = Re\left\{H_{mn}(l)Q_{mn}\hat{h}_{mn}(r) \otimes s(r)\right\} \tag{3.34}$$

where $H_{mn}(l)$ is the medium transfer function that is detailed in Section 3.2 and Q_{mn} is the total acoustic outflow by the object per unit force.

The technique has been experimentally tested on a gelatin phantom containing a spherical inclusion. Results demonstrated that the multifrequency VA could be used for vibrometry and VA applications. Several benefits of the multifrequency VA approach include the fact that it can provide very rich data sets which can be used to estimate the material properties of a tissue-like medium with artifact reduction. Compared to conventional VA systems, it can gain a large amount of information without increasing the scanning time.

3.4 Vibro-Acoustography Applications

The potential abilities of VA have been demonstrated in two main areas: medical applications and material evaluation. VA has been widely studied in the medical field with a particular focus on medical imaging and biomedicine. In medical imaging, VA has been investigated for imaging breast phantoms and human breasts, human vessels, in vivo imaging of animal arteries, image brachytherapy seeds positioned in the excised prostate tissue, liver tissue, and the detection of calcium deposits on the heart valves.

3.4.1 Breast Imaging Application

VA for detection of abnormal lesions in breast imaging has been investigated extensively both in ex vivo (Alizad et al. 2004a, 2005) and in vivo tissues (Alizad et al. 2006; Hosseini et al. 2007). Micro-calcifications in studies (Fatemi et al. 2002; Alizad et al. 2004b) were successfully detected in excised tissue samples. It is crucial to see micro-calcifications because benign and malignant lesions can be indicated by micro-calcifications. Results (Alizad et al. 2004b) showed that only 21.6% of micro-calcifications (false-negative) were not detected by VA. Experimental results demonstrate that VA may provide more accurate results for measuring the breast lesion size than ultrasound and mammography. Clinical studies (Robinson et al. 1968) on human subjects have indicated that VA can detect various types of breast lesions, including fibro-adenomas and calcifications. Results presented by Hosseini et al. combined the VA and mammography to maximize the clinical impact with two imaging modalities (Hosseini et al. 2007). However, further clinical studies on a larger number of human subjects are required to prove the significance and the advantages that VA has in determining lesion sizing.

3.4.2 Arteries Imaging Application

Calcifications in arteries in excised human breast tissue specimens in vitro and large vessels of pigs in vivo (Alizad et al. 2004b; Pislaru et al. 2010) were successfully detected in VA images. The calcifications in VA images often appeared as bright structures in comparison to X-ray images. Pislaru et al. (Pislaru et al. 2010) studied pig's arteries in VA, they have compared the measured plaque area in vivo with the measured plaque area in vitro VA and X-ray radiography, and results indicated that the sensitivity for calcification detection was 100% and the specificity for calcification detection was 86%.

3.4.3 Prostate Imaging Application

VA has also been applied in diagnostic and therapeutic applications for the prostate gland. Various studies (Mitri et al. 2004, 2009a, 2009b) suggested that VA can detect brachytherapy seeds and monitor cryotherapy. The traditional ultrasound image has limitations of low signal-to-noise ratio and low contrast caused by speckle noise. It is hard to detect small calcifications, and some of the tissue structures blend in with speckle noise. The VA images of prostate tissue were compared with ultrasound images (Alizad et al. 2008), in which the ultrasound focused at a depth of 15 mm from the top surface of the prostate tissue sample. Results showed that the detailed tissue structures include calcification (panel C) that are detected by VA image. Results (Alizad et al. 2008) further confirmed that VA is a speckle-free image that provides clear images of tissue, which can be used to identify tissue structures and detect small calcifications (<1 mm).

The VA detectability of brachytherapy seeds is enhanced by Mitri et al. They compared the integrated optical density with pulse-echo and VA imaging, and the results indicated that the former decreased significantly when the transducers were moved away from the normal incident with respect to the seeds; however, the latter maintained a near constant, integrated optical density when the transducer rotated from 20° to 90° (Mitri et al. 2009a).

3.4.4 Other Applications

Other biomedical applications of VA are including imaging lesions in excised human liver (Alizad et al. 2004c), imaging bone (Calle et al. 2003), imaging of thyroid (Alizad et al. 2013), fractures in metal heart valves, and monitoring ablation tissue (Konofagou et al. 2002). Calle et al. reported the variation of VA for monitoring bone fracture, and fracture healing (Calle et al. 2003; Alizad et al. 2006b). In a recent study, VA was applied in small partial thickness rotator cuff tears, and the experimental results showed that this imaging technique could become a valuable imaging tool for detecting them (Yamamoto et al. 2014). Material evaluation is another exciting application area of VA, including mechanical parameter evaluation, imaging, and nondestructive testing of materials.

3.5 General Remarks on VA

In this section, some general remarks will be given on the benefits and limitations for the methods discussed earlier. Also comparison with other methods is presented.

3.5.1 Benefits and Limitations of VA

Several beneficial features of VA can be summarised as follows:

1) **VA images are speckle-free images:** speckle noise (artifact) in ultrasound images is produced in a medium containing numerous sub-resolution scatterers. Speckles reduce the spatial resolution and contrast quality in the ultrasound images, and they also decrease the level of detail presented in the diagnostic accuracy of ultrasound. Scatterers sizes are of the order of the ultrasound wavelength. VA-based imaging techniques can produce speckle-free images because the tissue dimensions are much smaller than the required acoustic emission wavelength in VA, and interference of backscattered signals due to multiple scattering does not affect the acoustic emission.
2) **Evaluating the dynamic characteristics of the material:** it is well known that the dynamical properties of various objects can be measured by ARF-based technique. Resonances and different vibration modes can apply different modulation frequencies (from few Hz to 100 kHz). VA is an acoustic RF-based imaging technique, which may be helpful for material characterization and evaluation. The idea was applied in the detection of fractures in metal heart valves (Fatemi et al. 2000).
3) **High contrast between hard inclusion and soft tissue:** hard inclusion in a soft tissue is displayed in VA images because: (i) the hard inclusion can be a good ultrasound reflector due to the significant difference in the acoustic impedance compared to the soft tissue and (ii) hard inclusion can produce a much stronger acoustic emission than the surrounding tissue due to its better radiation source.
4) **High sensitivity:** VA can detect motion as small as a few nanometers. This high sensitivity is due to the fact that small vibrations of the object produce an acoustic emission pressure field that is easily detectable by a sensitive hydrophone while using a low transmittance power to vibrate the object. Using a hydrophone makes this technique simple to use in clinical settings; however, it requires an acoustically quiet environment for accurate signal detection.

3.5.2 Limitations of VA

VA limitations include

1) Scanning structures very deep (up to 4–5 cm in vivo breast imaging) into the body may reduce the clarity of VA images due to the increased signal absorption and attenuation at higher depths.
2) The scanning time may be too long for clinical applications. It is desirable to reduce the scanning time for clinical applications to avoid excess patient discomfort.
3) The VA system requires filling the matching coupling medium between the object tested and the transducer, which increases the difficulty in clinical applications of the technique and increases the examination cost.

3.5.3 Comparison of Vibro-Acoustography with Pulse-echo Systems

VA beamforming is similar to its counterpart in conventional ultrasound (B-mode). Table 3.2 summarizes the significant differences between pulse-echo systems (B-mode and C-mode) and VA systems.

3.5.4 Future Directions

Several groups have extensively studied the VA techniques since 1998, and VA is still a young imaging method compared to other imaging modalities. The theory behind the scenes of VA is still not fully understood. The following areas may be necessary for future study of VA:

Table 3.2 Differences between VA beamforming and ultrasound beamforming.

	Pulse-echo system (B-mode and C-mode)	**Vibro-acoustography system**
Beam number	B-mode ultrasound imaging uses one beam	Two ultrasound beams
High-depth-resolution achievement	Achieved by transmitting short wideband pulses	Broad bandwidth is not required, it is related to how one incident beam intersects the other spatially
System	Broadband transmitted pulses	Narrowband signals
Data collection	Ultrasound system collects data along A-line (one line in depth at a time)	Acquires data from one point at a time
Image represents	Represents object microstructure by displaying its ultrasonic reflectivity distribution	Represents both microstructure and macrostructure of the object
Incident	A linear function of the incident ultrasound pressure amplitude and the amplitude reflection coefficient of the object	The acoustic signal is proportional to the ultrasound power and power reflection coefficient of the object
Sensitive to medium	Not directly sensitive to medium absorption	The acoustic emission can be produced directly as a result of energy absorption by the medium, even in a homogeneous medium

- Quantitative VA for estimating viscoelastic material properties of an object. Analysis methods and modeling approaches should be refined to solve the spatial distribution of viscoelastic material properties.
- Developing a 3D VA system to produce 3D images of the object is another area to focus on. Most published VA works are related to 2D imaging, but there is limited study on 3D VA systems to produce 3D imaging of the object. Kamimura et al. (2013) recently created a 3D surface representation of an unknown shape of bones by combining VA and B-mode techniques. However, this approach needs further clinical validation.
- Optimization of electronic beamforming with array transducers to improve image resolution and sensitivity and reduce the scanning time can be another study area. Mayo Clinic developed linear and phased arrays to study VA clinically. Design and validation of random and spiral linear and nonlinear arrays should be helpful for this work.
- Clinical implementation of VA should be another study area. VA has been studied extensively in the laboratory environment; however, only very few examples of in vivo studies in pig arteries and human breast were conducted. Clinical studies on many patients are required to further prove the possibility and safety of VA in imaging biological tissues, such as heart, breast, prostate, and thyroid gland.
- Finally, developing a commercial VA system including more sensitive transducers is the natural pathway of this work.

Bibliography

Alizad, A., Fatemi, M., Whaley, D.H., and Greenleaf, J.F. (2004a) Application of vibro-acoustography for detection of calcified arteries in breast tissue. *Journal of Ultrasound in Medicine* 23, 267–273.

Alizad, A., Fatemi, M., Wold, L.E., and Greenleaf, J.F. (2004b) Performance of vibro-acoustography in detecting microcalcifications in excised human breast tissue: a study of 74 tissue samples. *IEEE Transactions on Medical Imaging* 23, 307–312.

Alizad, A., Wold, L.E., Greenleaf, J.F., and Fatemi, M. (2004c) Imaging mass lesions by vibro-acoustography: modeling and experiments. *IEEE Transactions on Medical Imaging* 23, 1087–1093.

Alizad, A., Whaley, D.H., Greenleaf, J.F., and Fatemi, M. (2005) Potential applications of vibro-acoustography in breast imaging. *Technology in Cancer Research & Treatment* 4, 151–158.

Alizad, A., Whaley, D.H., Greenleaf, J.F., and Fatemi, M. (2006a) Critical issues in breast imaging by vibroacoustography. *Ultrasonics* 44(Suppl 1): e217–e220.

Alizad, A., Walch, M., Greenleaf, J.F., and Fatemi, M. (2006b) Vibrational characteristics of bone fracture and fracture repair: application to excised rat femur. *Journal of Biomechanical Engineering* 128(3): 300–308.

Alizad, A., Whaley, D.H., Greenleaf, J.F., and Fatemi, M. (2008) Image features in medical vibro-acoustography: in vitro and in vivo results. *Ultrasonics* 48, 559–562.

Alizad, A., Urban, M.W., Morris, J.C. et al. (2013) In vivo thyroid vibro-acoustography: a pilot study. *BMC Medical Imaging* 13(1): 12.

Altberg, W. (1903) Uber die Druckkrafte der Schallwellen und die absolute Messung der Schallintensitat. *Annalen der Physik* 11, 405–420.

Bavu, É., Gennisson, J.L., Couade, M. et al. (2011) Noninvasive in vivo liver fibrosis evaluation using supersonic shear imaging: a clinical study on 113 hepatitis C virus patients. *Ultrasound in Medicine & Biology* 37(9): 1361–1373.

Bercoff, J., Pernot, M., Tanter, M., and Fink, M. (2004a) Monitoring thermally-induced lesions with supersonic shear imaging. *Ultrasonic Imaging* 26(2): 71–84.

Bercoff, J., Tanter, M., and Fink, M. (2004b) Supersonic shear imaging: a new technique for soft tissue elasticity mapping. *IEEE Transactions on Ultrasonic Ferroelectrics* 51, 396–409.

Beyer, R.T. (1978) Radiation pressure – the history of a mislabeled tensor. *The Journal of the Acoustical Society of America* 63, 1025–1030.

Bjerknes, V.F.K. (1909) *Die Kraftfelder*. Germany: Braunschweig. Vieweg und Sohn.

Borgnis, F.E. (1953) Acoustic radiation pressure of plane compressional waves. *Reviews of Modern Physics* 25(3): 653–664.

Brandt, E.H. (1989) Levitation in physics. *Science* 243, 349–355.

Brandt, E.H. (2001) Acoustic physics – suspended by sound. *Nature* 413, 474–475.

Calle, S., Remenieras, J.P., Bou Matar, O., and Patat, F. (2002) Presence of nonlinear interference effects as a source of low frequency excitation force in vibro-acoustography. *Ultrasonics* 40, 873–878.

Calle, S., Remenieras, J.P., Bou Matar, O. et al. (2003) Application of nonlinear phenomena induced by focused ultrasound to bone imaging. *Ultrasound in Medicine & Biology* 29, 465–472.

Carson, P.L., Fischella, P.R., and Oughton, T.V. (1978) Ultrasonic power and intensities produced by diagnostic ultrasound equipment. *Ultrasound in Medicine & Biology* 3, 341–350.

Carstensen, E.L. and Schwan, H.P. (1959) Acoustic properties of hemoglobin solutions. *The Journal of the Acoustical Society of America* 31, 305–311.

Chen, S., Fatemi, M., Kinnick, R., and Greenleaf, J.F. (2004a) Comparison of stress field forming methods for vibro-acoustography. *IEEE Transactions on Ultrasonics, Ferroelectrics, and Frequency Control* 51, 313–321.

Chen, S.G., Fatemi, M., and Greenleaf, J.F. (2004b) Quantifying elasticity and viscosity from measurement of shear wave speed dispersion. *The Journal of the Acoustical Society of America* 115(6): 2781–2785.

Chen, S., Silva, G.T., Kinnick, R.R. et al. (2005) Measurement of dynamic and static radiation force on a sphere. *Physical Review. E, Statistical, Nonlinear, and Soft Matter Physics* 71, 056618.

Chen, S., Urban, M.W., Pislaru, C. et al. (2009) Shearwave dispersion ultrasound vibrometry (SDUV) for measuring tissue elasticity and viscosity. *IEEE Transactions on Ultrasonics, Ferroelectrics, and Frequency Control* 56, 55–62. doi:10.1109/TUFFC.2009.1005.

Cho, S., Lee, J., Han, J.K., and Choi, B.I. (2010) Acoustic radiation force impulse elastography for the evaluation of focal solid hepatic lesions: preliminary findings. *Ultrasound in Medicine & Biology* 36, 202–208.

Chu, B.-T. and Apfel, R.E. (1982) Acoustic radiation pressure produced by a beam of sound. *The Journal of the Acoustical Society of America* 72, 1673–1687.

Clevert, D.A., Stock, K., Klein, B. et al. (2009) Evaluation of Acoustic Radiation Force Impulse (ARFI) imaging and contrast-enhanced ultrasound in renal tumors of unknown etiology in comparison to histological findings. *Clinical Hemorheology and Microcirculation* 43, 95–107.

Coakley, W.T., Hawkes, J.J., Sobanski, M.A. et al. (2000) Analytical scale ultrasonic standing wave manipulation of cells and microparticles. *Ultrasonics* 38, 638–641.

Crum, L.A. (1971) Acoustic force on a liquid droplet in an acoustic stationary wave. *The Journal of the Acoustical Society of America* 50, 157–163.

D'onofrio, M., Gallotti, A., Salvia, R. et al. (2011) Acoustic radiation force impulse (ARFI) ultrasound imaging of pancreatic cystic lesions. *European Journal of Radiology*. 80(2): 241–244.

Dayton, P., Klibanov, A., Brandenburger, G., and Ferrara, K. (1999) Acoustic radiation force in vivo: a mechanism to assist targeting of microbubbles. *Ultrasound in Medicine & Biology* 25(8): 1195–1201.

Dayton, P.A., Zhao, S., Bloch, S.H. et al. (2006) Application of ultrasound to selectively localize nanodroplets for targeted imaging and therapy. *Molecular Imaging* 5(3): 160–174.

Deffieux, T., Montaldo, G., Tanter, M., and Fink, M. (2009) Shear wave spectroscopy for in vivo quantification of human soft tissues viscoelasticity. *IEEE Transactions on Medical Imaging* 28 (3): 313–322.

Doherty, J.R., Trahey, G.E., Nightingale, K.R., and Palmeri, M.L. (2013) Acoustic radiation force elasticity imaging in diagnostic ultrasound. *IEEE Transactions on Ultrasonics, Ferroelectrics, and Frequency Control* 60(4): 685–701.

Duck, F. (1990) *Physical Properties of Tissue, A Comprehensive Reference Book*. New York, NY: Academic Press.

Dunn, F. (1962) Temperature and amplitude dependence of acoustic absorption in tissue. *The Journal of the Acoustical Society of America* 34, 1545–1547.

Dunn, F. and Fry, W.J. (1961) Ultrasonic absorption and reflection by lung tissue. *Physics in Medicine and Biology* 5, 401–410.

Dunn, F., Edmonds, P.D., and Fry, W.J. (1969) Absorption and dispersion of ultrasound in biological media. In: *Biological Engineering* (ed. H.P. Schwan), 205–332. New York: McGraw-Hill.

Dymling, S.O., Persson, H.W., Hertz, T.G., and Lindstrom, K. (1991) A new ultrasonic method for fluid property measurements. *Ultrasound in Medicine & Biology* 17(5): 497–500.

Dyson, M., Pond, J.B., Joseph, J., and Warwick, R. (1968) The stimulation of tissue regeneration by means of ultrasound. *Clinical Science* 35, 273–285.

Dyson, M., Woodward, B., and Pond, J.B. (1971) Flow of red blood cells stopped by ultrasound. *Nature* 232, 572–573.

Eckart, C. (1948) Vortices and streams caused by sound waves. *Physics Review* 73(2): 68–76.

Evans, K.T., McCarthy, C.F., Read, A.E., and Wells, P.N. (1966) Ultrasound in diagnosis of liver disease. *British Medical Journal* 2, 1368–1369.

Fahey, B.J., Nightingale, K.R., Stutz, D.L., and Trahey, G.E. (2004) Acoustic radiation force impulse imaging of thermally- and chemically-induced lesions in soft tissues: Preliminary ex vivo results. *Ultrasound in Medicine & Biology* 30, 321–328.

Fahey, B.J., Nightingale, K.R., McAleavey, S.A. et al. (2005a) Acoustic radiation force impulse imaging of myocardial radiofrequency ablation: initial in vivo results. *IEEE Transactions on Ultrasonics, Ferroelectrics, and Frequency Control* 52, 631–641.

Fahey, B.J., Nightingale, K.R., Nelson, R.C. et al. (2005b) Acoustic radiation force impulse imaging of the abdomen: demonstration of feasibility and utility. *Ultrasound in Medicine & Biology* 31, 1185–1198.

Fahey, B.J., Nelson, R.C., Bradway, D.P. et al. (2008a) In vivo visualization of abdominal malignancies with acoustic radiation force elastography. *Physics in Medicine and Biology* 53, 279–293.

Fahey, B.J., Nelson, R.C., Hsu, S.J. et al. (2008b) In vivo guidance and assessment of liver radio-frequency ablation with acoustic radiation force elastography. *Ultrasound in Medicine & Biology* 34, 1590–1603.

Faraday, M. (1831) On a peculiar class of acoustic figures. *Philosophical Transactions. Royal Society of London* 121, 299–327.

Fatemi, M. and Greenleaf, J.F. (1998) Ultrasound-stimulated vibro-acoustic spectrography. *Science* 280, 82–85.

Fatemi, M. and Greenleaf, J.F. (1999) Vibro-acoustography: an imaging modality based on ultrasound stimulated acoustic emission. *Proceedings of the National Academy of Sciences of the United States of America* 96, 6603–6608. [PubMed:10359758].

Fatemi, M., Rambod, E., Gharib, M., and Greenleaf, J.F. (2000) Nondestructive testing of mechanical heart valves by vibro-acoustography. In: *7th International Congress on Sound and Vibration, Garmisch-Partenkirchen, Germany*, 4–7.

Fatemi, M., Ogburn, P.L., and Greenleaf, J.F. (2001) Fetal stimulation by pulsed diagnostic ultrasound. *Journal of Ultrasound in Medicine* 20, 883–889.

Fatemi, M., Wold, L.E., Alizad, A., and Greenleaf, J.F. (2002) Vibro-acoustic tissue mammography. *IEEE Transactions on Medical Imaging* 21, 1–8.

Fatemi, M., Manduca, A., and Greenleaf, J.F. (2003) Imaging elastic properties of biological tissues by low-frequency harmonic vibration. *Proceedings of the IEEE* 91(10): 1503–1519.

Ferraioli, G., Tinelli, C., Dal Bello, B. et al. (2012) Accuracy of real-time shear wave elastography for assessing liver fibrosis in chronic hepatitis C: a pilot study. *Hepatology* 56(6): 2125–2133.

Fick, S.E. and Breckenridge, F.R. (1996) Ultrasonic power output measurement by pulsed radiation pressure. *Journal of Research of the National Institute of Standards and Technology* 101, 659–669.

Fierbinteanu-Braticevici, C., Andronescu, D., Usvat, R. et al. (2009) Acoustic radiation force imaging sonoelastography for noninvasive staging of liver fibrosis. *World Journal of Gastroenterology: WJG* 15(44): 5525.

Friedrich-Rust, M., Wunder, K., Kriener, S. et al. (2009) Liver fibrosis in viral hepatitis: noninvasive assessment with acoustic radiation force impulse imaging versus transient elastography 1. *Radiology* 252(2): 595–604.

Fry, W.J. (1952) Mechanism of acoustic absorption in tissue. *The Journal of the Acoustical Society of America* 24, 412–415.

Fry WJ, Fry FJ. Fundamental neurological research and human neurosurgery using intense ultrasound. *IRE Transactions on Medical Electronics*, 1960; 166–181. ME-7. doi: 10.1109/iret-me.1960.5008041.

Fung, Y.C. (1993) *Biomechanics: Mechanical Properties of Living Tissues*, 2ee. New York, NY: Springer.

Gallotti, A., D'Onofrio, M., and Pozzi, M.R. (2010) Acoustic radiation force impulse (ARFI) technique in the ultrasound study with Virtual Touch tissue quantification of the superior abdomen. *La Radiologia Medica* 115, 889–897.

Goertz, R.S., Amann, K., Heide, R. et al. (2011) An abdominal and thyroid status with Acoustic Radiation Force Impulse Elastometry – a feasibility study: Acoustic Radiation Force Impulse Elastometry of human organs. *European Journal of Radiology*. 80(3): e226–e230.

Gor'kov, L.P. (1962) On the forces acting on a small particle in an acoustic field in an ideal fluid. *Soviet Physics – Doklady* 6, 773–775.

Greenspan, M., Breckenridge, F.R., and Tschiegg, C.E. (1978) Ultrasonic transducer power output by modulated radiation pressure. *The Journal of the Acoustical Society of America* 63, 1031–1038.

Haake, A., Neild, A., Kim, D.H. et al. (2005) Manipulation of cells using an ultrasonic pressure field. *Ultrasound in Medicine & Biology* 31(6): 857–864.

Hah, Z., Hazard, C., Cho, Y.T. et al. (2010) Crawling waves from radiation force excitation. *Ultrasonic Imaging* 32(3): 177–189.

Haque, M., Robinson, C., Owen, D. et al. (2010) Comparison of acoustic radiation force impulse imaging (ARFI) to liver biopsy histologic scores in the evaluation of chronic liver disease: a pilot study. *Annals of Hepatology* 9(3): 289–293.

Hasegawa, T. and Yosioka, K. (1969) Acoustic-radiation force on a solid elastic sphere. *The Journal of the Acoustical Society of America* 46, 1139–1143.

Hawkes, J.J., Barber, R.W., Emerson, D.R., and Coakley, W.T. (2004) Continuous cell washing and mixing driven by an ultrasound standing wave within a microfluidic channel. *Lab on a Chip* 4, 446–452.

Heikkila, J. and Hynynen, K. (2006) Investigation of optimal method for inducing harmonic motion in tissue using a linear ultrasound phased array – a simulation study. *Ultrasonic Imaging* 28, 97–113.

Hertz, H.M. (1995) Standing-wave acoustic trap for nonintrusive positioning of microparticles. *Journal of Applied Physics* 78, 4845–4849.

Hosseini, H.G., Alizad, A., and Fatemi, M. (2007) Integration of vibro-acoustography imaging modality with the traditional mammography. *International Journal of Biomedical Imaging* 2007, 40980.

Hou, G.Y., Luo, J., Marquet, F. et al. (2011) Performance assessment of HIFU lesion detection by harmonic motion imaging for focused ultrasound (HMIFU): a 3-D finite-element-based framework with experimental validation. *Ultrasound in Medicine & Biology* 37(12): 2013–2027.

Hu, J. and Santoso, A.K. (2004) A pi-shaped ultrasonic tweezers concept for manipulation of small particles. *IEEE Transactions on Ultrasonics, Ferroelectrics, and Frequency Control* 51(11): 499–507.

Hueter, T.F. and Bolt, R.H. (1951) An ultrasonic method for outlining the cerebral ventricles. *The Journal of the Acoustical Society of America* 23, 160–167.

Hultström, J., Manneberg, O., Dopf, K. et al. (2007) Proliferation and viability of adherent cells manipulated by standing-wave ultrasound in a microfluidic chip. *Ultrasound in Medicine & Biology* 33, 145–151.

Jiang, Z.Y. and Greenleaf, J.F. (1996) Acoustic radiation pressure in a three-dimensional lossy medium. *The Journal of the Acoustical Society of America* 100, 741–747.

Joyner, C.R. Jr. and Reid, J.M. (1963) Applications of ultrasound in cardiology and cardiovascular physiology. *Progress in Cardiovascular Diseases* 5, 482–497.

Kamimura, H.A., Wang, L., Carneiro, A.A. et al. (2013) Vibroacoustography for the assessment of total hip arthroplasty. *Clinics* 68(4): 463–468.

Kasai, C., Koroku, N., Koyano, A., and Omoto, R. (1985) Real-time two-dimensional blood flow imaging using an autocorrelation technique. *IEEE Transactions on Ultrasonics, Ferroelectrics, and Frequency Control* SU-32, 458–463.

King, L.V. (1934) On the acoustic radiation pressure on spheres. *Proceedings of the Royal Society of America* 147, 212–240.

Konofagou, E.E. (2004) Quo vadis elasticity imaging? [review]. *Ultrasonics* 42, 331–336.

Konofagou, E.E. and Hynynen, K. (2003) Localized harmonic motion imaging: theory, simulations and experiments. *Ultrasound in Medicine & Biology* 29, 1405–1413.

Konofagou, E., Thierman, J., Karjalainen, T., and Hynynen, K. (2002) The temperature dependence of ultrasound-stimulated acoustic emission. *Ultrasound in Medicine & Biology* 28, 331–338.

Kossoff, G. (1966) Diagnostic applications of ultrasound in cardiology. *Australasian Radiology* 10, 101–106.

Kossoff, G., Wadsworth, J.R., and Dudley, P.F. (1967) The round window ultrasonic technique for treatment of Ménière's disease. *Archives of Otolaryngology* 86, 535–542.

Kossoff, G., Robinson, D.E., and Garrett, W.J. (1968) Ultrasonic two-dimensional visualization for medical diagnosis. *The Journal of the Acoustical Society of America* 44, 1310–1318.

Kundt, A. and Lehman, O. (1874) Ueber longitudinale Schwinungen und Klangfiguren in cylindrischen Flussigkeitssaulen. *Annalen der Physik und Chemie* 153, 1–12.

Kuznetsova, L.A. and Coakley, W.T. (2007) Applications of ultrasound streaming and radiation force in biosensors. *Biosensors & Bioelectronics* 22(8): 1567–1577.

Lai, W.M., Rubin, D.H., Rubin, D., and Krempl, E. (2009) *Introduction to Continuum Mechanics*. Butterworth-Heinemann.

Lee, J. and Shung, K.K. (2006) Radiation forces exerted on arbitrarily located sphere by acoustic tweezer. *The Journal of the Acoustical Society of America* 120, 1084–1094.

Lee, C.P. and Wang, T.G. (1993) Acoustic radiation pressure. *The Journal of the Acoustical Society of America* 94, 1099–1109.

Lee, J., Ha, K., and Shung, K.K. (2005) A theoretical study of the feasibility of acoustical tweezers: ray acoustics approach. *The Acoustical Society of America* 117(5): 3273–3280.

Lehmann, J.F. and Herrick, J.F. (1953) Biologic reactions to cavitation, a consideration for ultrasonic therapy. *Archives of Physical Medicine* 3, 86–98.

Lizzi, F.L., Burt, W.J., and Coleman, D.J. (1970) Effects of ocular structures on propagation of ultrasound in the eye. *Archives of Ophthalmology* 84, 635–640.

Lizzi, F.L., Muratore, R., Deng, C.X. et al. (2003) Radiation-force technique to monitor lesions during ultrasonic therapy. *Ultrasound in Medicine & Biology* 29(11): 1593–1605.

Malcolm, A.E., Reitich, F., Yang, J. et al. (2008a) A combined parabolic-integral equation approach to the acoustic simulation of vibro-acoustic imaging. *Ultrasonics* 48, 553–558.

Malcolm, A.E., Reitich, F., Yang, J. et al. (2008b) Numerical modeling for assessment and design of ultrasound vibroacoustography systems. In: *Biomedical Applications of Vibration and Acoustics in Imaging and Characterizations* (ed. M. Fatemi and A. Al-Jumaily), 21–40. New York: ASME Press.

Maleke, C. and Konofagou, E.E. (2008) Harmonic motion imaging for focused ultrasound (HMIFU): a fully integrated technique for sonication and monitoring of thermal ablation in tissues. *Physics in Medicine and Biology* 53(6): 1773–1793.

Maleke, C., Pernot, M., and Konofagou, E.E. (2006) Single-element focused ultrasound transducer method for harmonic motion imaging. *Ultrasonic Imaging* 28, 144–158.

Maleke, C., Luo, J.W., Gamarnik, V. et al. (2010) A simulation study of amplitude-modulated (AM) Harmonic Motion Imaging (HMI) for early detection and stiffness contrast quantification of tumors with experimental validation. *Ultrasonic Imaging* 32, 154–176.

Marston, P.L. (1980) Shape oscillation and static deformation of drops and bubbles driven by modulated radiation stresses – theory. *The Journal of the Acoustical Society of America* 67, 15–26.

Marston, P.L. and Apfel, R.E. (1980) Quadrupole resonance of drops driven by modulated acoustic radiation pressure – experimental properties. *The Journal of the Acoustical Society of America* 67, 27–37.

Mayer, A.G. and Woodword, R.S. (1916) Biographical memoir of Alfred Marshall Mayer. National Academy of Sciences of the United States of America. *Biographical Memoirs V.VIII.*

McAleavey, S.A., Menon, M., and Orszulak, J. (2007) Shear-modulus estimation by application of spatially-modulated impulsive acoustic radiation force. *Ultrasonic Imaging* 29, 87–104.

McLaughlin, J. and Renzi, D. (2006a) Shear wave speed recovery in transient elastography and supersonic imaging using propagating fronts. *Inverse Problems* 22(2): 681.

McLaughlin, J. and Renzi, D. (2006b) Using level set based inversion of arrival times to recover shear wave speed in transient elastography and supersonic imaging. *Inverse Problems* 22(2): 707.

McNamara, F.L. and Beyer, R.T. (1953) A variation of the radiation pressure method of measuring sound absorption in liquids. *The Journal of the Acoustical Society of America* 25, 259–262.

Mitri, F.G. (2005) Acoustic radiation force acting on elastic and viscoelastic spherical shells placed in a plane standing wave field. *Ultrasonics* 43, 681–691.

Mitri, F.G. and Chen, S. (2005) Theory of dynamic acoustic radiation force experienced by solid cylinders. *Physical Review. E, Statistical, Nonlinear, and Soft Matter Physics* 71, 016306.

Mitri, F.G. and Fatemi, M. (2005) Dynamic acoustic radiation force acting on cylindrical shells: theory and simulations. *Ultrasonics* 43, 435–445.

Mitri FG, Fellah ZE. Amplitude-modulated acoustic radiation force experienced by elastic and viscoelastic spherical shells in progressive waves. *Ultrasonics* 2006; 44:287–296. [PubMed:16677678]

Mitri, F.G., Trompette, P., and Chapelon, J.Y. (2004) Improving the use of vibro-acoustography for brachytherapy metal seed imaging: a feasibility study. *IEEE Transactions on Medical Imaging* 23, 1–6.

Mitri, F.G., Davis, B.J., Alizad, A. et al. (2008) Prostate cryotherapy monitoring using vibroacoustography: preliminary results of an ex vivo study and technical feasibility. *IEEE Transactions on Biomedical Engineering* 55, 2584–2592.

Mitri, F.G., Davis, B.J., Greenleaf, J.F., and Fatemi, M. (2009a) In vitro comparative study of vibro-acoustography versus pulse-echo ultrasound in imaging permanent prostate brachytherapy seeds. *Ultrasonics* 49(1): 31–38.

Mitri, F.G., Davis, B.J., Urban, M.W. et al. (2009b) Vibro-acoustography imaging of permanent prostate brachytherapy seeds in an excised human prostate – preliminary results and technical feasibility. *Ultrasonics* 49, 389–394. [PubMed:19062061].

Morse, P.M. (1981) *Vibration and Sound*, 3e. Woodbury, NY: The Acoustical Society of America.

Morse, P.M. and Ingard, K.U. (1968) *Theoretical Acoustics*. New York: McGraw–Hill.

Nightingale, K.R., Kornguth, P.J., Walker, W.F., and McDermott, B.A. (1995) A novel ultrasonic technique for differentiating cysts from solid lesions: preliminary results in the breast. *Ultrasound in Medicine & Biology* 21(6): 745–751.

Nightingale, K.R., Nightingale, R.W., Palmeri, M.L., and Trahey, G.E. (2000) A finite element model of remote palpation of breast lesions using radiation force: factors affecting tissue displacement. *Ultrasonic Imaging* 22, 35–54.

Nightingale, K.R., Palmeri, M.L., Nightingale, R.W., and Trahey, G.E. (2001) On the feasibility of remote palpation using acoustic radiation force. *The Journal of the Acoustical Society of America* 110, 625–634.

Nightingale, K., Soo, M.S., Nightingale, R., and Trahey, G. (2002) Acoustic radiation force impulse imaging: in vivo demonstration of clinical feasibility. *Ultrasound in Medicine & Biology* 28(2): 227–235.

Nightingale, K., McAleavey, S., and Trahey, G. (2003) Shear-wave generation using acoustic radiation force: in vivo and ex vivo results. *Ultrasound in Medicine & Biology* 29, 1715–1723.

Nyborg, W.L. (1965) *Acoustic Streaming* (ed. W.P. Mason). New York: Academic Press.

Nyborg, W.L. (1968) Mechanisms for nonthermal effects of sound. *The Journal of the Acoustical Society of America* 44, 1302–1309.

Nyborg, W.L. (1982) Ultrasonic microstreaming and related phenomena. *The British Journal of Cancer. Supplement* Suppl. 5, 156–160.

Ostrovsky, L., Sutin, A., Il'inskii, Y. et al. (2007) Non-dissipative mechanism of acoustic radiation force generation. *JASA* 121(3): 1324–1331.

Piersol, G.M., Schwan, H.P., Pennell, R.B., and Carstensen, E.L. (1952) Mechanism of absorption of ultrasonic energy in blood. *Archives of Physical Medicine and Rehabilitation* 33, 327–332.

Piscaglia, F., Salvatore, V., Di Donato, R. et al. (2011) Accuracy of Virtual Touch Acoustic Radiation Force Impulse (ARFI) imaging for the diagnosis of cirrhosis during liver ultrasonography. *Ultraschall in der Medizin-European Journal of Ultrasound* 32(2): 167–175.

Pislaru, C., Kantor, B., Kinnick, R.R. et al. (2008) In vivo vibroacoustography of large peripheral arteries. *Investigative Radiology* 43(4): 243–252.

Pislaru, C., Greenleaf, J.F., Kantor, B., and Fatemi, M. (2010) Vibro-acoustography of arteries. In: *Atherosclerosis Disease Management* (ed. J. Suri). Berlin: Springer.

Rayleigh. (1902) On the pressure of vibrations. *The Philosophical Magzine* 3, 338–346.

Robinson, D.E., Garrett, W.J., and Kossoff, G. (1968) Fetal anatomy displayed by ultrasound. *Investigative Radiology* 3, 442–449.

Sarvazyan, A. (2001) Elastic properties of soft tissue. In: *Handbook of Elastic Properties of Solids, Liquids and Gases* (eds. M. Levy, H.E. Bass and R.R. Stern), 107–127. London, UK: Academic Press.

Sarvazyan, A., Skovoroda, A., Emelianov, S. et al. (1995) Biophysical bases of elasticity imaging. *Acoustical Imaging* 21, 223–240.

Sarvazyan, A.P., Rudenko, O.V., Swanson, S.D. et al. (1998) Shear wave elasticity imaging: a new ultrasonic technology of medical diagnostics. *Ultrasound in Medicine and Biology* 24, 1419–1435.

Sarvazyan, A.P., Rudenko, O.V., and Nyborg, W.L. (2010) Biomedical applications of radiation force of ultrasound: historical roots and physical basis. *Ultrasound in Medicine & Biology* 36(9): 1379–1394.

Schwan, H.P., Carstensen, E.L., and Li, K. (1954) Comparative evaluation of electromagnetic and ultrasonic diathermy. *Archives of Physical Medicine and Rehabilitation* 35, 13–19.

Shou, W.D., Huang, X.W., Duan, S.M. et al. (2006) Acoustic power measurement of high intensity focused ultrasound in medicine based on radiation force. *Ultrasonics* 44, e17–e20.

Silva, G.T. (2006) Dynamic radiation force of acoustic waves on solid elastic spheres. *Physical Review E* 74, 026609.

Silva, G.T. (2010) Dynamic radiation force of acoustic waves on absorbing spheres. *Brazilian Journal of Physics* 40(2): 184–187.

Silva, G.T., Greenleaf, J.F., and Fatemi, M. (2004) Linear arrays for vibro-acoustography: a numerical simulation study. *Ultrasonic Imaging* 26, 1–17.

Silva, G.T., Chen, S., Frery, A.C. et al. (2005) Stress field forming of sector array transducers for vibro-acoustography. *IEEE Transactions on Ultrasonics, Ferroelectrics, and Frequency Control* 52, 1943–1951.

Silva, G.T., Chen, S., Greenleaf, J.F., and Fatemi, M. (2005) Dynamic ultrasound radiation force in fluids. *Physical Review. E, Statistical, Nonlinear, and Soft Matter Physics* 71, 056617.

Silva, G.T., Chen, S., and Viana, L.P. (2006a) Parametric amplification of the dynamic radiation force of acoustic waves in fluids. *Physical Review Letters* 96, 234301.

Silva, G.T., Frery, A.C., and Fatemi, M. (2006b) Image formation in vibro-acoustography with depth-of-field effects. *Computerized Medical Imaging and Graphics* 30, 321–327.

Silva, G.T., Urban, M.W., and Fatemi, M. (2007) Multifrequency radiation force of acoustic waves in fluids. *Physica D: Nonlinear Phenomena* 232(1): 48–53.

Sirli, R., Bota, S., Sporea, I. et al. (2013) Liver stiffness measurements by means of supersonic shear imaging in patients without known liver pathology. *Ultrasound in Medicine & Biology* 39(8): 1362–1367.

Stock, K.F., Klein, B.S., Vo Cong, M.T. et al. (2010) ARFI-based tissue elasticity quantification in comparison to histology for the diagnosis of renal transplant fibrosis. *Clinical Hemorheology and Microcirculation* 46(2): 139–148.

Tanter, M., Bercoff, J., Athanasiou, A. et al. (2008) Quantitative assessment of breast lesion viscoelasticity: initial clinical results using supersonic shear imaging. *Ultrasound in Medicine & Biology* 34(9): 1373–1386.

Torr, G.R. (1984) The acoustic radiation force. *American Journal of Physics* 52, 402–408.

Trahey, G.E., Palmeri, M.L., Bentley, R.C., and Nightingale, K.R. (2004) Acoustic radiation force impulse imaging of the mechanical properties of arteries: in vivo and ex vivo results. *Ultrasound in Medicine & Biology* 30, 1163–1171.

Urban, M.W., Silva, G.T., Fatemi, M., and Greenleaf, J.F. (2006) Multifrequency vibroacoustography. *IEEE Transactions on Medical Imaging* 25(10): 1284–1295.

Vappou, J., Maleke, C., and Konofagou, E.E. (2009) Quantitative viscoelastic parameters measured by harmonic motion imaging. *Physics in Medicine and Biology* 54, 3579–3594.

Viola, F., Kramer, M.D., Lawrence, M.B. et al. (2004 May). Sonorheometry: a noncontact method for the dynamic assessment of thrombosis. *Annals of Biomedical Engineering* 32(5): 696–705.

Walker, W. and Trahey, G. (1995) A fundamental limit on delay estimation using partially correlated speckle signals. *IEEE Transactions on Ultrasonics, Ferroelectrics, and Frequency Control* 42, 301–308.

Wall, P.D., Fry, W.J., Stephens, R. et al. (1951) Changes produced in the central nervous system by ultrasound. *Science* 114, 686–687.

Westervelt, P.J. (1951) The theory of steady forces caused by sound waves. *The Journal of the Acoustical Society of America* 23, 312–315.

Westervelt, P.J. (1957) Acoustic radiation pressure. *The Journal of the Acoustical Society of America* 29, 26–29.

Wiklund, M. and Hertz, H.M. (2006) Ultrasonic enhancement of bead-based bioaffinity assays. *Lab on a Chip* 6, 1279–1292.

Wild, J.J. and Reid, J.M. (1952a) Application of echo-ranging techniques to the determination of structure of biological tissues. *Science* 115, 226–230.

Wild, J.J. and Reid, J.M. (1952b) Further pilot echographic studies on the histologic structure of tumors of the living intact human breast. *The American Journal of Pathology* 28, 839–861.

Wild, J.J. and Reid, J.M. (1954) Echographic visualization of lesions of the living intact human breast. *Cancer Research* 14, 277–282.

Wood, R.W. and Loomis, A.L. (1927) The physical and biological effects of high frequency sound waves of great intensity. *Philosophical Magazine* 4, 417–436.

Wu, J.R. (1991) Acoustical tweezers. *The Acoustical Society of America* 89(5): 2140–2143.

Yamamoto, N., Kinnick, R.R., Fatemi, M. et al. (2014) Diagnosis of small partial-thickness rotator cuff tears using vibro-acoustography. *Journal of Medical Ultrasonics* 1–5.

Yosioka, K. and Kawasima, Y. (1955) Acoustic radiation pressure on a compressible sphere. *Acustica* 5, 167–173.

Zhang, X., Kinnick, R.R., Fatemi, M., and Greenleaf, J.F. (2005) Noninvasive method for estimation of complex elastic modulus of arterial vessels. *IEEE Transactions on Ultrasonics, Ferroelectrics, and Frequency Control* 52, 642–652. doi:10.1109/TUFFC.2005.1428047.

Zhao, S., Borden, M., Bloch, S.H. et al. (2004) Radiation force assisted targeting facilitates ultrasonic molecular imaging. *Molecular Imaging* 3(3): 135–148.

4

Human Respiratory System

4.1 Introduction

Since the following four chapters focus on using pressure oscillation (PO) as a diagnostic tool as well as a treatment tool for the upper, central, and lower airways, it is appropriate to have a short chapter on lung mechanics and how each part of the lung is associated with various diseases. The processes of how PO can target these parts and help in diagnostics or treatments of these diseases are well explained.

4.2 Respiratory System

Breathing is essentially a mechanical process in which the muscles of the thorax and abdomen work together controlled by the brain to produce the pressures required to expand the lung so that air is sucked into it from the environment. Breathing requires a pressure gradient from outside of the body to the alveoli and then to the ambient air. If air is to move out of the lungs, the intra-alveolar pressure must exceed the atmospheric pressure. Lung diseases can cause breathing difficulties and can even cause respiratory failure. When breathing becomes uncomfortable, it is usually perceived as a sense of breathlessness known as dyspnea. In other words, there is something wrong with the mechanical properties of the lungs. People may find it difficult to breathe when their lungs become encased in thick scar tissue, as occurs in pulmonary fibrosis.

Figure 4.1 shows the human respiratory system (RS) (Tortora et al. 2018), which consists of several specific organs and tissues for the process of respiration. The lungs are the main organs of the RS, where oxygen O_2 is taken into the body and CO_2 is breathed out. The red blood cells deliver the O_2 from the lungs to the organs and take away the CO_2 from the organs, and they also transport CO_2 back to the lungs, and then the CO_2 is breathed out during exhalation.

In a spontaneously breathing subject, negative pressure is generated around the outside of the lungs by the respiratory muscles. This produces a flow of gas along the pulmonary airways in the direction of decreasing stress. The airways can be divided into the upper and the lower airways. The conducting airways connect the atmospheric air with the gas exchange membrane of the lungs. These airways do not participate in gas exchange but provide the

Pressure Oscillation in Biomedical Diagnostics and Therapy, First Edition. Ahmed Al-Jumaily and Lulu Wang.

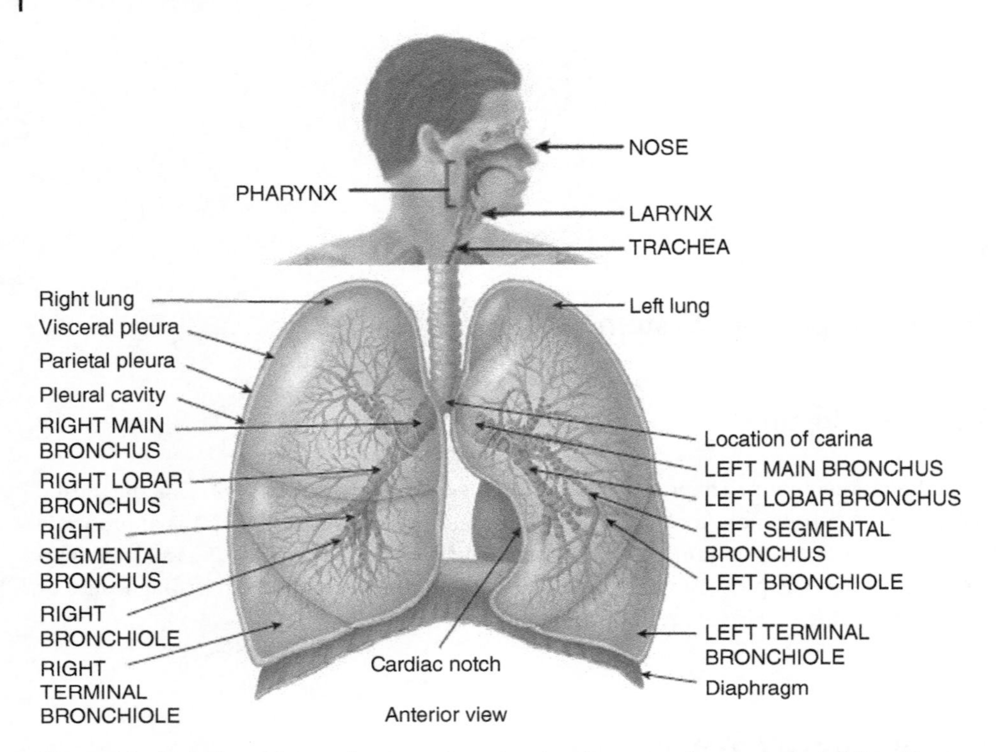

Figure 4.1 Majority of the respiratory system showing the upper airways separated from the lower airways for clarification. *Source:* Tortora et al. (2018)/with permission of John Wiley & Sons.

pathway by which inspired air reaches the gas exchange surface. Inspired air is warmed, humidified, and filtered by the upper airways in transit to this surface. The upper and lower airways share a common susceptibility to various agents, such as allergens, infectious agents, occupational sensitizers, and drugs, and respond to them in a similar fashion (Bruce et al. 1975; Townley et al. 1975; Fish et al. 1980; Naclerio et al. 1985; Iliopoulos et al. 1990; Muller et al. 1993; Rak et al. 1994; Bonavia et al. 1996). Nevertheless, each part of the airways has specialized functions: humidification, filtration of the air and the sense of smell in the nose, phonation in the larynx, and gas exchange in the lungs.

4.2.1 Upper Airways

The upper airways play an essential role in protecting the lower airways and the formation of the sound produced by the vocal cords (Corren et al. 2003). This conduit has a variety of functions, including the filtering and conditioning of air, olfaction, mastication and deglutition, phonation, coughing, and protection of the lower airways and lungs from large particulate material (Proctor 1977, 1986; Thach and Brouillette 1979; Cauna 1982; Isaacs and Sykes 2002; Doyle and Arellano 2002). The upper airways consist of several main structures, including the nasal cavity, oral cavity, pharynx, and larynx.

Nasal cavity: The nose supports and protects the anterior portion of the nasal cavity, which is a hollow space within the nose and skull. The nasal cavity helps to warm, moisturize, and filter the air entering the body before it reaches the lungs. Hairs and mucus lining in the nasal cavity help to trap dust, mold, pollen, and other environmental contaminants before they can reach the inner portions of the body. Air exiting the body through the nose returns moisture and heat to the nasal cavity before being exhaled into the environment.

Oral cavity: The oral cavity is the secondary external opening for the respiratory tract. It lacks the hairs and sticky mucus that filter air passing through the nasal cavity. Compared to breathing through the nose, the one advantage of breathing through the mouth is its shorter distance and larger diameter to allow more air to quickly enter the body. However, oral cavity neither warms nor moisturizes the air entering the lungs.

Pharynx: The pharynx is divided into the nasopharynx, oropharynx, and laryngopharynx regions. The nasopharynx is the superior region of the pharynx found in the posterior of the nasal cavity. Inhaled air from the nasal cavity passes into the nasopharynx and descends through the oropharynx, located in the posterior of the oral cavity. Air inhaled through the oral cavity enters the pharynx at the oropharynx. The inhaled air then descends into the laryngopharynx, where it is diverted into the opening of the larynx by the epiglottis. The epiglottis is a flap of elastic cartilage that acts as a switch between the trachea and the esophagus.

Larynx: The larynx is made up of the thyroid, cricoid, and arytenoid cartilages. It contains the true and false vocal cords. Its rostral aperture can be closed by the epiglottis, for example, during swallowing or reflex due to strong stimulation of irritant receptors in the lower airways. It is innervated by the superior laryngeal nerve and the recurrent vagal nerve. One of the main functions of the larynx is phonation. Pathological narrowing of the larynx (in vocal cord dysfunction, enlarged thyroid) gives rise to an increased inspiratory resistance and stridor. This is in contrast with a narrowing of the intrathoracic airways, which causes a predominantly expiratory resistance. Increased upper airway resistance, in vocal cord dysfunction or in athletes during stressful performances, can mimic exercise-induced asthma.

4.2.2 Lower Airways

The structures of upper airways are different from lower airways, but mucosal histology is similar. The main difference is the presence of sinusoids in the nose and bronchial smooth muscle in the bronchi. However, both upper and lower airways respond to various stimuli by narrowing the lumen, mucous secretion, and inflammation. Inflammatory cellular infiltration and neurogenic inflammation are similar in the nose and bronchi, but mechanisms causing airway narrowing differ. The lower airways which from the trachea to alveolar ducts consist of the following main structures.

Trachea: The trachea, also called the windpipe, connects the larynx to the bronchi and allows air to pass through the neck and into the thorax. The main function of the trachea is to provide a clear airway for air to enter and exit the lungs. In addition, the epithelium lining of the trachea produces mucus that traps dust and other contaminants and prevents it from reaching the lungs.

Bronchi and bronchioles: The airway splits into the tree-like branches of the bronchi and bronchioles. The primary bronchi contains many C-shaped cartilage rings that firmly hold the airway open and give the bronchi a cross-sectional shape like a flattened circle or a letter D. The main function of the bronchi and bronchioles is to carry air from the trachea into the lungs. The bronchioles differ from the structure of the bronchi in that they do not contain any cartilage at all. The presence of smooth muscles and elastin allow the smaller bronchi and bronchioles to be more flexible and contractile. Smooth muscle tissue in their walls helps to regulate airflow into the lungs. When more volumes of air are required by the body, such as during exercise, the smooth muscle relaxes to dilate the bronchi and bronchioles. The dilated airway provides less resistance to airflow and allows more air to pass into and out of the lungs. The smooth muscle fibers can contract during rest to prevent hyperventilation. The bronchi and bronchioles also use the mucus and cilia of their epithelial lining to trap and move dust and other contaminants away from the lungs.

Lungs: The lungs are the main organs of the RS, where the O_2 is taken into the body and CO_2 is breathed out. The red blood cells deliver the O_2 from the lungs to the organs and take away the CO_2 back to the lungs, and is then breathed out when human exhales.

Muscles: The muscles are surrounding the lungs. They are able to cause air to go into or out from the lungs. They can be divided into the internal and the external intercostal muscles. The inner intercostal muscles are the deeper set of muscles that depress the ribs to compress the thoracic cavity and force air to be exhaled from the lungs. The external intercostal muscles are found superficial to the internal intercostal muscles and function to elevate the ribs, expanding the volume of the thoracic cavity and causing air to be inhaled into the lungs.

Thoracic diaphragm: It is a dome-shaped structure of muscle and fibrous tissue that separates the thoracic cavity containing the heart and lungs from the abdominal cavity. As this diaphragm contracts, the volume of the thoracic cavity increases and the air is drawn into the lungs.

4.3 Lung Development

The lungs grow during childhood, subsequently, they deteriorate with aging and intercurrent illnesses. Smoking causes additional damage. Lung development is partitioned into five phases with the context of gestational age: embryonic phase (3–6 weeks), pseudo-glandular phase (6–16 weeks), canalicular phase (16–26 weeks), saccular phase (26–36 weeks), and alveolar phase (36 weeks–3 years) (Gray 2000; Goldsmith and Karokin 2003). The expected date of delivery is calculated at 40 weeks of pregnancy. Normal pregnancy is considered to occur between 37 and 42 weeks. Infants born before 37 weeks of age are considered preterm, and infants are termed as neonates up to 28 days after birth.

Surfactant is produced by the alveolar type II cells in lamellar bodies. It performs the function of decreasing the surface tension forces at the air–liquid interface in the alveoli and is therefore a prerequisite to breathing. Once secreted by the lamellar bodies, the surfactant is transported to the air–liquid interface where it spreads as a monolayer and acts to reduce the surface tension. This process starts around week 26 postconception and is not yet fully completed in the saccular phase (26–36 weeks). Infants born in this period have lungs with insufficient surfactant and alveoli to facilitate adequate ventilation. A lack of surfactant in preterm infants means that the surface tension forces in their alveoli are much higher and act to decrease the surface area of the alveoli to such an extent that they collapse, marking the onset of respiratory diseases.

4.4 Gas Exchange and Control

Respiration is a gas exchange procedure that includes ventilation, external respiration, internal respiration, and cellular respiration.

Ventilation as demonstrated in Figure 4.2 contains inspiration and expiration phases. Inspiration (inhalation) is an active process in which the air enters the lungs and requires the diaphragm and, in some cases, the intercostal muscles to work. The downward movement of the diaphragm enlarges the thorax, and when necessary, the intercostal muscles increase the thorax by drawing the ribs upward and outward. Expiration (exhalation) is a passive process because no muscular contractions are involved, and it is simply driven by the elastic recoil of the lungs in healthy individuals. Forced expiration occurs in certain diseases as well as during exercise (Tortora et al. 2018).

The average levels of lung volume and other indices of functions are defined in terms of age and height for the two sexes separately. For optimal accuracy, allowance should also be made for ethnic groups and body composition. Figure 4.3 and Table 4.1 show various lung volumes and capacities in a respiratory system (Polin and Fox 1997; West 1999).

Various guidelines for measuring lung volumes and airflows have been published (American Thoracic Society 1991; Quanjer et al. 1993), which will be detailed in the following chapters. Briefly, the participant should rest 15 minutes prior to the test. A mouthpiece

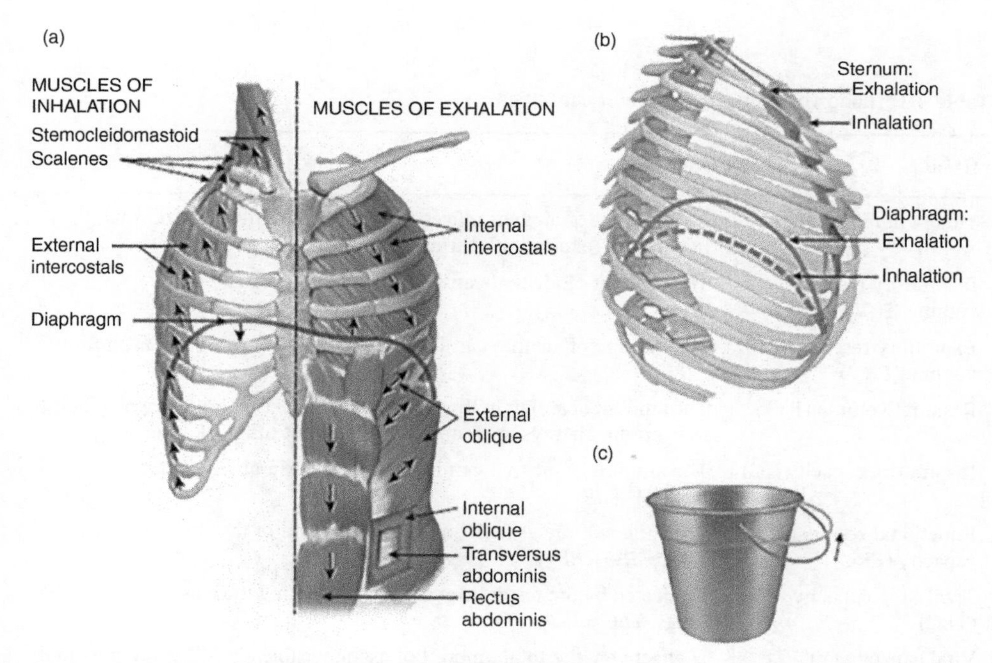

Figure 4.2 Inspiration and expiration processes: (a) muscles of inhalation (left); muscles of exhalation (right); (b) changes in size of thoracic cavity during inhalation and exhalation arrows indicate the direction of muscle contraction; and (c) during inhalation, the lower ribs (7–10) move upward and outward like the handle on a bucket. *Source:* Tortora et al. (2018)/with permission of John Wiley & Sons.

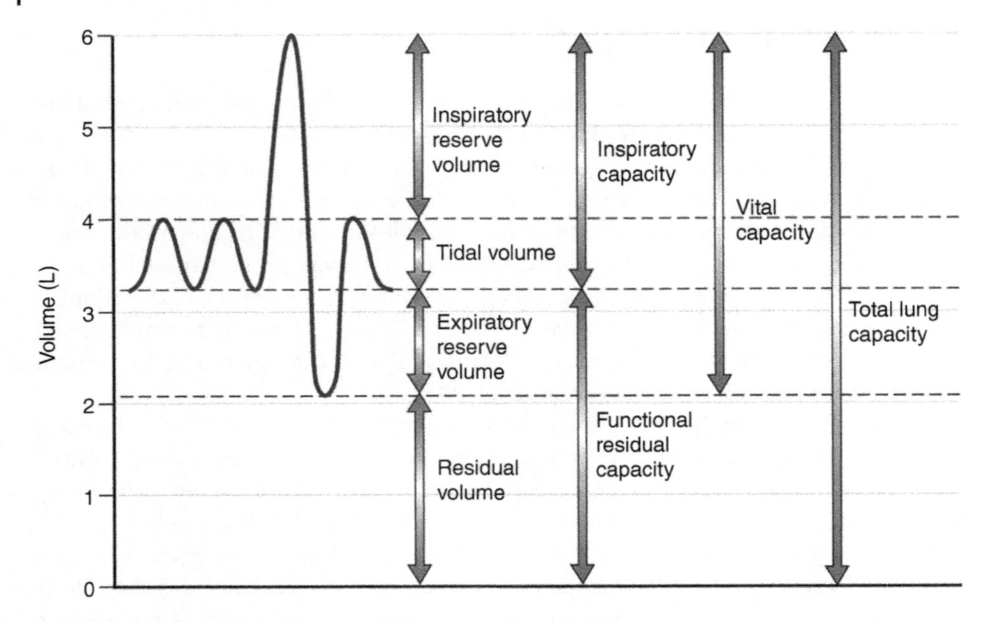

Figure 4.3 Respiratory volumes and capacities. *Source:* Based on Polin and Fox (1997) and West (1999).

Table 4.1 Lung volumes and ventilator capacities.

Name	Function
Tidal volume (TV)	The amount of air that moves into (out) the lungs with normal inspiration (expiration) during quiet breathing.
Inspiratory reserve volume (IRV)	The amount of air that can be inspired beyond the TV.
Expiratory reserve volume (ERV)	The volume of air that can be forcibly exhaled after quiet expiration.
Residual volume (RV)	The amount of air left in the lungs after a forced expiration, which helps to keep the airways open and hence prevent lung collapse.
Inspiratory capacity (IC)	The amount of air that can be inspired after a quiet expiration, $IC = TV + IRV$.
Functional residual capacity (FRC)	The lung volume at the equilibrium position of the thoracic cage, $FRC = RV + ERV$.
Total lung capacity (TLC)	Is affected by the respiratory muscles stretching the elastic tissue of the lungs, $TLC = IC + FRC$.
Vital capacity (VC)	Is effectively the total amount of exchangeable air. VC is the maximal tidal excursion from TLC to RV, or vice versa.
FEV_1	This forced expiratory volume in 1 second reflects the maximal ventilatory capacity of the lung

Source: Based on Polin and Fox (1997) and West (1999).

is inserted between the teeth and lips, and the nose is clipped. Preliminary explanation and forceful coaching during the performance are required. At least three satisfactory blows should be performed, with a maximum of eight blows if performance is faulty. Criteria of satisfactory performance of forced ventilatory maneuvers are detailed in the ATS and ERS recommendations. For clinical use, the largest values of forced vital capacity (FVC) and forced expiratory volume at the first second (FEV_1) values can be recorded by using spirometer.

4.5 Respiratory System Mechanics

The mechanical properties of the lungs determine how muscular pressures, airway flows, and lung volumes are related. The study of lung mechanics is about linking structure to function, and the experimental measurements of mechanical function are made in the laboratory and then used to infer something about the structure of the lung itself.

4.5.1 Mechanical Properties

Respiratory muscle forces are expressed as pressure and are measured with a pressure transducer calibrated with a U-tube manometer containing water or mercury. The greatest pressure a person can exert against such a manometer without using the cheek muscles is about 15 kPa above atmospheric pressure on expiratory effort or 10 kPa below atmospheric pressure during inspiratory effort. Clinical experience shows that patients who exert less than 2 kPa are weak and may need assisted ventilation.

To expand the lungs during inspiration, the muscular force has to overcome (i) the force of elastic recoil, a restorative force after stretching the tissue of the lungs, chest wall, and abdomen (proportional to the volume of expansion), (ii) force of friction during airflow (proportional to the rate of inspiratory airflow), and (iii) force to overcome the inertia of the respiratory system (proportional to mass acceleration). The pressure difference across the lung can be expressed as:

$$P = \frac{V}{C} + \frac{RdV}{dt} + \frac{Id^2V}{dt^2} \tag{4.1}$$

where V is the volume, C is the compliance, R is the resistance to airflow, and I is the inertia.

4.5.2 Airway Resistance

The resistance of the airway tree is determined by the internal dimensions of its various branches. The conducting airways are lined with a delicate epithelium that partakes in numerous metabolic activities. Some of the cells in the epithelium continually secrete protective mucus that, being sticky, acts to trap inhaled particles of potentially noxious materials. The mucus and its particle prisoners are then swept up to the tracheal opening by tiny hair-like projections known as cilia. The cilia project into the airway lumen from specialized epithelial cells and beat in the direction of the tracheal opening. The walls of the airways also contain significant amounts of smooth muscle. In the trachea, this smooth muscle exists as a continuous band along its posterior aspect, linking the open ends of cartilage horseshoes that give the trachea its mechanical stability. Contraction of the tracheal smooth muscle causes the loose ends of the horseshoes to approach each other, thereby decreasing the cross-sectional area of the tracheal lumen. Smooth muscles are wrapped circumferentially around the airways distal to the trachea and extend as far down as the alveolar ducts that serve as the entrance to the gas-exchanging zone of the lung. Whether or not there is any survival advantage to having smooth muscles in our lungs is still debated, but a disease such as asthma leaves little doubt that its presence can have adverse consequences.

The airway resistance (R_{aw}) is the resistance of the respiratory tract to airflow during inspiration and expiration:

$$R_{aw} = \frac{\Delta P}{\dot{V}} = \frac{P_{atm} - P_t}{\dot{V}} \tag{4.2}$$

where ΔP is the pressure difference driving airflow, P_{atm} is the atmospheric pressure, P_t is the alveolar pressure, and $\dot{V}$ is the flow rate.

Airflow size and airflow types (laminar or turbulent) are the most critical parameters to determine R_{aw}. Generally, narrow airways have a large resistance and vice versa. In fluid dynamics, assuming the flow is laminar, viscous, incompressible, and moving through a constant circular cross section that is substantially longer than its diameter, then R_{aw} can be obtained by Hagen–Poiseuille law (Leff and Schumacker 1993; West 1999):

$$R_{aw} = \frac{8\mu l}{\pi r^4} \tag{4.3}$$

where l is the length of the pipe, μ is the dynamic viscosity, and r is the radius of the pipe. However, equation (4.3) cannot accurately determine airway resistance, particularly the first few generations of the branches, because the actual airway bronchi are larger than the assumptions in the Hagen–Poiseuille equation.

Flow in a laminar manner has less resistance compared with a turbulent flow. A large increase in pressure difference is required to maintain flow if it becomes turbulent. Reynolds number (R_e) is introduced to determine flow type:

$$R_e = \frac{\rho v r}{2\mu} \tag{4.4}$$

where ρ is the density, v is the mean velocity, r is the radius of the pipe, and μ is the density viscosity. Generally, flow within a pipe is laminar if $R_e < 2300$ (West 1999).

The **lung tissue resistance** (R_t) cannot be measured but can be estimated. It normally represents approximately 10% of the total pulmonary resistance, and it increases in pulmonary fibrosis and other conditions where the quantity of interstitial lung tissue is increased. However, the increase can be simulated in different parts of the lung, and it varies widely in their dispensability (Tortora et al. 2018).

Total pulmonary resistance can be obtained by:

$$R_t = R_1 - R_{aw} \tag{4.5}$$

where R_1 is the pulmonary flow resistance in kPa (or cm H_2O/ml/s). To secure a velocity of flow of 1 l/s, the pressure difference must be applied between the pleural surface of the lungs and the lips.

Total thoracic resistance is the sum of the components attributable to the rib cage, the diaphragm, the abdominal wall and contents, the lung tissue, and the gas in the lung airways:

$$R_{TTR} = \frac{P_{TTR}}{\dot{V}^{n1}} = \frac{P_{th} + P_{aw} + P_t}{\dot{V}^{n1}} \tag{4.6}$$

where P is the force required to overcome the frictional resistance and thoracic, tissue, and airway, respectively. The value of $n1$ is 1.0–1.1.

The total resistance of the airway tree (Figure 4.4) is defined as:

$$R_{tot} = \frac{\Delta P}{\dot{V}} = R_P + \frac{R_{d1} R_{d2}}{R_{d1} + R_{d2}} \tag{4.7}$$

where ΔR_P is the pressure across the parent, ΔR_d is the pressure across each daughter, and R_{d1} and R_{d2} are resistances at the two daughters airways.

Several common methods to measure the lung resistance include plethysmography, pressure–flow method, and interrupter method:

- Plethysmography is the best way of measuring airway resistance because it is accurate and acceptable to most subjects, and in addition, it yields thoracic gas volume; the latter is necessary for interpreting resistance measurements. Knowledge of thoracic gas volume also has other uses.
- The classical pressure–flow method measures pulmonary resistance. It requires esophageal intubation but can be applied in the absence of a plethysmograph. The method does not measure thoracic gas volume.
- The interrupter method is a less sensitive guide to airways obstruction, and the advantage of this method is that it is readily portable for surveys (Verbanck et al. 1998).

4.5.3 Surface Tension

Water molecules align themselves so that positively and negatively charged regions of adjacent molecules are juxtaposed. This gives rise to a horizontally directed force holding the molecules

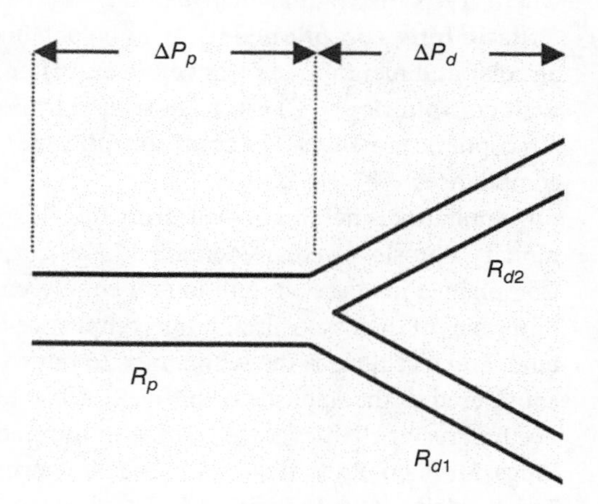

Figure 4.4 Resistance of the airway tree.

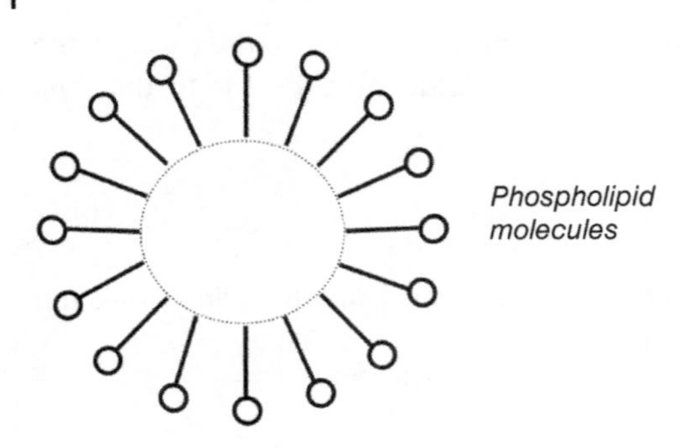

Figure 4.5 Phospholipid molecules at the air–liquid interface.

together at the alveoli surface, known as **surfaces tension**. The importance of surface tension for the lung arises from the fact that an alveolus is a very delicate membranous sac lined with liquid and acts somewhat like a bubble of fluid with an air–liquid interface at its inner surface, Figure 4.5. Surface tension (T) on the inner surface of the bubble can be expressed as:

$$T = \frac{Pr}{2} \tag{4.8}$$

where P is the positive pressure across the air–liquid interface and r is the radius of the bubble.

4.5.4 Elastance and Compliance

Elastance (E) measures the difficulty of the stretch of tissue to increase the volume it surrounds. It is determined by the lung size, tissue type, and tissue stiffness (Christie 1934; Kelly et al. 1994). On the other hand, the compliance reflects the elastic properties of the RS, and the reciprocal of the lung compliance is expressed as:

$$C = \frac{\Delta V}{\Delta P} = \frac{1}{E} \tag{4.9}$$

where ΔV is the volume change and ΔP is the pressure in the RS.

Static lung compliance (C_L) can be obtained by recording the difference between the alveolar and pleural pressures before and after inflation with a known volume. The **chest wall compliance** (C_{cw}) can be measured by recording the difference between pleural and atmospheric pressures. The **total compliance** (C_T) is the sum of the lung and the chest wall compliances.

Compliance is helpful in measuring the diseased states. It is reduced in conditions of pulmonary fibrosis, alveolar edema, and atelectasis, which prevent the inflation of the lung. Compliance increases in pulmonary emphysema and in the aging lungs where the elastic properties of the tissue alter. Lung compliance describes the willingness of the lungs to distend, and elastance is the willingness to return to the resting position. Compliance can be considered as the opposite of stiffness. A low lung compliance would mean that the lungs need a greater than average change in intrapleural pressure to change the volume of the lungs. More energy is required to breathe normally in a person with low lung compliance. The elasticity of the lung tissue and surface tension are the two major things that determine the lung compliance. People with low lung compliance due to disease, therefore, tend to take shallow breaths and breathe more frequently.

4.5.5 Impedance

Lung impedance is an empirical characterization of how the organ functions mechanically and is an expression of the overall impediment to flow in the RS (Kaczka and Dellacá 2011):

$$Z(\omega) = P(\omega)/\dot{V}(\omega) \tag{4.10}$$

where P is the pressure at airway opening, $\dot{V}$ is the airflow, $\omega = 2\pi f$ is an oscillation frequency, and f is the excitation frequency.

Using Cartesian coordinates, impedance can be expressed as (Kaczka and Dellacá 2011):

$$Z(\omega) = R(\omega) + jX(\omega) \tag{4.11}$$

where R is the respiratory resistance and X is the respiratory reactance, $j = \sqrt{-1}$.

4.6 Respiratory System Models

Various lung models based on single-compartment linear and general nonlinear models have been developed to determine the interaction of the RS with respiratory support equipment and ventilation techniques.

Figure 4.6a shows the simplest lung model which consists of single compliance (C) and a single resistance (R) (Woo et al. 2004). This RC model is suitable for simple tidal breathing predictions but is inadequate for higher frequency analysis due to the lack of inertance (I). The RIC model, Figure 4.6b, adds inertance that is useful in the study of simple maturation studies due to the small number of required model parameters (Schmidt et al. 1998). Figure 4.6c displays a two-compartment model (Crooke and Head 1996). This model can predict tidal volumes, flows, and pressures in each compartment and is well suited to investigate the uneven ventilation. However, airway and chest wall compliance are not included in this model (Bates and Milic-Emili 1993). Figure 4.6d shows the seven-parameter model (Mead 1961). Such a model contains more properties and can simulate the influences of mask leaks and compliance on the RS as well as the effects of bronchiole obstruction; however, it cannot simulate uneven ventilation.

4.7 Measurement Methods

In this section, some of the most important commonly used measurement methods are presented.

4.7.1 Lung Function Tests

It is essential to understand the lung function in diagnosing lung diseases. Lung function tests measure how well the lungs work and include breathing tests and tests that measure

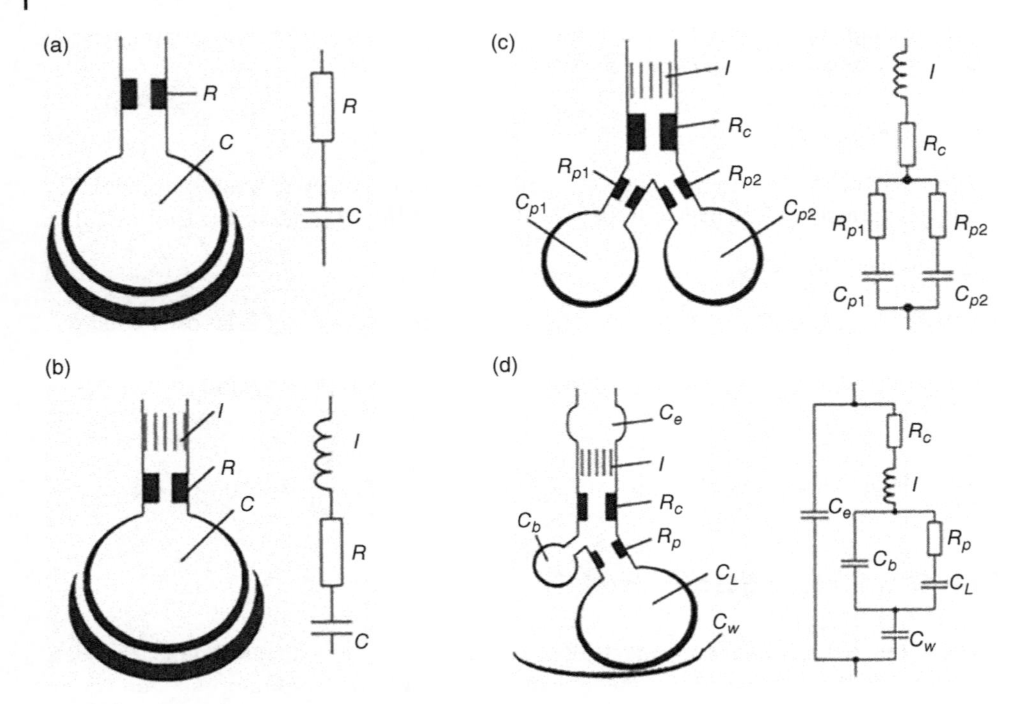

Figure 4.6 Linear lung models for simulation of input impedance: (a) RC model, (b) RIC model, (c) two-compartment model, and (d) Mead's model, inertance *I*, central and peripheral resistances (*Rc* and *Rp*) and lung, chest wall, bronchial and extrathoracic compliances (*Cl*, *Cw*, *Cb*, and *Ce*). *Source:* Schmidt et al. (1998)/with permission of Elsevier.

the oxygen level in the blood. In pulmonary function testing, a person blows air forcefully through a mouthpiece. As the person performs various breathing maneuvers, a machine records the volume and flow of the air through the lungs. Pulmonary function testing can identify the presence of obstructive lung disease or restrictive lung disease, as well as their severity. Common lung function tests include measuring the lung volumes and air flows. In adults, lung volume depends on age, gender, height, and ethnicity. Therefore, reference values for normal volumes will be different based on the factors mentioned.

4.7.2 Spirometry

Spirometry is one of the most common ways to measure lung function tests, specifically the volume and/or flow of air that is inhaled and exhaled. A spirometer is a well-known device in performing a spirometer test, and most spirometers display spirograms, Figure 4.7. Table 4.2 lists the most common lung function values measured with spirometry.

Controlled perturbations in these variables can be easily applied as probes to investigate the lung's internal workings. Nevertheless, most events inside the lung that influence its mechanical function are not accessible by direct observation. For example, a sharp increase in peak airway pressure during mechanical ventilation likely indicates that something has suddenly impeded the flow of air into the lungs. Even though an elevated peak pressure on

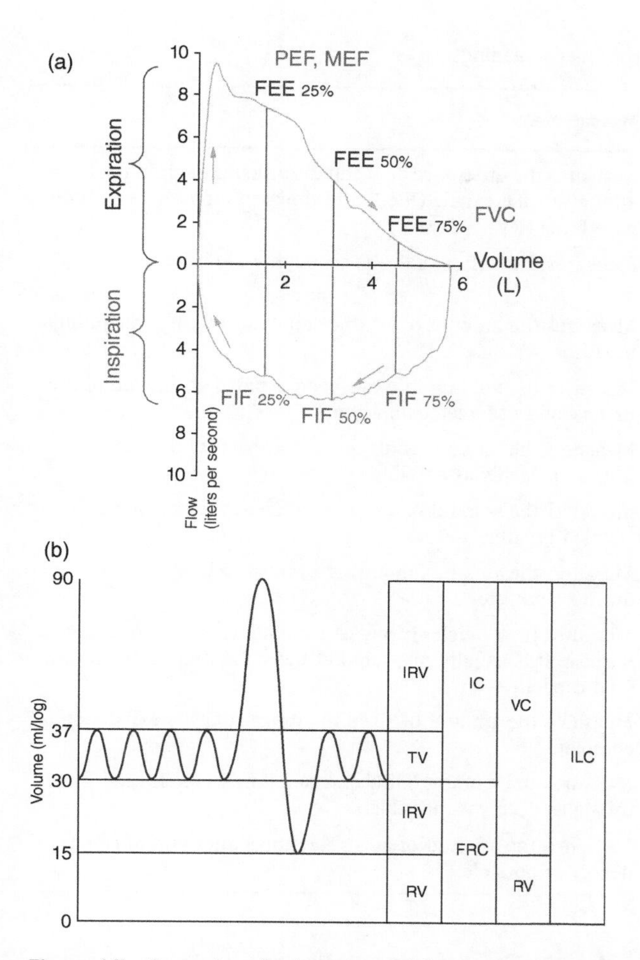

Figure 4.7 Examples of spirograms: (a) flow–volume loop and (b) volume–time curve.

its own gives no information as to where the airway impediment might be, it may still prompt an anesthesiologist to check the endotracheal tube and determine if it is blocked with mucus – clearly a helpful outcome. Empirical quantities such as peak airway pressure, FEV1, and FVC that are derived from measured variables may thus have great utility despite being of limited specificity.

4.7.3 Forced Oscillation Technique

The forced oscillation technique (FOT) is a noninvasive method to measure the respiratory mechanics. FOT employs small amplitude POs superimposed on normal breathing and therefore has the advantage over conventional lung function techniques as it does not require the performance of respiratory maneuvers.

The FOT measures the dynamic mechanical properties of the total respiratory system by analyzing its response to an externally applied excitation (DuBois et al. 1956). More

Table 4.2 Most common values of spirometry readings.

Function	Measurement
Expiratory reserve volume (ERV)	Measures the difference between the amount of air in the lungs after a normal exhale (FRC) and the amount after exhalation with force (RV).
Forced expiratory flow 25–75% (FEF_{25-75})	Measures the airflow halfway through an exhale.
Forced expiratory volume (FEV)	Measures the amount of air one can exhale with force in one breath.
Forced expiratory volume at the first second (FEV_1)	Measures the volume that has been exhaled at the end of the first second of forced expiration.
Forced vital capacity (FVC)	Measures the amount of air one can exhale with force after inhale as deeply as possible.
Functional residual capacity (FRC)	Measures the amount of air in the lungs at the end of a normal exhaled breath.
Maximum voluntary ventilation (MVV)	Measures the greatest amount of air one can breathe in and out during 1 minute.
Peak expiratory flow (PEF)	Measures how much air one can exhale when one tries the hardest. It is usually measured at the same time as the forced vital capacity (FVC).
Residual volume (RV)	Measures the amount of air in the lungs after the exhalation is complete.
Slow vital capacity (SVC)	Measures the amount of air one can slowly exhale after one inhale as deeply as possible.
Total lung capacity (TLC)	Measures the amount of air in the lungs after one inhale as deeply as possible.

commonly, FOT measures the input impedance (Z) of the respiratory system. The input impedance contains the respiratory resistance (R_{rs}, the real part of Z) and the respiratory reactance (X_{rs}, the imaginary part of Z), and both R_{rs} and X_{rs} values are highly dependent on the frequency of the applied oscillations. Generally, these measurements (magnitude and shape) over a low-frequency range (0.1–10 Hz) can be considered as sensitive indicators for the changes of the lungs mechanical properties and can be used to predict lung diseases but do not discriminate between obstructive and restrictive lung disorders. There is no consensus regarding the sensitivity of FOT for bronchodilation testing in adults. Values of respiratory resistance have proved sensitive to bronchodilation in children, although the reported cutoff levels remain to be confirmed in future studies.

FOT is a reliable method in the assessment of bronchial hyperresponsiveness in adults and children. FOT has been shown to be as sensitive as spirometry in detecting impairments of lung function due to smoking or exposure to occupational hazards. Modifying the airway smooth muscle tone is not required in FOT compared to spirometry where a deep inspiration is needed. Together with the minimal requirement for the subject's cooperation, this makes FOT an ideal lung function test for epidemiological and field studies.

4.8 Respiratory System Diseases

Human respiratory system diseases can occur from intrinsic causes as well as the inhalation of foreign bodies. The respiratory tract is constantly exposed to microbes due to the extensive surface area, which is why the respiratory system includes many mechanisms to defend itself and prevent pathogens from entering the body. Generally, respiratory diseases can be divided into two categories: obstructive and restrictive.

Figure 4.8 shows a simple view of obstructive and restrictive lung diseases that are separable based on the kinds of expiratory flow–volume curves they produce. Both pathologies have the normal PEF. However, the TLC is commensurately only reduced in the restrictive case. The RV is frequently increased in obstructive disease due to air becoming trapped behind airways that close as lung volume decreases.

4.8.1 Obstructive Lung Diseases

The most common obstructive lung diseases that can affect the respiratory system include asthma, chronic obstructive pulmonary disease (CPOD), bronchiectasis, and cystic fibrosis.

Asthma is the condition in subjects with widespread narrowing of the bronchial airways, which changes in severity over short periods of time and leads to cough, wheezing, and difficulty in breathing. A wide range of stimuli can cause asthma, including allergens, drugs, exertion, emotion, infections, and air pollution. Asthma can be controlled by medication or may subside spontaneously. A stepwise form of therapy with inhaled corticosteroid drugs and inhaled bronchodilators is an effective treatment in most patients. The onset of asthma is usually early in life, and in atopic subjects may be accompanied by other manifestations of hypersensitivity, such as hay – fever and dermatitis; however, the onset may be delayed into

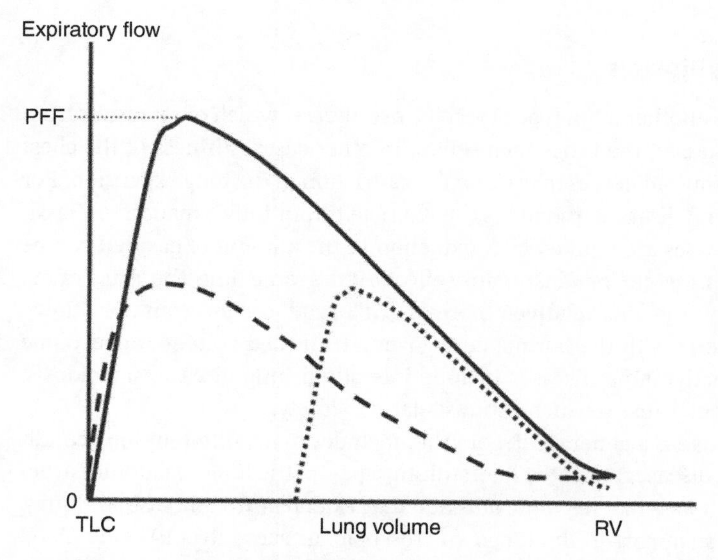

Figure 4.8 Forced expiratory flow–volume curves for a normal lung (solid line), an obstructed lung (dashed line), and a restricted lung (dotted line).

adulthood or even middle or old age. Asthma normally can be treated with bronchodilators, with or without corticosteroids, usually administered via aerosol or dry powder inhalers, or if the condition is severe via a nebulizer. Oral corticosteroids are reserved for patients who fail to respond adequately to these measures. Severe asthmatic attacks may need large doses of corticosteroids. Avoidance of known allergens, especially the house dust mites, allergens arising from domestic pets and food additives, will help to reduce the frequency of attacks, as will the discouragement of smoking.

COPD includes two main diseases: chronic bronchitis and emphysema. Chronic bronchitis is a form of COPD characterized by a chronic productive cough, which is particularly prevalent in Britain and associated with cigarette smoking and air pollution, and emphysema. The patient coughs up excessive mucus secreted by enlarged bronchial mucus glands. The bronchospasm cannot always be relieved by bronchodilator drugs. Acute bronchitis is normally caused by viruses or bacteria. Emphysema is normally caused by cigarette smoke which makes it hard to breathe if suffering from this. Pulmonary emphysema is when the air sacs of the lungs are enlarged and damaged, which reduces the surface area for the exchange of O_2 and CO_2. COPD can be treated with inhaled medications called anticholinergics, but it does not tend to resolve with therapy. Treatment tends to suppress the disease, but it continues to progress throughout the patient's life. Its progression is slowed by the cessation of cigarette smoking.

Cystic fibrosis is a severe disease that changes how the body makes mucus and sweat. It affects how well the lungs, digestive system, and some other body parts work. It is caused by a flawed gene. People with cystic fibrosis have difficulty in breathing and may have too thick mucus or too salty sweat. Most people diagnosed with cystic fibrosis today are babies. More than 75% of children with cystic fibrosis get a diagnosis by the age of 2 years. Cystic fibrosis can be life-threatening. However, with early treatment nowadays most people with cystic fibrosis live twice as long as they did 30 years ago.

4.8.2 Restrictive Lung Diseases

Restrictive lung disease is another main type of respiratory disease, which often results from a condition causing stiffness in the lungs themselves. In other cases, stiffness of the chest wall, weak muscles, or damaged nerves may cause the restriction in the lung expansion. For people with restrictive lung diseases, the lungs are restricted from fully expanding. Classically, restrictive lung diseases are typified by a reduction in the amount of gas that can be drawn into the lungs during a maximal inspiratory effort. At the same time, the shape of the maximum expiratory flow remains relatively normal. This produces an expiratory flow–volume curve that intersects with the normal curve over a truncated volume range. Some conditions causing restrictive lung diseases include interstitial lung disease, sarcoidosis, scoliosis, and neuromuscular disease such as muscular dystrophy.

Interstitial lung disease is a general category that includes many different lung conditions. All interstitial lung diseases affect the interstitium, a part of the lungs' anatomic structure. The interstitium is a lace-like network of tissue that extends throughout both lungs. The interstitium provides support to the lungs' microscopic air sacs (alveoli). Tiny blood vessels travel through the interstitium, allowing gas exchange between blood and the air in the lungs. Normally, X-rays or CT scans cannot identify the interstitium due to its thin shape.

Sarcoidosis is an inflammatory disease that affects multiple organs in the body, but primarily the lungs and lymph glands. In people with sarcoidosis, abnormal masses or nodules consisting of inflamed tissues form in specific organs of the body. These granulomas may alter the normal structure and possibly the function of the affected organs.

Scoliosis is a lateral curvature in the normally straight vertical line of the spine. When viewed from the side, the spine should show a mild roundness in the upper back and shows a degree of swayback in the lower back. When a person with a normal spine is viewed from the front or back, the spine appears to be straight. When a person with scoliosis is viewed from the front or back, the spine appears to be curved.

Muscular dystrophy is a group of diseases that makes muscles weaker and less flexible over time. It is caused by a problem in the genes that control how the body keeps muscles healthy. Many people get the disease in their childhood. Most people's condition will get worse over time, and some people may lose the ability to walk, talk, or care for themselves. People with mild symptoms normally can have longer lives.

4.9 Diagnosis of Lung Diseases

People with lung diseases normally visit a doctor when they find it hard to breathe. A doctor's interview (including smoking history), physical examination, and lab tests may provide additional clues to the cause of the lung disease. To diagnose lung diseases, physicians regularly assess mechanical abnormalities in the lung. Assessing lung mechanical function is vital to areas of basic science such as pulmonary pharmacology and immunology. Imaging tests include chest X-ray and computed tomography (CT scan) of the chest, which are almost always part of the diagnosis of lung disease. A bronchoscopy may also be recommended to diagnose the lung condition causing lung disease. In bronchoscopy, a doctor uses an endoscope to look inside the airways and take samples of lung tissue.

4.10 Respiratory Diseases Treatment

Insufficient surfactant production in the alveoli and underdeveloped lung structure normally cause respiratory distress syndrome (RDS) (Avery and Mead 1959). RDS can be treated by using traditional surfactant therapies and ventilation techniques.

4.10.1 Surfactant Therapy

Natural surfactants that are currently used in the treatment of RDS are lipid extracts of bovine lung mince, extracts of bovine lung washes, and extracts of porcine lung mince. Surfactant treatments have been shown to improve oxygenation and lung mechanics as well as lower the mean airway pressures and the percentage concentration of oxygen in the gas (fraction of inspired oxygen – FiO_2) delivered by ventilators to premature infants with RDS. A surfactant can be administered either prophylactically (within ten minutes of birth), as an early treatment (within two hours of birth), or as a rescue treatment (when respiratory system disease has been established). Although there are no trials comparing prophylactic treatment with early treatment, clinical trials show that early treatment is more effective than delaying treatment until RDS has been established (Morley and Davis 2004; Been and Zimmeman 2007).

4.10.2 Ventilation Treatments

Mechanical ventilation techniques have improved the treatment of neonates with RDS since the 1970s. Traditional modes of mechanical ventilation include intermittent mandatory ventilation (IMV) and synchronized intermittent mandatory ventilation (SIMV) with pressure and volume control modes. Early techniques of IMV are functioned by taking over the role of the respiratory muscles by facilitating gas exchange. The ventilator could be preset to deliver "mechanical breaths" at regular breath cycles. The disadvantage is that infants often breathe asynchronously with the mechanical breaths, which leads to inefficient gas exchange, gas trapping, and air leaks (Schreiner and Kisling 1982; Donn and Sinha 2003; Greenough and Sharma 2007).

Clinicians have the option within a ventilatory mode to choose between pressure-controlled ventilation (PCV) and volume-controlled ventilation (VCV). Controlling the applied pressure works in such a way that if a consistent peak inspiratory pressure (PIP) is set, the volume of air delivered to the patient would depend on lung compliance. This means that if the compliance is poor, the infant would receive a lower volume than required and if the compliance improved, the infant would receive a higher volume. Controlling the tidal volume, however, does not present much improvement of the undesirable possibility of generating high pressures in the lung. A reduced compliance increases resistance, or active exhalation increases the risk of ventilator-induced lung injury such as BPD (Schreiner and Kisling 1982; Donn and Sinha 2003; Greenough and Sharma 2007).

Mechanical ventilation causes damage and inflammation in the delicate premature lungs and impairs the lungs' ability to repair itself and develop further. Clinical investigations using mechanical ventilation performed on several species showed that lungs that were exposed to high levels of oxygen in an attempt to achieve higher levels of oxygen saturation in the blood developed significant persistent lung disease (Slutsky and Tremblay 1998). Studies on preterm animals also showed that mechanical ventilation damaged lungs by causing an inflammatory response (Coalson et al. 1995; Naik et al. 2000; Jobe and Bancalari 2001).

Due to the high risks involved with mechanical ventilation, other "gentler" techniques have been developed, such as high-frequency ventilation (HFV), biologically variable ventilation (BVV), and CPAP to avoid neonates from progressing to respiratory failure.

4.10.3 Ventilation Techniques using Pressure Oscillations

The use of POs in ventilation before the introduction of Bubble CPAP was not entirely unknown. This section presents a review of the literature focusing on describing the ventilation techniques that use POs, their clinical benefits, and proposed mechanisms of improvement.

4.10.4 High-Frequency Ventilation

HFV is a radical departure from mechanical ventilation in that it delivers gas at small tidal volumes (less than the dead space of the lung) at frequencies up to 15 Hz. The pulses of small gas volumes at rapid rates create lower pressures in the alveoli, decreasing the incidence of lung tissue injury (Kesler and Durand 2001; Owen and Lewis 2001; Ritacca and Stewart 2003). It has been suggested that the vibratory or pulsing nature of the airflow causes a “sloshing motion” of air in the lungs, known as pendelluft, and that this feature of HFV may offer better gas transport to the respiratory zone of the lung (Chang 1984; High et al. 1991). Clinical results (Rimensberger et al. 2000; Courtney et al. 2002) have shown that early, exclusive use of HFV decreases the incidence of BPD in premature infants with RDS when compared to mechanical ventilation. Infants also required a much shorter period on ventilator support.

Several mechanisms of gas transport during HFV have been suggested. During normal breathing frequencies, bulk convection and molecular diffusion take place. These mechanisms also occur during HFV, but other mechanisms have also been suggested. Chang (Chang 1984) proposed six modes of gas transport that are collectively responsible during HFV, including direct alveolar ventilation, pendelluft, convective gas transport, facilitated diffusion, lateral diffusion, and molecular diffusion. Although all six modes of gas transport may exist during HFV, it is possible that certain modes are more dominant in certain regions of the respiratory system. It has been suggested that direct ventilation and facilitated diffusion dominate ventilation in the trachea, while pendelluft or convective gas transport dominates in the medium-sized airways. Smaller airways are ventilated by pendelluft and alveoli by molecular diffusion. There are many variants of the HFV technique, but they all contain three basic elements, a high-pressure flow generator, a valve for flow interruption, and a breathing circuit connected to the patient. Variants include high-frequency jet ventilation (HFJV), high-frequency flow interruption (HFFI), high-frequency (push–pull) oscillation (HFO), high-frequency positive pressure ventilation (HFPPV), and high-frequency percussive ventilation (HFPV).

4.10.5 Continuous Positive Airway Pressure (CPAP) with Pressure Oscillations

Nasal continuous positive airway pressure (nCPAP) is a noninvasive type of respiratory support compared to endotracheal ventilation. It applies a continuous positive pressure to the alveoli (through nasal prongs or tubes) throughout the respiratory cycle and has been increasingly used in hospitals to treat neonates with RDS (Locke et al. 1991; Gitterman et al. 1997). A variable resistance creates the positive pressure applied to the lung at exhalation. It produces a more regular breathing pattern when compared to mechanical ventilation, since it allows the infant to breathe spontaneously. The continuous distending pressure provided by nCPAP increases lung volume and promotes better gas exchange in the alveoli by keeping them open (Donn and Sinha 2003). Studies have shown that nCPAP reduced the need for mechanical ventilation in neonates with moderate RDS and proved to be an acceptable form of ventilation by improving oxygenation without posing any additional harmful side effects (De Klerk and De Klerk 1987, 2001a; Verder et al. 1999; Mazzella et al. 2001).

The Bubble CPAP System was designed to offer respiratory support to spontaneously breathing neonates suffering from RDS. The Bubble CPAP System operates under the same principles but has been improved with additional features with more accuracy in maintaining required mean pressures and flows and other features that consider patient comfort.

Hospital and clinical studies have identified several benefits of Bubble CPAP making it a feasible device for managing RDS in neonates. Hospital studies on preterm and deficient birth weight infants concluded that the use of Bubble CPAP reduces the possibility of lung injury by reducing the need and number of days required for mechanical ventilation (De Klerk and De Klerk. 2001; Narendran et al. 2003). Studies also showed that hospitals that used Bubble CPAP to a greater extent on preterm infants with RDS reported a significantly lower rate of BPD than hospitals that did not (Van Marter et al. 2000).

To date, no clinical studies have ventured so far as to determine the optimum amplitude and frequencies of POs in the treatment of RDS. Some analytical studies (Manilal 2004; Manilal et al. 2004) have been done that demonstrate the ability to use the Bubble CPAP System to optimize the amplitudes and frequencies of POs produced by the system, but these remain to be validated by clinical studies.

4.10.6 Noisy Ventilation

Noisy ventilation uses a computer controller to mimic the normal variability in a spontaneously breathing lung by producing random variations in tidal volume and breath rate. Previous studies on porcine (Mutch et al. 2000a, 2000b) and rodent (Arold et al. 2002) models showed that noisy ventilation techniques improve the oxygenation of arterial blood and have enhanced the performance of mechanical ventilators. The studies on mathematical models (Suki et al. 1998) also suggest that mechanical ventilation accompanied by randomly varying breath patterns improved alveolar recruitment by opening collapsed alveoli and increasing the net lung volume without causing increases in mean airway pressures.

The mechanisms of alveolar recruitment for noisy ventilation techniques are not fully understood, although theories do exist on the mechanisms of gas transport. These are mostly associated with the benefits due to the better mixing of gases in noisy ventilation (Otis et al. 1956). Various authors (Suki et al. 1998; Arold et al. 2002) suggest that noisy ventilation is an example of stochastic resonance. Studies on small animals (Arold et al. 2002) have shown that respiratory support systems can optimize lung recruitment by tuning parameters such as the timing and amplitude of the pressure fluctuations. Whether this applies to larger animals and humans remains to be proven.

4.10.7 The Role of Vibration

The POs produced in Bubble CPAP facilitate gas transport and vibrate the lung and chest wall. These can be felt if one's hand is placed on the infant's chest. Investigations on the mechanical response of the lung walls to high frequencies are a field of study, and the relationship is not fully developed. Simulations of normal adult lungs (Fredberg and Hoenig 1978) and experiments on canine lungs (Fredberg et al. 1978) showed that the geometry and mechanical properties of the lungs are essential factors in the mechanical response of the lung. Furthermore, it is suggested that vibrations caused by POs improve the elastic condition of the lung walls (Fredberg and Hoenig 1978; Fredberg et al. 1978), and that the behavior of the lung is susceptible to the frequency of those POs (Grimal et al. 2002). However, any effect that POs have on respiratory parameters due to the mechanical properties and dynamics of the respiratory system are unknown.

Longitudinal vibrations (up to 37 Hz) to canine trachea smooth muscles (Dhanaraj and Pidaparti 2002) showed a decrease in muscle stiffness with an increase in vibration frequency. A threefold reduction in stiffness was recorded for frequencies around 35 Hz. It was postulated that mechanically inducing vibrations to lung tissue disrupted the cohesive mechanical interactions between protein filaments during contraction, resulting in a lengthening or relaxation of the tissue. However, a different mechanism (Homma 1980; Manning et al. 1991; Binks et al. 2001) related to lung relaxation found it is conceivable that ventilators such as BVV, oscillatory CPAP, and HFV which induce lung vibration at frequencies of 8–28 Hz positively affect breathing by stimulating pulmonary receptors which send information to the brain.

4.11 Closure

The airways play an important role in the respiratory function of the body. Evaluation of the structure and function of the airways should be included in all diagnostic processes in patients with respiratory problems. In many lower respiratory tract diseases, treatment cannot be successful unless a concomitant pathology of the upper airways is under control.

Monitoring of respiratory mechanics can be done widely in pulmonary medicine and in intensive care units. Mechanical ventilation is the most commonly used method to measure the respiratory mechanics of patients. In mechanically ventilated patients, mechanics measurements can provide information about the severity of disease, the response to treatment, and the safety of ventilator discontinuation. Mechanics have also become a treatment modality because measuring plateau pressures and making appropriate ventilator adjustments can lead to improved outcomes in selected patients who are receiving mechanical ventilation. Clinical studies have shown that there are physiological benefits to adding POs to respiratory treatment. However, the mechanisms for gas transport and the effect of the mechanical and surfactant properties on the dynamic response of the lungs still require further study.

To summarize, it is apparent that vibrations or POs affect different elements of the respiratory system in different ways. In the context of neonatal respiratory mechanics, lung maturity and surfactant function are the major contributors to lung disease, and the current chapter focuses on the effect that POs have on respiratory performance due to mechanical and surface tension effects of the lung.

Bibliography

American Thoracic Society (1991) Lung function testing: selection of reference values and interpretative strategies. *The American Review of Respiratory Disease* 144, 1202–1218.

Andrea, A. and Antonio, P. (2014) *Mechanics of Breathing*. New York Dordrecht London: Springer Milan Heidelberg.

Arold, S.P., Mora, R., Lutchen, K. et al. (2002) Variable tidal volume ventilation improves lung mechanics and gas exchange in a rodent model of acute lung injury. *American Journal of Respiratory and Critical Care Medicine* 165, 366–371.

Avery, M.E. and Mead, J. (1959) Surface properties in relation to atelectasis and hyaline membrane disease. *The American Journal of Disease of Children* 97, 517–523.

Bates, J.H.T. and Milic-Emili, J. (1993) Influence of the viscoelastic properties of the respiratory system on the energetically optimum breathing frequency. *Annals of Biomedical Engineering* 21, 489–499.

Been, J.V. and Zimmerman, L.J.I. (2007) What's new in surfactant? *European Journal of Pedriatrics* 166, 889–899.

Binks, A.P., Bloch-Salisbury, E., Banzett, R.B. et al. (2001) Oscillation of the lung by chest-wall vibration. *Respiration Physiology* 126(3): 245–249.

Bonavia, M., Crimi, E., Quaglia, A., and Brusasco, V. (1996) Comparison of early and late responses between patients with allergic rhinitis and mild asthma. *The European Respiratory Journal* 9, 905–909.

Bruce, C.A., Rosenthal, R.R., Lichtenstein, L.M., and Norman, P.S. (1975) Quantative inhalation bronchial challenge in ragweed hay fever patients: a comparison with ragweed-allergic asthmatics. *The Journal of Allergy and Clinical Immunology* 56, 331–337.

Cauna, N. (1982) Blood and nerve supply of the nasal lining. In: *The Nose* (eds. D.F. Proctor and I. B. Anderson), 44–69. Oxford: Elsevier Biomedical Press.

Chang, H.K. (1984) Mechanisms of gas transport during ventilation by high frequency oscillation. *Journal of Applied Physiology* 56(3): 553–563.

Christie, R.V. (1934) The elastic properties of the emphysematous lung and their clinical significance. *The Journal of Clinical Investigation* 13(2): 295–321.

Coalson, J.J., Winter, V., and deLemos, R.A. (1995) Decreased alveolarisation in baboon survivors with bronchopulmonary dysplasia. *American Journal of Respiratory and Critical Care Medicine* 152, 640–646.

Corren, J., Togias, A., and Bousquet, J. (eds.) (2003) *Upper and Lower Respiratory Disease.* CRC Press.

Courtney, S. E., Durand, D. J., Asselin, J. M. et al. (2002) High-frequency oscillatory ventilation versus conventional mechanical ventilation for very-low-birth-weight infants. *New England Journal of Medicine* 347(9): 643–652.

Crooke, P.S. and Head, J.D. (1996) A general two-compartment model for mechanical ventilation. *Mathematical and Computer Modeling* 24(7): 1–18.

De Klerk, A.M. and De Klerk, R.K. (1987) Nasal continuous positive airway pressure and outcomes of preterm infants. *Journal of Pediatrics and Child Health* 79(1): 26–29.

De Klerk, A.M. and De Klerk, R.K. (2001a) Nasal positive airway pressure and outcomes of preterm infants. *Journal of Pediatrics and Child Health* 37, 161–167.

De Klerk, A.M. and De Klerk, R.K. (2001b) Use of continuous positive airways pressure in pretern infants: comments and experience from New Zealand. *Pediatrics* 108(3): 761–763.

Dhanaraj, N. and Pidaparti, R.M. (2002) Smooth muscle tissue reponse to applied vibration following extreme isotonic shortening. *Proceedings. ASME International Mechnical Engineering Congress and Exposition*, New Orleans (17–22 November).

Donn, S.M. and Sinha, S.K. (2003) *Invasive and noninvasive neonatal ventilation. Respiratory Care* 48, 426–441.

Doyle, D.J. and Arellano, R. (2002) Upper airway diseases and airway management: a synopsis. *Anesthesiology Clinics of North America* 20(767–87): vi.

DuBois, A.B., Brody, A.W., Lewis, D.H., and Burgess, B.F. Jr. (1956) Oscillation mechanics of lungs and chest in man. *Journal of Applied Physiology* 8, 587–594.

Fish, J.E., Ankin, M.G., Kelly, J.F., and Peterman, V.I. (1980) Comparison of responses to pollen extract in subjects with allergic asthma and non-asthmatic subjects with allergic rhinitis. *The Journal of Allergy and Clinical Immunology* 65, 154–161.

Fredberg, J.J. and Hoenig, A. (1978) Mechanical response of the lungs at high frequencies. *Journal of Biomechanical Engineering* 100(2): 57–65. doi: https://doi.org/10.1115.

Fredberg, J.J., Sidell, R., and DeJong, R. (1978) Canine pulmonary input impedance measured by transient forced oscillations. *Journal of Biomechanical Engineering-Transactions of the ASME* 100(2): 57–65. doi: 10.1115/1.3426194.

Gittermann, M.K., Fusch, C., Gittermann, A.R. et al. (1997) Early nasal continuous positive airway pressure treatment reduces the need for intubation in very low birth weight infants. *European Journal of Pediatrics* 156(5):384–388.

Goldsmith, J.P. and Karotkin, E.H. (2003) *Assisted Ventilation of the Neonate.* Philadelphia: Saunders.

Gray, H. (2000) *Gray's Anatomy of the Human Body.* Philedelphia: Lea and Febiger, 1918.

Greenough, A. and Sharma, A. (2007) What is new in ventilation strategies for the neonate? *European Journal of Pedriatrics* 166, 991–996.

Grimal, Q., Watzky, A., and Naili, S. (2002) A one-dimensional model for the propagation of transient pressure waves through the lung. *Journal of Biomechanics* 35(8): 1081–1089.

High, K.C., Ultham, J.S., and Karl, S.R. (1991) Mechanically induced Pendelluft flow in a model airway bifurcation during high frequency oscillation. *Transactions of the ASME* 113, 342–347.

Homma, I. (1980) Inspiratory inhibitory reflex caused by the chest-wall vibration in man. *Respiration Physiology* 39, 345–353.

Iliopoulos, O., Proud, D., Adkinson Franklin Jr, N. et al. (1990) Relationship between the early, late, and rechallenge reaction to nasal challenge with antigen: observations on the role of inflammatory mediators and cells. *The Journal of Allergy and Clinical Immunology* 86, 851–861.

Isaacs, R.S. and Sykes, J.M. (2002) Anatomy and physiology of the upper airway. *Anesthesiology Clinics of North America* 20(733–45): v.

Jobe, A.H. and Bancalari, E. (2001) Bronchopulmonary dysplasia. *American Journal of Respiratory and Critical Care Medicine* 163, 1723–1729.

Kaczka, D.W. and Dellacá, R.L. (2011) Oscillation mechanics of the respiratory system: applications to lung disease. *Critical Reviews in Biomedical Engineering* 39(4): 337–359.

Kelly, S.M., Bates, J.H., and Michel, R.P. (1994) Altered mechanical properties of lung parenchyma in postobstructive pulmonary vasculopathy. *Journal of Applied Physiology* 77(6): 2543–2551.

Kesler, M. and Durand, D.J. (2001) Neonatal high-frequency ventilation. Past, present and future. *Clinical Perinatology* 28(3): 579–607.

Leff, A. and Schumacker, P. (1993) *Respiratory Physiology. Basics and Applications.* Philadelphia, PA: Saunders.

Locke, R., Greenspan, J.S., and Shaffer, T.H. (1991) Effect of Nasal CPAP on thoracoabdominal motion in neonates with respiratory insufficiency. *Pediatric Pulmonology* 11(3): 259–264.

Manilal, P. (2004) The effect of pressure oscillations on neonatal breathing. Master of Engineering Thesis, Diagnostics and Control Research Centre, Auckland University of Technology, Auckland, p. 134.

Manilal, P., Al-Jumaily, A., and Prime, N. (2004) The effect of pressure oscillations on neonatal breathing. *2004 ASME International Mechanical Engineering Congress & Exposition*, Anaheim, California (13–19 November 2004).

Manning, H.L., Basner, R., Ringler, J. et al. (1991) Effect of chest wall vibration on breathlessness in normal subjects. *Journal of Applied Physiology* 71(1): 175–181.

Mazzella, M., Bellini, C., Calevo, M.G. et al. (2001) A randomised control study comparing the infant flow driver with nasal continuous positive airway pressure in preterm infants. *Archives of Disease in Childhood Fetal & Neonatal Edition* 85(2), F86.

Mead, J. (1961) Mechanical properties of the lung. *Physiological Reviews* 41, 281–320.

Morley, C. and Davis, P. (2004) Surfactant treatment for premature lung disorders: a review of best practices in 2002. *Paediatric Respiratory Reviews* 5(Suppl A): S299–S304.

Muller, A.B., Cheryl, A.L., Smith, R.M. et al. (1993) Comparisons of specific and non-specific bronchoprovocation in subjects with asthma, rhinitis, and healthy subjects. *The Journal of Allergy and Clinical Immunology* 91, 758–772.

Mutch, W., Stefan, H., Ruth, G.M. et al. (2000a) Biologically variable or naturally noisy mechanical ventilation recruits atelactic lung. *American Journal of Respiratory and Critical Care Medicine* 162, 319–323.

Mutch, W., Harms, S., Lefevre, G.R. et al. (2000b) Biologically variable ventilation increases arterial oxygenation over that seen with positive end-expiratory pressure alone in a porcine model of acute respiratory distress syndrome. *Critical Care Medicine* 28(7):2457–2464.

Naclerio, R.M., Proud, D., Togias, A.G. et al. (1985) Inflammatory mediators in late antigen-induced rhinitis. *The New England Journal of Medicine* 313, 65–70.

Naik, A., Kallapur, S., Bachurski, C. et al. (2001) Effects of ventilation with different positive end-expiratory pressures on cytokine expression in the preterm lamb lung. *American Journal of Respiratory & Critical Care Medicine* 164(3): 494.

Narendran,V., Donovan, E.F., Hoath, S.B. et.al. (2003) Early bubble CPAP and outcomes in ELBW preterm infants. *Journal of Perinatology* 23(3): 195–199.

Otis, A.B., Mckerrow, C.B., Bartlett, R.A. et al. (1956) Mechanical factors in distribution of pulmonasry ventilation. *Journal of Applied Physiology* 8(4): 427–443.

Owen, M.R. and Lewis, M.A. (2001) The mechanics of lung tissue under high-frequency ventilation. *SIAM Journal on Applied Mathematics* 61(5): 1731–1761.

Polin, R.A. and Fox, W.W. (1997) *Fetal and Neonatal Physiology*, 2ee. Philadelphia: WB Saunders Co.

Proctor, D.F. (1977) The upper airways. II. The larynx and trachea. *The American Review of Respiratory Disease* 115, 315–342.

Proctor, D. (1986) Form and function of the upper airways and the larynx. In: *Handbook of Physiology. Section 3. The Respiratory System, Mechanics of Breathing* (eds. P. Macklem and J. Mead), 63–74. Bethesda, MD: American Physiological Society.

Quanjer, P.H., Tammeling, G.J., Cotes, J.E. et al. (1993) Lung volumes and forced ventilatory flows. Report working party. Standardization of lung function tests. European Community for Steel and Coal. *European Respiratory Journal* 6(Suppl. 16): 5–40.

Rak, S., Jacobson, M.R., Sudderick, R.M. et al. (1994) Influence of prolonged treatment with topical corticosteroid (fluticasone propionate) on early and late phase nasal responses and cellular infiltration in the nasal mucosa after allergen challenge. *Clinical and Experimental Allergy* 24, 930–939.

Rimensberger, P.C., Beghetti, M., Hanquinet, S. et al. (2000) First intention high-frequency oscillation with early lung volume optimization improves pulmonary outcome in very low birth weight infants with respiratory distress syndrome. *Pediatrics* 105(6): 1202–1208.

Ritacca, F.V. and Stewart, T.E. (2003) Clinical review: high frequency osillatory ventilation in adults – a review of the literature and practical applications. *Critical Care* 7, 385–390.

Schmidt, M., Foitzik, B., Hochmuth, O. et al. (1998) Computer simulation of the measured respiratory impedance in newborn infants and the effect of the measurement equipment. Medical Engineering & Physics 20(3): 220–228.

Schreiner, R.L. and Kisling, J.A. (1982) *Practical Neonatal Respiratory Care*. New York: Raven Press.

Slutsky, A.S. and Tremblay, L.N. (1998) Multiple system organ failure. Is mechanical ventilation a contributing factor? *American Journal of Respiratory and Critical Care Medicine* 157, 1721–1725.

Suki, B., Alencar, A.M., Sujeer, M.K. et al. (1998) Life-support system benefits from noise. *Nature* 393, 127–128.

Thach, B. and Brouillette, R. (1979) The respiratory function of pharyngeal musculature: relevance to clinical obstructive apnea. In: *Central Nervous Control Mechanisms in Breathing* (eds. C. Euler and H. Lagercrantz), 483–494. Oxford: Pergamon Press.

Tortora, G.J., Derrickson, B., Burkett, B. et al. (2018) *Principles of Anatomy and Physiology*, 2nd Asia-Pacifice. Wiley. ProQuest Ebook Central. https://ebookcentral.proquest.com/lib/AUT/detail.action?docID=5561269.

Townley, R.G., Ryo, U.K., Kolotzin, M.B., and Kang, B. (1975) Bronchial sensitivity to methacholine in current and former asthmatic and allergic rhinitis patients and control subjects. *The Journal of Allergy and Clinical Immunology* 56, 429–442.

Van Marter, L.J., Allred, E.N., Pagano, M. et al. (2000) Do clinical markers of Barotrauma and oxygen toxicity explain interhospital variation in rates of chronic lung disease? *Pediatrics* 105 (6): 1194–1201.

Verbanck, S., Schuermans, D., Van Muylem, A. et al. (1998) Conductive and acinar lung-zone contributions to ventilation inhomogeneity in COPD. *American Journal of Respiratory and Critical Care Medicine* 157, 1573–1577.

Verder, H., Albertsen, P., Ebbesen, F. et al. (1999) Nasal continuous positive airway pressure and early surfactant therapy for respiratory distress syndrome in newborns of less than 30 weeks' gestation. *Pediatrics*
103(2): E24.

West, J.B. (1999) *Respiratory Physiology. The Essentials*, 6ee. Philadelphia, PA: Lippincott Williams & Wilkins.

Woo, T., Diong, B., Mansfield, L. et al. (2004) A comparison of various respiratory system models based on parameter estimates from impulse oscillometry data. *Proceedings of the 26th Annual Conference IEEE EMBS*, San Francisco. 5, 3828–3831.

5

Forced Oscillation Technique

5.1 Introduction

The forced oscillation technique (FOT) is another application of the pressure oscillation method for lung diagnostics. It is based on the measurement of the respiratory impedance, which has evolved into powerful tools for assessing various mechanical phenomena in the mammalian lung during health and disease. Although FOT was introduced in the previous chapter, it is presented in detail in this chapter, including working principle, instrumentation, measurement arrangement, and impedance measurement methods. Finally, some clinical applications of FOT are also described.

Application of cyclic muscular pressure to the chest wall produces normal tidal breathing. The breathing flow generated by the driving pressure is determined by the mechanical properties of the respiratory system. Therefore, information on the mechanical properties of the airways and lung and chest wall tissues can be derived from the relationship between the driving pressure and the resulting flow. Airflow and volume changes can be easily recorded with a pneumotachograph placed at the airway opening. However, muscular pressure cannot be directly measured. An alternative procedure is to establish an esophageal balloon and regard this recording as an estimate of pleural pressure. A noninvasive approach is to apply low-amplitude pressure oscillation with a loudspeaker connected to the mouth during spontaneous breathing. Using external excitation makes it possible to measure both the oscillatory driving pressure and the resulting oscillatory flow. This approach is known as the FOT (Marchal and Loos 1997; Navajas and Farré 1999).

The current applications of FOT can be divided into two main groups based on the frequency range. The first group studies the frequency range outside of spontaneous breathing (0.1–1 Hz). The respiratory impedance can be obtained at frequencies above (2–32 Hz) (Daróczy and Hantos 1990; Zerah et al. 1995; Pasker et al. 1996; de Melo et al. 1998; Rigau et al. 2002) or below the breathing frequency (0.01–0.1 Hz) (Suki et al. 1989). In the first case, the pressure oscillations are generated by using a piston or loudspeaker, while in the latter, a body chamber is used. In both cases, the breathing effect on the pressure and flow signals are deemed to be residual out-of-band noise and can easily be filtered out. This has the main advantage that the patient can breathe normally, and no special cooperation is required. The second FOT techniques measuring the respiratory impedance at frequency bands partly or entirely overlap with the frequency band of spontaneous breathing. In several applications, low-frequency respiratory impedance measurements are performed on patients who are

Pressure Oscillation in Biomedical Diagnostics and Therapy, First Edition. Ahmed Al-Jumaily and Lulu Wang.

maintained on artificial mechanical ventilation. The respiratory impedance is then assessed during end-expiratory pauses to avoid breathing disturbances. For these purposes, servo-controllers are often used in combination with pistons or loudspeakers to generate the desired pressure oscillations (Farré et al. 1995; Kaczka et al. 1997; Kaczka et al. 1999a; Kaczka and Lutchen 2004).

Apart from mechanical ventilation, other specific circumstances are created to measure the respiratory impedance at low frequencies without the introduction of breathing disturbances. The respiratory impedance was measured at frequencies 0.25–5 Hz on subjects (Hantos et al. 1986) familiar with lung function measurements. Modeling the low-frequency respiratory impedance in dogs (Hantos et al. 1990) was performed using oscillations 0.125–5 Hz generated by a loudspeaker. Low-frequency respiratory mechanics (Sly et al. 1996) were obtained from sedated infants by exploiting an apneic phase produced by triggering an inflation reflex.

5.2 Forced Oscillation Technique

FOT measures the impedance of the respiratory system to "forced" pressure oscillations produced by a loudspeaker. Respiratory impedance is a complex quantity that includes resistance and reactance. Resistance is the more familiar parameter, comprising the pressure–flow relationship of that portion of the pressure oscillation, which is "in phase" with airflow.

5.2.1 FOT Development History

In 1951 Dubois et al. first studied FOT (DuBois et al. 1956), by applying small sinusoidal pressures to stimulate the RS at frequencies higher than the normal breathing frequency and measured the flow response to demonstrate the human RS. Since then, a large number of research groups have investigated FOT techniques. Mead (Mead 1960a) studied the FOT theory and found that the forced oscillations could be superimposed on the normal breathing pattern to measure the flow resistance of the RS (Mead 1960b).

Fisher et al. investigated the FOT to evaluate the total respiratory resistance in human subjects and provided an evaluation of the tissue resistance in subjects with various thoracic diseases (Fisher et al. 1968). Grimby et al. conducted clinical trials of FOT on human subjects with chronic airflow obstruction (CAO) at a frequency range of 3–9 Hz, and the research findings demonstrated that the frequency dependences of resistance and compliance are interpreted as effects of uneven distribution of the mechanical properties in the lungs (Otis et al. 1956; Grimby et al. 1968; Ingram and O'Cain 1971). Figure 5.1 shows a typical graphic representation of FOT data in normal subjects and patients with CAO. Figure 5.1 indicates that for normal subjects, the resistance R1 is constant at all analyzed frequencies. However, the reactance X1 is slightly negative at the lowest frequency (4 Hz) and rises to cross zero (resonant frequency, f_{res}) at 8 Hz, and is positive at all higher frequencies. On the other hand, for subjects with CAO the resistance R2 is elevated, most at lower frequencies, and decreases as the frequency increases. However, the reactance X2 is markedly negative at low frequencies and does not reach zero (f_{res}) until 24 Hz.

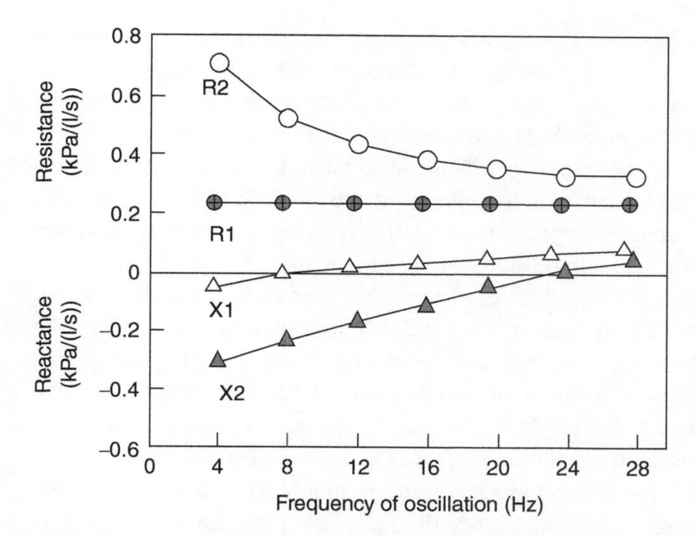

Figure 5.1 Schematic representation of resistance (R_{rs}) and reactance (X_{rs}) at different frequencies of forced oscillation: R1 and X1 Normal adult; R2 and X2 Patients with CAO. *Source:* Based on Otis et al. (1956), Grimby et al. (1968), and Ingram and O'Cain (1971).

Spectral analysis method to improve FOT applications was proposed by Michaelson et al. who studied the FOT over a wide frequency range (2–32 Hz) (Michaelson et al. 1975). Landser et al. further tested this method on healthy nonsmokers and subjects with CAO (Làndsér et al. 1976). The results demonstrated that the frequency-dependent resistance in subjects with CAO dropped substantially after bronchodilator inhalation.

Kjeldgaard et al. (Kjeldgaard et al. 1976) investigated the frequency-dependent resistance in asymptomatic smokers. Hayes et al. further studied the differences between normal nonsmokers and asymptomatic smokers for frequency-dependent resistance (Hayes et al. 1979). The expected values of FOT resistance and frequencies for normal subjects were not established until the middle of the 1980s. Many researchers have investigated FOT in normal children and children with airflow obstruction (Stanescu et al. 1979; Clement et al. 1987; Bisgaard and Klug 1995; Klug and Bisgaard 1996, 1997, 1998, 1999; Ducharme and Davis 1997, 1998; Ducharme et al. 1998; Delacourt et al. 2000; Klug et al. 2000; Nielsen and Bisgaard 2000a, 2000b).

5.2.2 Forced Oscillation Technique Types

Many researchers have studied the FOT in terms of measurement configuration, oscillation frequencies, and evaluation principles (Stanescu et al. 1979; Clement et al. 1987; Bisgaard and Klug 1995; Klug and Bisgaard 1996, 1997, 1998, 1999; Ducharme and Davis 1997, 1998; Ducharme et al. 1998; Delacourt et al. 2000; Klug et al. 2000; Nielsen and Bisgaard 2000a, 2000b). FOT can be divided into two groups: mono-frequency and multifrequencies.

The mono-frequency FOT measures R_{rs}, and oscillations are applied at a frequency close to the resonant frequency ω_o. It has been applied to assess upper airway resistance in sleep-disordered breathing and to assess instantaneous respiratory impedance at different times within the respiratory cycle. Mono-frequency FOT is easier for clinicians to use because there is only one resistance at each frequency. Multifrequency FOT applies frequencies below and above ω_o and measures R_{rs} and X_{rs} at different frequencies. It has been used in clinical practice to assess for the presence and severity of airflow obstruction.

Previous studies indicated that the impedance could be obtained below the breathing frequencies (0.01–0.1 Hz) (Suki et al. 1989) or at frequencies above (2–32 Hz) (Zerah et al. 1995; Pasker et al. 1996; Rigau et al. 2002). The pressure oscillations can be generated using a body chamber in the first case, while a piston or loudspeaker can be applied in the second case. Generally, measurements are performed with the frequency range of 2–32 Hz, and the signal is made up of a fundamental frequency of 2 Hz and its harmonics: 4, 6, 8· · ·32 Hz. The low-frequency (0.1–10 Hz) oscillations include frequencies of spontaneous breathing and, accordingly, can be applied only in apneic condition. In contrast, the high-frequency range contains oscillations up to several 100 Hz. Most commonly, the oscillation frequency scale utilized for multifrequency oscillation methods includes about 5–30 Hz.

The low-frequency (2–4 Hz) oscillations are required for spontaneous breathing. However, the characteristic rheology of the respiratory tissues below 2 Hz can be revealed by investigation during voluntary apnea according to previous studies (Hantos et al. 1986; Navajas et al. 1990). The different frequency dependences of the airway and tissue impedance allow the model-based separate estimation of their parameters (Hantos et al. 1990). These parameters are more relevant to the mechanical properties manifested during spontaneous breathing than those estimated with higher frequency oscillations. This is one of the advantages of the low-frequency oscillations.

The elevation of oscillation frequencies above 100 Hz reveals new patterns of frequency dependence of Z_{rs}, with the potential of estimating additional mechanical parameters (Frey et al. 1998a, 1998b, 1998c; MacLeod and Birth 2001). In particular, at frequencies above f_{res}, X_{rs} crosses zero in the negative direction, and this is called the first antiresonant frequency ($f_{ar,1}$). Healthy subjects and patients with airway obstruction have been differentiated based on both f_{res} and $f_{ar,1}$, and the forced spirometry indices have been shown to correlate better with $f_{ar,1}$ than with the medium-frequency oscillation parameters (MacLeod and Birth 2001). The high-frequency Z_{rs} is dominated by wave propagation processes in the airway tree and much less affected by tissue properties. Although the model-based analysis of these phenomena remains far from complete, the current work suggests that $f_{ar,1}$ carries information about airway wall compliance, which may be an important descriptor in the understanding of airway instability occurring in wheezing disorders in infancy (Frey et al. 1998c).

Although the low-frequency and high-frequency oscillations reveal different mechanical characteristics of the respiratory system and require different modeling approaches, studies have also shown that combined studies involving both oscillation ranges in the same subjects would undoubtedly be helpful, particularly in facilitating the interpretation of the medium frequency Z_{rs} data.

5.2.3 FOT Setup

To measure the low-frequency respiratory impedance in humans and animals, various measurement approaches have been undertaken. Mechanical ventilation and specific circumstances are the two main approaches to measure respiratory impedance. Compared to

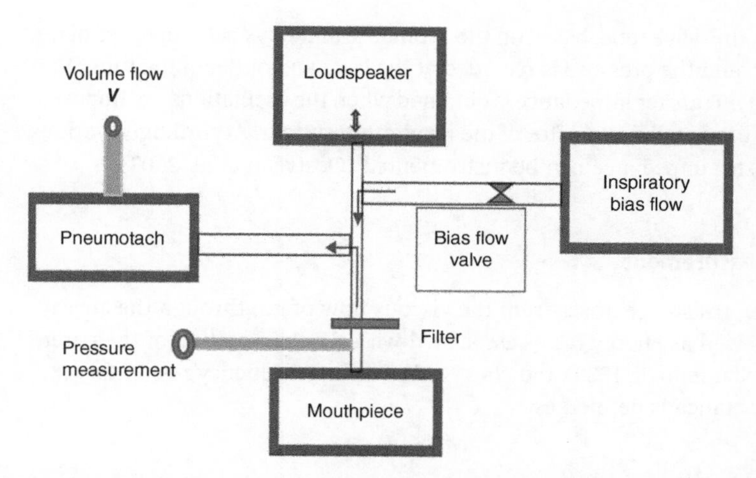

Figure 5.2 Schematic arrangement of the forced oscillatory respiratory impedance measurement.

circumstances, mechanical ventilation can cause breathing disturbances. To reduce breathing disturbances, a servo-controller was used in combination with a piston or loudspeaker to generate the desired pressure oscillations (Kaczka et al. 1999a; Kaczka and Lutchen 2004).

Figure 5.2 shows an example of low-frequency FOT measurement setup. The system requires the subject to breathe through a pneumotachograph via a mouthpiece, and pneumotachograph is connected to a loudspeaker. Oscillation signals include pressure (P) that is measured by a suitable transducer next to the mouthpiece and airflow ($\dot{V}$) measured at the airway opening with a pneumotachograph. To avoid losing forcing pressure during breathing, a low-resistance high-inertance tube is placed in parallel with the loudspeaker. The pneumotachograph is continuously flushed with air to minimize accumulation of expired gas. A bias flow is introduced between the loudspeaker and the pneumotachograph to reduce the equipment dead space.

5.3 Measurement Arrangement

Most commonly, the forced oscillations are applied at the airway opening, and the central airflow (V'_{ao}) is measured with a pneumotachograph attached to the mouthpiece, face mask, or endotracheal tube. Pressure is measured at the airway opening (P_{ao}) with reference to body surface pressure (P_{bs}). The input impedance of the respiratory system ($Z_{rs,in}(f) = P_{rs}(f)/V'_{ao}(f)$) is determined with the ratio of the transrespiratory pressure ($P_{rs} = P_{ao} - P_{atm}$) and the airflow. Based on the measurement of intraesophageally pressure (P_{es}), the lung impedance ($Z_L = (P_{ao} - P_{es})/V'_{ao}$) and the chest wall impedance ($Z_{cw} = (P_{es} - P_{bs})/V'_{ao}$) can be obtained when the impedance of the respiratory system (Z_{rs}) is partitioned into pulmonary and chest walls.

FOT with the generator where P_{ao} is applied around the head to minimize upper airway wall shunting is a special version of the FOT technique. A wave tube is placed at the loudspeaker to connect with the source of forced oscillation and the subject. $Z_{rs,in}$ is measured as the load impedance on the wave tube, based on the geometric and physical properties of the tube and the inside air, and the pressure is recorded at the inlet and outlet of the tube (Van de Woestijne et al. 1981). Transfer impedance is obtained when the oscillations are imposed, and P and V' are measured at different sites of the respiratory system. Accordingly, various measurements of transfer impedance can be instrumented (Oostveen et al. 2003).

5.3.1 Resistance Measurement

In a respiratory system, resistance arises from the viscous flow of gas through the airways (Pedley et al. 1970), as well as energy losses associated with the deformation of the parenchyma (Fredberg and Stamenovic 1989) and chest wall. For low-frequency respiratory system (below 10 Hz), reactance is defined as:

$$X(\omega) = \omega I + \frac{E}{\omega} \tag{5.1}$$

where I is an inertial component and E is an elastic component.

Moreover, the resonant frequency (ω_o) can be obtained:

$$\omega_o = \sqrt{\frac{E}{I}} \tag{5.2}$$

where the magnitude of the impedance is achieved at the minimum value of ω_o, and the inertia has a negligible contribution to the impedance at low-frequency range ($\omega \ll \omega_o$). Reactance can be expressed as:

$$X(\omega) = -\frac{E(\omega)}{\omega} \tag{5.3}$$

Generally, impedance can be obtained by calculating the transfer impedance of the respiratory system from the Fourier domain relationship between pressure oscillations applied to the body surface and flow measured at the mouth or vice versa (Navajas and Farré 1999). Transfer impedance (Z_{tf}) is defined as:

$$Z_{tf} = \frac{P_{bs}}{V_{ao}} \tag{5.4}$$

where P_{bs} is the pressure generated around the body surface and V_{ao} is the oscillatory flow at the airway opening.

Alternatively, oscillatory flows may be presented to the airway opening while simultaneously measuring the resulting airway opening pressure (P_{ao}) relative to atmosphere.

If the chest wall is intact, then:

$$Z = \frac{P_{ao}}{V_{ao}} \tag{5.5}$$

where P_{ao} is the pressure generated at airway opening.

If P_{ao} is measured relative to the pressure at the pleural surface (P_{al}), then:

$$Z = \frac{(P_{ao} - P_{al})}{\dot{V}} \tag{5.6}$$

5.3.2 Impedance Measurement Methods

There are several approaches to calculate impedance from the measured oscillatory pressure and flow. One of the simplest ways is to excite the respiratory system at one discrete frequency when the subject remains apneic and then to obtain the impedance from the corresponding pressure and flow at this frequency. Impedances over a particular frequency range can be obtained by continually forcing the respiratory system with different discrete frequencies (Barnas et al. 1991, 1993, 1997). Time-consuming is the major limitation of this method (Bates et al. 1994).

A more efficient method involves exciting the system with broadband waveforms that simultaneously contain all frequencies of interest. This can be done using multiple sinusoids or small-amplitude random noise (Kaczka and Lutchen 2004). The impedance can be determined at each particular frequency using Fourier analysis (Cooley and Tuckey 1965; Welch 1967; Kaczka et al. 1995). For multifrequency, the impedance is:

$$Z(\omega) = \frac{G_{P\dot{V}}(\omega)}{G_{\dot{V}\dot{V}}(\omega)} \tag{5.7}$$

where $G_{P\dot{V}}(\omega)$ is the cross-power spectrum between pressure and flow, and $G_{\dot{V}\dot{V}}(\omega)$ is the auto-power spectrum of flow, $0 < \omega \leq \omega_{max}$.

The coherence function is introduced to measure the linearity:

$$\gamma^2(\omega) = \frac{|G^2_{P\dot{V}}(\omega)|}{G_{\dot{V}\dot{V}}(\omega)G_{PP}(\omega)} \tag{5.8}$$

The coherence is used as an index of causality between the respiratory system's input flow and output pressure and ranging between 0 and 1. Values smaller than 0.95 are generally discarded to assess the quality of impedance data (Kaczka et al. 1999a).

Optimal ventilator waveform (OVW) was proposed (Lutchen et al. 1993) as a more practical and efficient method for measuring low-frequency impedance in humans. OVW is not recommended for patients with severe obstructions. It is useful for patients with mild-to-moderate obstructions who can be trained to relax their chest muscles (Kaczka et al. 1999b). Another drawback of OVW is that it deliveries via a closed system, and a bias flow of fresh gas must be delivered into the breathing circuit to ensure sufficient gas exchange (Kaczka et al. 1997). Within-breath analysis of the impedance is potentially valuable for a wide range of clinical applications, such as detecting obstructive sleep apnea, expiratory flow limitation, (Vassiliou et al. 1996; Dellacà et al. 2004, 2006), and the evaluation of lung mechanics during positive pressure.

5.4 Clinical Applications

The impulse oscillation system (Hellinckx et al. 2001) and MostGraph are two main commercial FOT devices used in clinical practice. Reports have demonstrated that the respiratory impedance can be a sensitive indicator of serial and parallel airway heterogeneity when applying FOT over low frequencies, which may be useful in the prediction of respiratory diseases (Lutchen and Gillis 1997; Lutchen et al. 1988).

5.4.1 FOT in Responsiveness Tests

FOT is particularly useful in responsiveness tests. In such tests, post- and pre-challenge respiratory resistance and the respiratory reactance values are compared. The reference value is not critical to evaluate the effect of the inhaled agent because the patient is self-controlled. Previous studies (Decramer et al. 1984; Rodriguez-Roisin et al. 1991; Snashall et al. 1991; Wesseling and Wouters 1992; Holmgren et al. 1993; Wesseling et al. 1993; Lorino et al. 1994; Weersink et al. 1995; Wilson et al. 1995; Pennings and Wouters 1997; Schmekel and Smith 1997; Ducharme and Davis 1998; Van Noord et al. 1998; Echazarreta et al. 2001) investigated the FOT in assessing the responsiveness tests to increase the airway obstruction by bronchial challenges with histamine, methacholine, cold air hyperventilation, carbachol, and glutathione. Several research groups (Wouters et al. 1992; Van Noord et al. 1994; Pauwels et al. 1997; Zerah et al. 1995; Hellinckx et al. 1998; Delacourt et al. 2000) used FOT to measure the changes in respiratory impedance caused by bronchodilation drugs. Many researchers reported that FOT is useful in dose–response tests due to the high time resolution provided by FOT in the assessment of respiratory mechanics (Duiverman et al. 1986; Sekizawa et al. 1986; Chinet et al. 1988; Mazurek et al. 1995; Cauberghs and Van de Woestine 1989; Peslin et al. 1993; Farré et al. 2001). FOT has a sensitivity and specificity similar to that of conventional spirometric and plethysmographic indices. Several publications (Cauberghs and Van de Woestine 1989; Peslin et al. 1993; Farré et al. 2001) reported that modifying the traditional setup and data analysis may improve the capability of FOT in detecting changes after bronchial challenge.

5.4.2 FOT for Detecting Asthma Phenotypes

More recently, Shirai et al. demonstrated the usefulness of color 3D imaging of respiratory impedance in asthma using the FOT device, MostGraph (Shirai et al. 2013). Their study involved 78 patients with asthma, and whole-breath and within-breath respiratory system resistance and reactance were measured for each patient. Images were grouped as chronic obstructive pulmonary disease (COPD)-like patterns, asthma patterns, and normal-like patterns. Three researchers who were unaware of the clinical information performed this classification. Their results showed that subjects with the COPD-like pattern were predominately female with a higher body mass index, lower forced expiratory volume in 1 second (FEV_1), and forced vital capacity (FVC) values, and higher R_{rs} and X_{rs} values (whole-breath and within-breath variation). Subjects with the regular pattern had higher FEV_1 and FVC values and a lower single-breath nitrogen washout slope. There were no differences in asthma control or exhaled nitric oxide levels among these groups. Their research findings suggested that color 3D imaging of respiratory impedance may show asthma phenotypes.

5.4.3 FOT in Patients Subjected to Ventilator Support

FOT has been applied in patients with sleep apnea-hypopnea syndromes (SAHS) subjected to continuous positive airway pressure (CPAP). The technique has been used during invasive and noninvasive mechanical ventilation of patients with acute or chronic respiratory failure (Farré et al. 2000). The time resolution of the FOT technique feature makes this technique particularly useful for quantifying the degree of airway obstruction during sleep.

Indeed, when applying a single-frequency oscillation, the signal-to-noise ratio achieved is high enough to allow a time resolution capable of accurately tracking the changes in R_{rs} and X_{rs} along the breathing cycle (Badia et al. 1998; Lorino et al. 1998). This possibility and the fact that FOT does not interfere with sleep make it useful in the diagnosis of SAHS (Badia et al. 1998) and in the CPAP titration value. Further, it has been shown that FOT can be easily implemented to control automatic CPAP treatment adapted to the degree of patient impedance.

5.4.4 Monitoring of Respiratory Mechanics

Clinical trials (Reddel et al. 2009; Frey et al. 2011) have suggested that FOT is useful in the management of obstructive lung diseases, including COPD and asthma. The measured respiratory reactance provides an index to assess expiratory flow limitation, which is of particular interest in COPD, since these patients are prone to this respiratory alteration mainly in supine body posture (Dellacà et al. 2006, 2009). Applying FOT to monitor respiratory mechanics has the potential in the treatment of bronchodilators or CPAP. It is suggested that this technique could play an essential role in monitoring the respiratory status of the patient at home (Rigau et al. 2002, 2003; Gulotta et al. 2012).

5.5 Concluding Remarks

FOT is a noninvasive method to monitor respiratory mechanics without patient's cooperation during mechanical ventilation. It is a helpful tool for characterizing the mechanics of the respiratory system to investigate the mechanisms of respiratory diseases (Ducharme and Davis 1998). Published reports demonstrated that FOT is easily implemented and valuable in many medical applications.

Drawbacks of low-frequency FOT include high-performance subwoofer speaker relatively free of harmonic distortion is required, only nonphysiological flows can be generated (typically less than 0.2 l/s) (Farré et al. 1995; de Melo et al. 1998), and cooperation and training are required from subjects (Hantos et al. 1986; Krishman and Brower 2000).

The main advantages of FOT are no forced spirometry, no particular breathing maneuver, or noticeable interference with respiration, and it is a noninvasive and versatile method. These make FOT particularly useful in assessing the changes in respiratory mechanics induced by the inhalation of bronchodilator and bronchoconstrictor drugs.

Bibliography

Badia, J.R., Farré, R., Montserrat, J.M. et al. (1998). Forced oscillation technique for the evaluation of severe sleep apnoea/hypopnoea syndrome: a pilot study. *The European Respiratory Journal* 11, 1128–1134.

Badia, R., Farré, R., Rigau, J. et al. (2001). Forced oscillation measurements do not affect upper airway muscle tone or sleep in clinical studies. *The European Respiratory Journal* 18, 335–339.

Barnas, G.M., Yoshino, K., Loring, S.H., and Mead, J. (1987). Impedance and relative displacements of relaxed chest wall up to 4 Hz. *Journal of Applied Physiology* 62(1): 71–81.

Barnas, G.M., Mills, P.J., Mackenzie, C.F. et al. (1991). Dependencies of respiratory system resistance and elastance on amplitude and frequency in the normal range of breathing. *The American Review of Respiratory Disease* 143, 240–244.

Barnas, G.M., Sprung, J., Craft, T.M. et al. (1993). Effect of lung volume on lung resistance and elastance in awake subjects measured sinusoidal forcing. *Anesthesiology* 78(6): 1082–1090.

Barnas, G.M., Delaney, P.A., Gheorghiu, I. et al. (1997). Respiratory impedances and acinar gas transfer in a canine model for emphysema. *Journal of Applied Physiology* 83(1): 179–188.

Bates, J.H.T., Lauzon, A.-M., Dechman, G.S. et al. (1994). Temporal dynamics of pulmonary response to intravenous histamine in dogs: effects of dose and lung volume. *Journal of Applied Physiology* 76(2): 616–626.

Bellardine Black, C.L., Hoffman, A.M., Tsai, L.W. et al. (2008). Impact of positive end-expiratory pressure during heterogeneous lung injury: insights from computed tomographic image functional modeling. *Annals of Biomedical Engineering* 36(6): 980–991.

Bisgaard, H. and Klug, B. (1995). Lung function measurements in awake young children. *The European Respiratory Journal* 8, 2067–2075.

Cauberghs, M. and Van de Woestijne, K. (1989). Effect of upper airway shunt and series properties on respiratory impedance measurements. *Journal of Applied Physiology* 66, 2274–2279.

Chinet, T., Pelle, G., Macquin-Mavier, I. et al. (1988). Comparison of the doseresponse curves obtained by forced oscillation and plethysmography during carbachol inhalation. *The European Respiratory Journal* 1, 600–605.

Clement, J., Dumoulin, B., Gubbelmans, R. et al. (1987). Reference values of total respiratory resistance 1, although generally slightly lar- and reactance between 4 and 26 Hz in children and adolescents aged 4–20 years. *Bulletin Européen de Physiopathologie Respiratoire* 23, 441–448.

Cooley, J.W. and Tukey, J.W. (1965). An algorithm for the machine calculation of complex Fourier series. *Mathematics of Computation* 19, 297–301.

Daróczy, B. and Hantos, Z. (1990). Generation of optimum pseudorandom signals for respiratory impedance measurements. *International Journal of Bio-Medical Computing* 25(1): 21–31.

Decramer, M., Demedts, M., and Van de Woestijne, K.P. (1984). Isocapnic hyperventilation with cold air in healthy non-smokers, smokers and asthmatic subjects. *Bulletin Européen de Physiopathologie Respiratoire* 20, 237–243.

Delacourt, C., Lorino, H., Herve-Guillot, M. et al. (2000). Use of the forced oscillation technique to assess airway obstruction and reversibility in children. *American Journal of Respiratory Data. Critical Care Medicine* 161, 730–736.

Dellacà, R.L., Santus, P., Aliverti, A. et al. (2004). Detection of expiratory flow limitation in COPD using the forced oscillation technique. *The European Respiratory Journal* 23(2): 232–240.

Dellacà, R.L., Rotger, M., Aliverti, A. et al. (2006). Noninvasive detection of expiratory flow limitation in COPD patients during nasal CPAP. *The European Respiratory Journal* 27(5): 983–991.

Dellacà, R.L., Pompilio, P.P., Walker, P.P. et al. (2009). Effect of bronchodilation on expiratory flow limitation and resting lung mechanics in COPD. *The European Respiratory Journal* 33, 1329–1337.

DuBois, A.B., Brody, A.W., Lewis, D.H., and Burgess, B.F. Jr. (1956). Oscillation mechanics of lungs and chest in man. *Journal of Applied Physiology* 8, 587–594.

Ducharme, F. and Davis, G. (1997). Measurement of respiratory resistance in the emergency department: feasibility in young children with acute asthma. *Chest* 111, 1519–1525.

Ducharme, F. and Davis, G. (1998). Respiratory resistance in the emergency department: a reproducible and responsive measure of asthma severity. *Chest* 113, 1566–1572.

Ducharme, F., Davis, G., and Ducharme, G. (1998). Pediatric reference values for respiratory resistance measured by forced oscillation. *Chest* 113, 1322–1328.

Duiverman, E., Clément, J., Van De Woestijne, K. et al. (1985). Forced oscillation technique. Reference values for resistance and reactance over a frequency spectrum of 2–26 Hz in healthy children aged 2.3–12.5 years. *Bulletin Européen de Physiopathologie Respiratoire* 21, 171–178.

Duiverman, E.J., Neijens, H.J., Van der Snee-van, S.M., and Kerrebijn, K.F. (1986). Comparison of forced oscillometry and forced expirations for measuring dose-related responses to inhaled methacholine in asthmatic children. *Bulletin Européen de Physiopathologie Respiratoire* 22, 433–436.

Echazarreta, A.L., Gomez, F.P., Ribas, J. et al. (2001). Pulmonary gas exchange responses to histamine and methacholine challenges in mild asthma. *The European Respiratory Journal* 17, 609–614.

Farré, R., Ferrer, M., Rotger, M., and Navajas, D. (1995). Servocontrolled generator to measure respiratory impedance from 0.25 to 26 Hz in ventilated patients at different PEEP levels. *European Respiratory Journal* 8(7): 1222–1227.

Farré, R., Gavela, E., Rotger, M. et al. (2000). Noninvasive assessment of respiratory resistance in severe chronic respiratory patients with nasal CPAP. *The European Respiratory Journal* 15, 314–319.

Farré, R., Manzini, M., Rorger, M. et al. (2001). Oscillatory resístanse measured during noninvasive proportional assist ventilation. *American Journal of Respiratory and Critical Care Medicine* 164, 790–794.

Fisher, A., Dubois, A., and Hyde, R. (1968). Evaluation of the forced oscillation technique for the determination of the resistance to breathing. *The Journal of Clinical Investigation* 47, 2045–2057.

Fredberg, J.J. and Stamenovic, D. (1989). On the imperfect elasticity of lung tissue. *Journal of Applied Physiology* 67(6): 2408–2419.

Frey, U., Suki, B., Kraemer, R., and Jackson, A.C. (1997). Human respiratory input impedance between 32 and 800 Hz, measured by interrupter technique and forced oscillations. *Journal of Applied Physiology* 82(3): 1018–1023.

Frey, U., Jackson, A.C., and Silverman, M. (1998a). Differences in airway wall compliance as a possible mechanism for wheezing disorders in infants. *European Respiratory Journal* 12(1): 136–142.

Frey, U., Silverman, M., Kraemer, R., and Jackson, A.C. (1998b). High-frequency respiratory impedance measured by forced-oscillation technique in infants. *American Journal of Respiratory and Critical Care Medicine* 158(2): 363–370.

Frey, U., Silverman, M., Kraemer, R., and Jackson, A.C. (1998c). High-frequency respiratory input impedance measurements in infants assessed by the high speed interrupter technique. *European Respiratory Journal* 12(1): 148–158.

Frey, U., Maksym, G., and Suki, B. (2011). Temporal complexity in clinical manifestations of lung disease. *Journal of Applied Physiology* 110, 1723–1731.

Goldman, M.D. (2001). Clinical application of forced oscillation. *Pulmonary Pharmacology & Therapeutics* 14(5): 341–350.

Grimby, G., Takishima, T., Graham, W. et al. (1968). Frequency dependence of flow resistance in patients with obstructive lung disease. *The Journal of Clinical Investigation* 47, 1455–1465.

Gulotta, C., Suki, B., Brusasco, V. et al. (2012). Monitoring the temporal changes of respiratory resistance: a novel test for the management of asthma. *American Journal of Respiratory and Critical Care Medicine* 185, 1330–1331.

Hantos, Z., Daroczy, B., Suki, B. et al. (1986). Forced oscillatory impedance of the respiratory system at low frequencies. *Journal of Applied Physiology* 60(1): 123–132.

Hantos, Z., Daroczy, B., Csendes, T. et al. (1990). Modeling of low-frequency pulmonary impedance in dogs. *Journal of Applied Physiology* 68(3): 849–860.

Hayes, D., Pimmel, R., Fullton, J., and Bromberg, P. (1979). Detection of respiratory mechanical dysfunction by forced random noise impedance parameters. *The American Review of Respiratory Disease* 120, 1095–1100.

Hellinckx, J., De, B.K., and Demedts, M. (1998). No paradoxical bronchodilator response with forced oscillation technique in children with cystic fi brosis. *Chest* 113, 55–59.

Hellinckx, J., Cauberghs, M., De Boeck, K., and Demedts, M. (2001). Evaluation of impulse oscillation system: comparison with forced oscillation technique and body plethysmography. *The European Respiratory Journal* 18, 564–570.

Holmgren, D., Engstrom, I., Bjure, J. et al. (1993). Respiratory resistance and transcutaneous PO2 during histamine provocation in children with bronchial asthma. *Pediatric Pulmonology* 15, 168–174.

Ingram, R. and O'Cain, C. (1971). Frequency dependence of compliance in apparently healthy smokers versus non-smokers. *Bulletin Européen de Physiopathologie Respiratoire* 7, 195–212.

Kaczka, D. and Lutchen, K. (2004). Servo-controlled pneumatic pressure oscillator for respiratory impedance measurements and high-frequency ventilation. *Annals of Biomedical Engineering* 32(4): 596–608.

Kaczka, D.W., Barnas, G.M., Suki, B., and Lutchen, K.R. (1995). Assessment of time-domain analyses for estimation of low-frequency respiratory mechanical properties and impedance spectra. *Annals of Biomedical Engineering* 23, 135–151.

Kaczka, D.W., Ingenito, E.P., Suki, B., and Lutchen, K.R. (1997). Partitioning airway and lung tissue resistances in humans: effects of bronchoconstriction. *Journal of Applied Physiology* 82, 1531–1541.

Kaczka, D., Ingenito, E., and Lutchen, K. (1999a). Technique to determine inspiratory impedance during mechanical ventilation: implications for flow limited patients. *Annals of Biomedical Engineering* 27(3): 340–355.

Kaczka, D.W., Ingenito, E.P., Israel, E., and Lutchen, K.R. (1999b). Airway and lung tissue mechanics in asthma: effects of albuterol. *American Journal of Respiratory and Critical Care Medicine* 159(1): 169–178.

Kaczka, D.W., Ingenito, E.P., Body, S.C. et al. (2001). Inspiratory lung impedance in COPD: effects of PEEP and immediate impact of lung volume reduction surgery. *Journal of Applied Physiology* 90, 1833–1841.

Kjeldgaard, J., Hyde, R., Speers, D., and Reichert, W. (1976). Dependence of total respiratory resistance in early airway disease. *The American Review of Respiratory Disease* 144, 501–508.

Klug, B. and Bisgaard, H. (1996). Measurement of lung function in awake 2–4-year-old asthmatic children during methacholine challenge and acute asthma. *Pediatric Pulmonology* 21, 290–300.

Klug, B. and Bisgaard, H. (1997). Repeatability of methacholine challenges in 2 to 4 year-old children with asthma, using a new technique for quantitative delivery of aerosol. *Pulmonology* 23, 278–286.

Klug, B. and Bisgaard, H. (1998). Specific airway resistance, interrupter resistance, and respiratory impedance in healthy children aged 2–7 years. *Pediatric Pulmonology* 25, 322–331.

Klug, B. and Bisgaard, H. (1999). Lung function and short-term outcome in young asthmatic children. *The European Respiratory Journal* 14, 1185–1189.

Klug, B., Nielsen, K.G., and Bisgaard, H. (2000). Observer variability of lung function measurements in 2–6-yr-old children. *The European Respiratory Journal* 16, 472–475.

Krishman, J.A. and Brower, R.G. (2000). High-frequency ventilation for acute lung injury and ARDS. *Chest* 118, 795–807.

Kurosawa, H., Ohishi, J., Shimizu, Y. et al. (2010). A new method to assess lung volume dependency of respiratory system resistance using forced oscillation. *American Journal of Respiratory and Critical Care Medicine* 181, A1240.

Làndsér, F., Nagels, J., Demedts, M. et al. (1976). A new method to determine frequency characteristics of the respiratory system. *Journal of Applied Physiology* 41, 101–106.

Lorino, A.M., Lofaso, F., Lorino, H., and Harf, A. (1994). Changes in respiratory resistance to low dose carbachol inhalation and to pneumatic trouser inflation are correlated. *The European Respiratory Journal* 7, 2000–2004.

Lorino, A.M., Lofaso, F., Duizabo, D. et al. (1998). Respiratory resistive impedance as an index of airway obstruction during nasal continuous positive airway pressure titration. *American Journal of Respiratory and Critical Care Medicine* 158, 1465–1470.

Lutchen, K.R. and Gillis, H. (1997). Relationship between heterogeneous changes in airway morphometry and lung resistance and elastance. *Journal of Applied Physiology* 83, 1192–1201.

Lutchen, K.R., Hantos, Z., and Jackson, A.C. (1988). Importance of low-frequency impedance data for reliably quantifying parallel inhomogeneities of respiratory mechanics. *IEEE Transactions on Biomedical Engineering* 35, 472–481.

Lutchen, K.R., Yang, K., Kaczka, D.W., and Suki, B. (1993). Optimal ventilation waveforms for estimating low-frequency respiratory impedance. *Journal of Applied Physiology* 75(1): 478–488.

MacLeod, D. and Birch, M. (2001). Respiratory input impedance measurement: forced oscillation methods. *Medical and Biological Engineering and Computing* 39(5): 505–516.

Maki, B.E. (1986). Interpretation of the coherence function when using pseudorandom inputs to identify nonlinear systems. *IEEE Transactions on Biomedical Engineering* 33(8): 775–779.

Marchal, F. and Loos, N. (1997). Respiratory oscillation mechanics in infants and preschool children. *European Respiratory Monograph* 5, 58–87.

Mazurek, H.K., Marchal, F., Derelle, J. et al. (1995). Specifi city and sensitivity of respiratory impedance in assessing reversibility of airway obstruction in children. *Chest* 107, 996–1002.

Mead, J. (1960a). Control of respiratory frequency. *Journal of Applied Physiology* 15, 325.

Mead, J. (1960b). Volume displacement body plethysmographfor respiratory measurements in human subjects. *Journal of Applied Physiology* 15, 736.

de Melo, P.L., Werneck, M.M., and Giannella-Neto, A. (1998). Linear servo-controlled pressure generator for forced oscillation measurements. *Medical & Biological Engineering & Computing* 36, 11–16.

Michaelson, E.D., Grassman, E.D., and Peters, W. (1975). Pulmonary mechanics by spectral analysis of forced random noise. *The Journal of Clinical Investigation* 56, 1210–1230.

Montserrat, J.M., Badia, J.R., Farré, R. et al. (1999). Routine application of the forced oscillation technique (FOT) for CPAP titration in the sleep apnea/hypopnea syndrome. *American Journal of Respiratory and Critical Care Medicine* 160, 1550–1554.

Navajas, D. and Farré, R. (1999). Oscillation mechanics. *European Respiratory Monograph* 12, 112–140.

Navajas, D., Farre, R., Canet, J. et al. (1990). Respiratory input impedance in anesthetized paralyzed patients. *Journal of Applied Physiology* 69, 1372–1379.

Navajas, D., Farré, R., Rotger, M. et al. (1998). Assessment of airflow obstruction during CPAP by means of forced oscillation in patients with sleep apnea. *American Journal of Respiratory and Critical Care Medicine* 157, 1526–1530.

Nielsen, K. and Bisgaard, H. (2000a). Lung function response to cold air challenge in asthmatic and healthy children of 2–5 years of age. *American Journal of Respiratory and Critical Care Medicine* 161, 1805–1809.

Nielsen, K. and Bisgaard, H. (2000b). The effect of inhaled budesonide on symptoms, lung function, and cold air and methacholine responsiveness in 2- to 5-year-old asthmatic children. *American Journal of Respiratory and Critical Care Medicine* 162, 1500–1506.

Oostveen, E., MacLeod, D., Lorino, H. et al. (2003). The forced oscillation technique in clinical practice: methodology, recommendations and future developments. *European Respiratory Journal* 22(6): 1026–1041.

Otis, A., McKerrow, C., Bartlett, R. et al. (1956). Mechanical factors in distribution of pulmonary ventilation. *Journal of Applied Physiology* 8, 427–443.

Pasker, H., Schepers, R., Clement, J., and Van de Woestijne, K. (1996). Total respiratory impedance measured by means of the forced oscillation technique in subjects with and without respiratory complaints. *The European Respiratory Journal* 9(1): 131–139.

Pauwels, J.H., Desager, K.N., Creten, W.L. et al. (1997). Study of the bronchodilating effect of three doses of nebulized oxitropium bromide in asthmatic preschool children using the forced oscillation technique. *European Journal of Pediatrics* 156, 329–332.

Pedley, T.J., Schroter, R.C., and Sudlow, M.F. (1970). The prediction of pressure drop and variation of resistance within the human bronchial airways. *Respiration Physiology* 9, 387–405.

Pennings, H.J. and Wouters, E.F. (1997). Effect of inhaled beclomethasone dipropionate on isocapnic hyperventilation with cold air in asthmatics, measured with forced oscillation technique. *The European Respiratory Journal* 10, 665–671.

Peslin, R., Duvivier, C., Didelon, J., and Gallina, C. (1985). Respiratory impedance measured with head generator to minimize upper airway shunt. *Journal of Applied Physiology* 59(6): 1790–1795.

Peslin, R., Felicio-da, S.J., Duvivier, C., and Chabot, F. (1993). Respiratory mechanics studied by forced oscillations during artificial ventilation. *The European Respiratory Journal* 6, 772–784.

Randerath, W.J., Parys, K., Feldmeyer, F., and Sanner, B. (1999). Self adjusting nasal continuous positive airway pressure therapy base on measurement of impedance: a comparison of two different maximum pressure levels. *Chest* 116, 991–999.

Reddel, H.K., Taylor, D.R., Bateman, E.D. et al. (2009). American Thoracic Society/European Respiratory Society Task Force on Asthma Control and Exacerbations. An official American Thoracic Society/European Respiratory Society statement: asthma control and exacerbations: standardizing endpoints for clinical asthma trials and clinical practice. *American Journal of Respiratory and Critical Care Medicine* 180, 59–99.

Rigau, J., Farré, R., Roca, J. et al. (2002). A portable forced oscillation device for respiratory home monitoring. *The European Respiratory Journal* 19(1): 146–150.

Rigau, J., Burgos, F., Hernández, C. et al. (2003). Unsupervised self-testing of airway obstruction by forced oscillation at the patient's home. *The European Respiratory Journal* 22, 668–671.

Rodriguez-Roisin, R., Ferrer, A., Navajas, D. et al. (1991). Ventilationperfusion mismatch after methacholine challenge in patients with mild bronchial asthma. *The American Review of Respiratory Disease* 144, 88–94.

Schmekel, B. and Smith, H.J. (1997). The diagnostic capacity of forced oscillation and forced expiration techniques in identifying asthma by isocapnic hyperpnoea of cold air. *The European Respiratory Journal* 10, 2243–2249.

Sekizawa, K., Sasaki, H., Shimizu, Y., and Takishima, T. (1986). Dose-response effects of methacholine in normal and in asthmatic subjects. Relationship between the site of airway response and overall airway hyperresponsiveness. *The American Review of Respiratory Disease* 133, 593–599.

Shirai, T., Mori, K., Mikamo, M. et al. (2013). Usefulness of colored 3D imaging of respiratory impedance in asthma. *Allergy, Asthma & Immunology Research* 5(5): 322–328.

Sly, P.D., Hayden, M.J., Peták, F., and Hantos, Z. (1996). Measurement of low-frequency respiratory impedance in infants. *American Journal of Respiratory and Critical Care Medicine* 154(1): 161–166.

Snashall, P.D., Parker, S., Phil, M. et al. (1991). Use of an impedance meter for measuring airways responsiveness to histamine. *Chest* 99, 1183–1185.

Stanescu, D., Moavero, N., Veriter, C., and Brasseur, L. (1979). Frequency dependence of respiratory resistance in healthy children. *Journal of Applied Physiology* 47, 268–272.

Suki, B., Peslin, R., Duvivier, C., and Farre, R. (1989). Lung impedance in healthy humans measured by forced oscillations from 0.01 to 0.1 Hz. *Journal of Applied Physiology* 67(4): 1623–1629.

Van de Woestijne, K.P., Franken, H., Cauberghs, M. et al. (1981). A modification of the forced oscillation technique. In: *Advances in Physiological Sciences Respiration, Proceedings of the 28th International Congress of Physiological Sciences*, 655–660. Oxford: Pergamon.

Van Noord, J.A., Smeets, J., Clement, J. et al. (1994). Assessment of reversibility of airfl ow obstruction. *American Journal of Respiratory and Critical Care Medicine* 150, 551–554.

Van Noord, J.A., Clement, J., Van de Woestijne, K.P., and Demedts, M. (1998). Total respiratory resistance and Reactance as a measurement of response to bronchial challenge with histamine. *The American Review of Respiratory Disease* 139, 921–926.

Vassiliou, M., Peslin, R., Saunier, C., and Duvivier, C. (1996). Expiratory flow limitation during mechanical ventilation detected by the forced oscillation method. *The European Respiratory Journal* 9, 779–786.

Weersink, E.J., Elshout, F.J., Van Herwaarden, C., and Folgering, H. (1995). Bronchial responsiveness to histamine and methacholine measured with forced expirations and with the forced oscillation technique. *Respiratory Medicine* 89, 351–356.

Welch, P.D. (1967). The use of fast Fourier transform for the estimation of power spectra: a method based on time averaging over short, modified periodograms. *IEEE Transactions on Audio and Electroacoustics* 15(2): 70–73.

Wesseling, G.J. and Wouters, E.F.M. (1992). Respiratory impedance measurements in a dose-response study of isocapnic hyperventilation with cold air. *Respiration* 59, 259–264.

Wesseling, G.J., Vanderhoven, A.I., and Wouters, E.F. (1993). Forced oscillation technique and spirometry in cold air provocation tests. *Thorax* 48, 254–259.

Wilson, N.M., Bridge, P., Phagoo, S.B., and Silverman, M. (1995). The measurement of methacholine responsiveness in 5 year old children: three methods compared. *The European Respiratory Journal* 8, 364–370.

Wouters, E.F., Landser, F.J., Polko, A.H., and Visser, B.F. (1992). Impedance measurement during air and helium-oxygen breathing before and after salbutamol in COPD patients. *Clinical and Experimental Pharmacology & Physiology* 19, 95–101.

Zerah, F., Lorino, A.M., Lorino, H. et al. (1995). Forced oscillation technique vs spirometry to assess bronchodilatation in patients with asthma and COPD. *Chest Journal* 108(1): 41–47.

Part II

Lung Therapies

6

Obstructive Sleep Apnea

6.1 Introduction

This chapter briefly describes the obstructive sleep apnea (OSA) syndromes, diagnostic methods using the combined clinical evaluation and objective sleep research findings, most available current treatment methods with their advantages and disadvantages, and finally the significance of pressure oscillation (PO) in the treatment of the disease. In the diagnostic sense, polysomnography is the best way to confirm the clinical suspicion of OSA syndromes, and it assesses severity and guides treatments. However, in the treatment scene, the continuous positive airway pressure (CPAP) is still the most recommended gold standard therapeutical method for the OSA treatment. This method will be explained in detail, including how competitive this technology is with other available therapy methods. The chapter will end with various improvement methods on CPAP, including the superimposed PO method. Technical and clinical explanations will be given to highlight the significance of PO in improving this therapy.

6.2 Obstructive Sleep Apnea

OSA is considered as the most common upper respiratory system disorder affecting millions of people worldwide with approximately 4% of men and 2% of women. OSA occurs due to the obstruction (narrowing) in the upper airway (UA) region. It is characterized by repeated obstruction and/or collapse of the pharyngeal airway during sleep, though the breathing effort still exists. UA obstruction may occur partially (known as hypopnea) or completely (known as apnea). It occurs due to anatomic obstructions and/or collapse, which will be explained in detail in Section 6.2.1. This, in turn, results in many recurrent and consecutive episodes of airflow cessation, blood oxygen desaturation (known as hypoxia or hypoxemia), blood carbon dioxide saturation increase (known as hypercapnia or hypercarbia), and arousal (sleep disruption) to restore the airway potency. Moderate-to-severe OSA is normally associated with a significantly increased risk of cardiovascular morbidity and mortality. OSA may be referred to by many terms, such as sleep apnea (SA), sleep apnea syndrome (SAS), sleep apnea–hypopnea syndrome (SAHS), obstructive sleep apnea syndrome (OSAS), obstructive sleep apnea–hypopnea syndrome (OSAHS) (Young et al. 1993; Findley et al. 1998; Bassiri and Guilleminault 2000; Emsellem and Murtagh 2005; Health 2006).

Pressure Oscillation in Biomedical Diagnostics and Therapy, First Edition. Ahmed Al-Jumaily and Lulu Wang.

Table 6.1 RDI and AHI severities.

OSA severity	RDI	AHI
Normal (undiagnosed)	$RDI \leq 5$	$AHI \leq 5$
Mild	$5 < RDI \leq 20$	$5 < AHI \leq 15$
Moderate	$20 < RDI \leq 40$	$15 < AHI \leq 30$
Sever	$RDI > 40$	$AHI > 30$

OSA severity can be determined by two indices, the respiratory disturbance index (RDI) and/or the apnea–hypopnea index (AHI). The first one is defined as:

$$RDI = \frac{Total\ number\ of\ apnea\ and\ hypopnea\ events + arousals}{Actual\ Sleeping\ time\ in\ Hours\ per\ night} \tag{6.1}$$

RDI is generally used to identify OSA severity. OSA syndromes are disregarded or not recognized if $RDI < 5$. For RDI of 5 or larger with choking, snoring, gasping, and daytime sleepiness symptoms, OSA is seriously considered. On the other hand, AHI is defined as:

$$AHI = \frac{\text{Total number of apnea and hypopnea events}}{Actual\ Sleep\ time\ in\ Hours\ per\ night} \tag{6.2}$$

Severity is assessed by *AHI*, which ranges from average to severe based on the number of chokes and sleep interruptions per hour. See Table 6.1 for the ranges of RDI and AHI. These two indices are interchangeable and normally determined by a polysomnogram. More details are in the following sections.

Figure 6.1 shows the sleep-disordered breathing that can be affected by many factors. Subjects with simple snoring might move into airway resistance, sleep hypopnea, or SA if their weights are significantly increased. Subjects who drink alcohol can also move up the scale toward Pickwickian syndrome (obesity hypoventilation syndrome), which consists of the

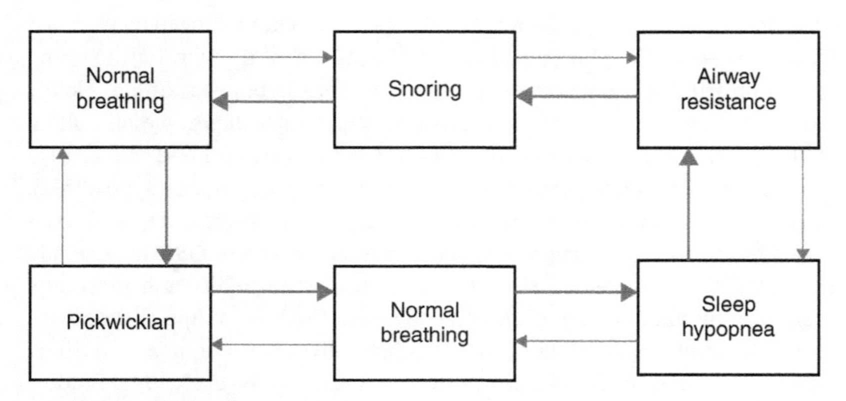

Figure 6.1 Sleep-disordered breathing occurance.

triad of obesity, sleep-disordered breathing, and chronic hypercapnia during wakefulness in the absence of other known causes of hypercapnia. Similarly, patients with SA may move to sleep hypopnea, airway resistance, simple snoring, or a normal state of breathing at night if they lose weight.

6.2.1 Anatomic Contributors to OSA

UA anatomy plays an essential role in OSA developments. As stated earlier, UA obstruction may occur partially (hypopnea) or completely (apnea), and the following anatomic factors contribute significantly to these conditions: (i) large tonsils size and/or adenoids (hypertrophy), (ii) large tongue size (known as macroglossia), (iii) large size of soft palate tip (uvula), (iv) lowered soft palate, (v) deviation in nasal septum, (vi) retrognathia, and (vii) micrognathia. These anatomic obstructions can reduce the cross-sectional area (CSA) of the pharyngeal airway which in turn increases the tendency of the UA to collapse. Other non-anatomical contributors such as drinking alcohol, smoking, and heavy weight also contribute significantly to triggering OSA symptoms.

Airway collapse occurs when the forces of the airway wall muscle (known as dilating forces) are lesser than the forces generated from the airway negative (pharyngeal) pressures (known as collapse forces) during the respiration process. It may occur at one or more of the following pharyngeal sites (Hudgel and Hendricks 1988; Morrison et al. 1993; Mylavarapu et al. 2010):

i) **Nasopharynx**, sometimes known as velopharynx or even retropalatal pharynx: extending from the rear of the nose to the bottom of the soft palate tip.
ii) **Oropharynx**, sometimes known as retroglossal pharynx: extending from the rear of the soft palate tip to the rear of epiglottis tip.
iii) **Hypopharynx**, sometimes known as laryngopharynx or retro-epiglottic pharynx: extending from the rear of epiglottis tip to the beginning of the larynx. These are the most collapsible sites due to the absence of bone support as illustrated in Figure 6.2 (Morrison et al. 1993).

Apnea occurs during sleep when there is an airflow cessation (airflow reduces by 90% or more) for an interval of 10 seconds or more associated with 4% or more of blood oxygen desaturation, or the air gap between the soft palate tip and the airway wall reduces by 75% (partial obstruction) (Susarla et al. 2010). Collapse results in reducing the diameter of the pharyngeal airway, which in turn increases the resistance of the airway to airflow (Isono et al. 1997; Obstructive Sleep Apnea 2015). Collapse increases during sleeping in a supine position due to the gravitational force which pulls the tongue and the soft palate tip toward the airway walls; see Figure 6.2.

Hypopnea is a reduction in the airflow associated with an electroencephalogram (EEG) arousal or a decrease in oxygen saturation (American Academy of Sleep Medicine Task Force 1999). It is defined as a reduction in the nasal pressure (or airflow) by 30% or more for an interval of 10 seconds or more, associated with 4% or more blood oxygen desaturation. However, the flow of air still exists (but with a lower airflow rate). It is sometimes recognized when the airflow is reduced by 50–90%. It also occurs when the air gap between the uvula and the airway walls reduces by 75%, accompanied by minimum wall static

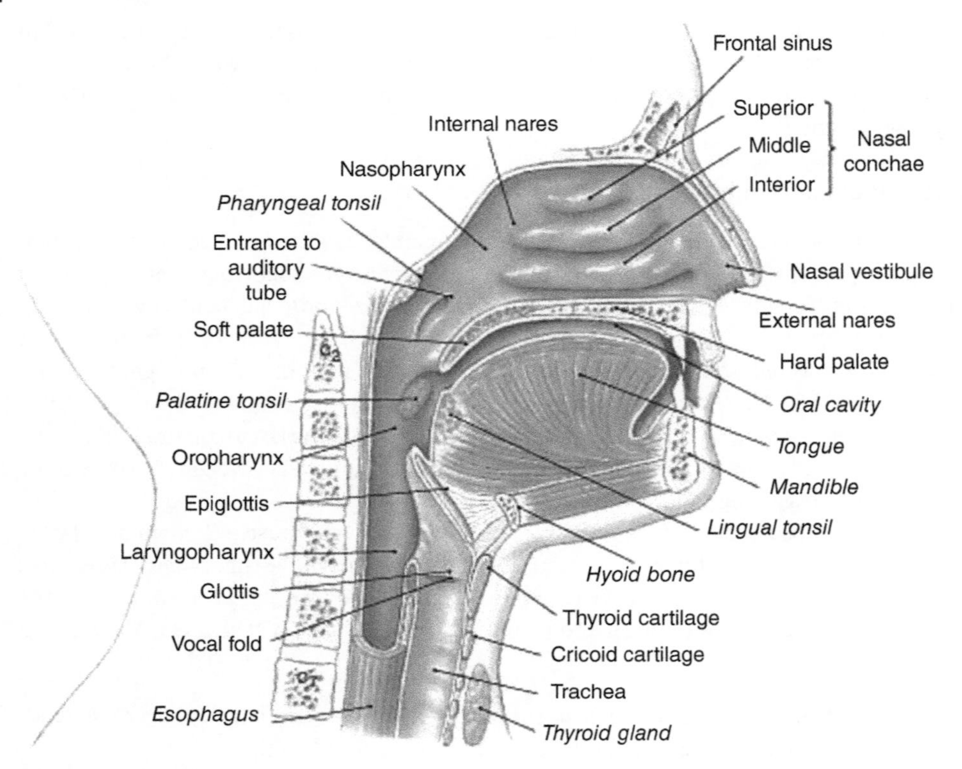

Figure 6.2 UA anatomy and obstruction sites. *Source:* Tortora et al. (2018)/with permission of John Wiley & Sons.

pressure (Susarla et al. 2010). It is usually associated with the chest and abdominal movements and snoring; see Figure 6.3.

As stated earlier, obesity significantly contributes to OSA syndromes. Overweight or underweight is usually assessed by the body mass index (BMI), which is defined as the mass of the body (kg) divided by the squared height of the body (m^2), and it is the measuring scale of obesity, see Table 6.2, which can be noted as:

$$BMI = \frac{Mass}{(height)^2} \tag{6.3}$$

Hypoxia (hypoxemia): It is defined as the reduction in blood oxygen saturation during sleep.

Hypercapnia (hypercarbia): It is an increase in carbon dioxide saturation in blood during sleep.

Apneic event: When a person starts sleeping, the muscles of the pharyngeal airway relax, causing the airway lumen to narrow, increasing the resistance to airflow, and causing airway obstruction. Also, due to gravity, the soft tissues move toward the airway walls narrowing the airway further. The event differs during inspiration and expiration as explained:

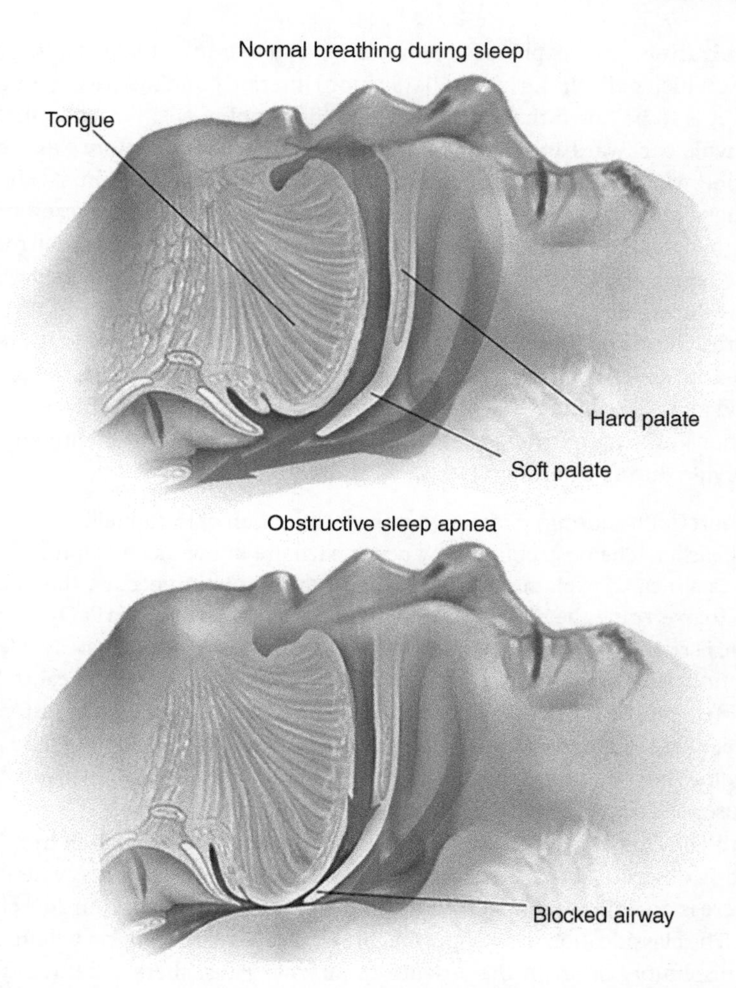

Figure 6.3 Normal and obstructed airways. *Source:* Used with permission of Mayo Foundation for Medical Education and Research.

Table 6.2 BMI and obesity.

Weight categories	Body mass index (BMI)
Underweight	<18.5
Normal	18.5–24.9
Overweight	25–29.9
Obese	30–34.9
Morbid obese	≥35

1) **During inspiration**: The diaphragm contracts causing a drop in the negative pharyngeal pressure, which pulls the airway walls (suction) interiorly and increasing the airway obstruction. It is to be noted that during sleeping in a supine position, the inhaled air pushes the uvula backward and away from the airway wall. It is therefore expected that the obstruction will not occur at the rear of the uvula during inspiration. At the same time, the tongue moves toward the airway wall due to pressure difference resulting from the diaphragm contraction, muscle relaxation, and its own weight, which narrows the airway at the rear of the tongue; therefore, the obstruction would commonly occur rear of the tongue during inspiration.
2) **During expiration**: The diaphragm expands causing a positive pharyngeal pressure, which pushes the airway walls anteriorly opening the airway; however, the outgoing airflow (exhaled air) pushes the uvula forward toward the airway walls (when sleeping in supine position) causing obstruction; therefore, the obstruction will commonly occur rear of the uvula during expiration (Menn et al. 1996).

A **gasp** or a snort (loud snoring) may occur at the termination of the apneic event for both cases described earlier. The obstruction may occur partially known as hypopnea, or completely known as apnea as explained earlier. When the obstruction occurs, the breathing effort increases to overcome the obstruction, the blood oxygen saturation (PaO_2) decreases (known as hypoxia or hypoxemia), and the blood carbon dioxide ($PaCO_2$) increases (known as hypercapnia or hypercarbia) (Qureshi et al. 2003) This in turn activates the central nervous system (CNS) to arouse the person in order to allow airway wall muscles to restore their potency accompanied by a snort or a gasp. The airflow is then resumed allowing the person to fall asleep again. The obstruction then occurs again, followed by arousal, then an obstruction, which causes sleep fragmentation and consequently restless sleep.

A patient with OSA may have hundreds of apneic events per night, ranging from 200 to 500 events, and is observed to have consecutive silence and snoring events, where silent events mean there is no airflow (apnea), and snoring events suggest that the airflow obstruction is relieved. The classification of an abnormal breathing event depends on whether there is evidence of respiratory effort in the absence of airflow. Thus, there are three types of apnea: (i) obstructive – events with the absence of airflow but with continued respiratory effort; (ii) central – events with the absence of airflow without respiratory effort; and (iii) mixed – events with mixed characteristics of obstructive and central events exist. These events typically start with a period that meets the central criteria but with respiratory effort without the flow (Stierer and Punjabi 2005).

6.2.2 OSA Risks and Symptoms

There are many risk factors that can significantly contribute or lead to OSA symptoms. These include but are not limited to obesity, aging, acromegaly (it was reported that 13–39% of individuals diagnosed with OSA have acromegaly), retrognathia, smoking, alcohol, sedatives, nasal obstructions, neck size, cardiovascular diseases, hypertension, large uvula, large tonsils and/or adenoids (hypertrophy), large tongue, lowered soft palate, airway tumor, and micrognathia.

There are four classifications of OSA symptoms: daytime, nighttime, psychological, and physiological symptoms. These symptoms are summarized:

a) **Daytime symptoms** include daytime tiredness (fatigue), sleep attacks while driving, eating or even talking, headache (cephalgia), especially in the morning, and excessive daytime sleepiness (hypersomnolence).
b) **Nighttime symptoms** include loud snoring, nocturnal chocking (angina) and pauses in breathing, the body moves excessively during sleeping, somnambulism (walking asleep), long sleep duration without rest, fragmented sleep, nocturia and enuresis (bed wetting), gastroesophageal reflux, excessive night sweating (diaphoresis), sore throat, and insomnia.
c) **Psychological symptoms** include less concentration, rapid mood changes, depression (especially in women), distress, memory loss, and irritability.
d) **Physiological symptoms** include excessive unexplained weight gain, hypertension (increased blood pressure), arrhythmias or heart rate instability (bradycardia and tachycardia), erythrocytosis, low sexual drive (libido), and erectile ability.

With the earlier discussed symptoms, OSA has many health implications such as blood oxygen desaturation during sleep (hypoxia or hypoxemia), increase in carbon dioxide in the blood during sleep, arrhythmias (bradycardia and tachycardia), resistance to insulin (diabetes), stroke, hypertension, depression, sudden death, cardiovascular diseases, nocturnal enuresis, atherosclerosis (atherogenesis), cerebrovascular diseases, low sexual drive, and memory loss.

6.2.3 OSA Diagnostic Methods

Medical history and physical examination are the cornerstones of the clinical diagnosis of OSA. Several studies (Strohl et al. 1978; Manon-Espaillat et al. 1988; el Bayadi et al. 1990) have suggested that family history may influence the risk of OSA. Familial clustering was initially demonstrated in several families with multiple members affected and a high prevalence of OSA symptoms in the first-degree relatives of affected individuals (Redline et al. 1992). Subsequent studies using either abbreviated methods for monitoring respiration during sleep (Guilleminault et al. 1995; Redline et al. 1995) or complete montage polysomnography (Douglas et al. 1993; Mathur and Douglas 1995) have confirmed the familial aggregation of OSA (Mathur and Douglas 1995). The increased risk in family members appears to be independent of other inherited factors such as obesity. A study using whole-genome scans in a Cleveland Family Study (Palmer et al. 2003, 2004) has shown that there may, in fact, be specific chromosomal linkages that may increase the susceptibility to OSA. A familial tendency toward OSA could be attributable to the inheritable factors such as craniofacial features or respiratory neural control mechanisms that decrease UA collapsibility during sleep. However, presently, the specific determinants that increase the OSA risk in families are not known.

Physical examination is based on using a polysomnography (Michaelson et al. 2006) which is a noninvasive method of diagnosis, whereby multiple function monitoring instruments are placed on the patient's skin during normal and deep sleep (Figure 6.4). The instruments are used to monitor (i) the heart rhythm (electrocardiography [ECG]), (ii) brain waves (EEG), (iii) muscle activity (electromyography [EMG]), and (iv) eye movement (electrooculography (EOG) (Rodrigues et al. 2017) with another instrument to monitor (v) oxygen in the blood supply (i.e. oximetry), (vi) airflow using a nasal cannula that is

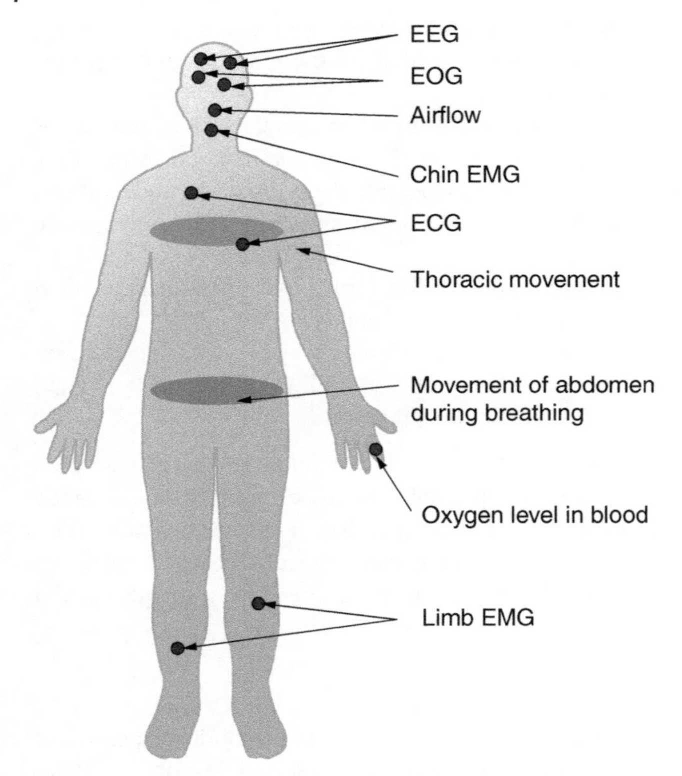

Figure 6.4 Polysomnography leads (polysomnography – sleep apnea, online, http://bozwell.co.uk/poly.html).

placed in the nostrils, and (vii) snoring (sound recorder). This process is usually conducted overnight in a controlled environment and requires six hours of monitoring to obtain viable results (Douglas et al. 1992; Chesson et al. 1997). Unfortunately, this process can sometimes take longer, as a typical patient with OSA struggles to complete full sleep cycles for the polysomnography process to establish results. OSA is highly prevalent in rapid eye movement (REM) sleep (Findley et al. 1985). REM sleep is a stage in which body muscles are relaxed with no body movement, but closed eyes move side to side. These movements are diagnosed with EOG; therefore, the patient usually needs to come back for follow-ups to piece together the data recordings and obtain results.

6.3 Treatment Option For OSA

The treatment of OSA depends on the severity of the disease. The purposes of any treatment method remain the same, to prevent the apneic events (Mathur and Douglas 1995) and to improve (reduce) the flow parameters to keep the airway open during sleep to prevent airway collapse. The values of these parameters usually change due to the narrowness of the

UA due to obstruction. These parameters include but are not limited to wall shear stress, airway resistance, maximum airflow velocity, and the minimum wall static pressure.

There are four principal options available to treat patients with OSA: lifestyle changing, drug therapy, nonsurgical oral appliances, and noninvasive lung supportive devices.

1) **Lifestyle changing:** This is considered as one of the treatment modalities for patients diagnosed with OSA; weight loss is an easy method for treating obese patients diagnosed with only mild OSA. This results in (i) reduced hypersomnolence, (ii) less apneic time during sleep, and (iii) decreased blood oxygen desaturation (Marti et al. 2002). If a patient is diagnosed with moderate-to-severe OSA, losing weight is not an effective practical option. Also, not all patients can lose weight (Oostveen et al. 2003). Changing lifestyle may involve stopping drinking alcohol, stopping taking sedatives (which relax muscles), and quitting smoking (Vgontzas et al. 1994). The patient also must avoid sleeping in the supine position to reduce the tendency for apneic events. The patient is advised to sleep at an incline of 30–60° (to the horizontal) (Qureshi et al. 2003) to minimize the apneic events that occur during sleeping in a supine position due to the gravitational force which pulls the tongue and the uvula toward the airway walls (Kiely et al. 1999).
2) **Drug Therapy:** Over many decades, no drug was proven to be an effective treatment for OSA patients. However, the drug therapy is restricted only for patients with mild to severe OSA. It is associated with harmful side effects such as urine retention, sore throat, constipation, and low erectile ability (Conway et al. 1982).
3) **Oral Appliances:** Various oral appliance models have been developed; however, mandibular repositioning device (MRD) and tongue retaining device (TRD) are the two most widely used oral appliances. MRD, as shown in Figure 6.5, anteriorizes the mandible position and the tongue base to increase the area of the hypopharynx. Several models have been applied for this purpose (Schmidt-Nowara et al. 1991; Menn et al. 1996; Hans et al. 1997; Deane et al. 2009). The most common complication related to MRD is discomfort and excessive salivation. One end of the MRD appliance tightly fits the patient's upper teeth, and the other end pulls the lower jaw forward by holding the lower teeth to the closest position to the upper teeth. The soft tissue is moved away from the airway opening by

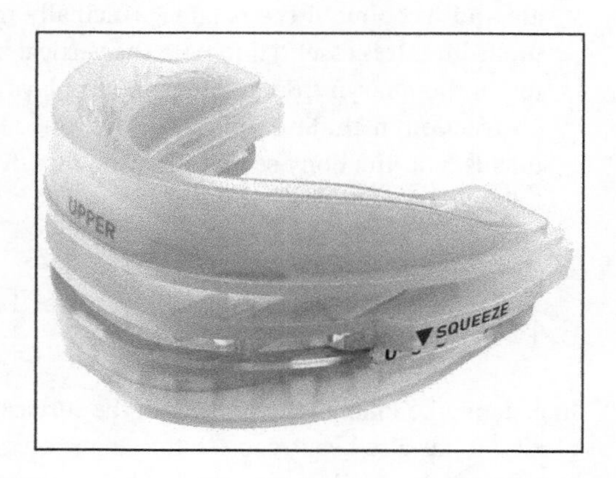

Figure 6.5 Mandibular repositioning device (Henke et al. 2000). *Source:* Courtesy of snorerx.com.

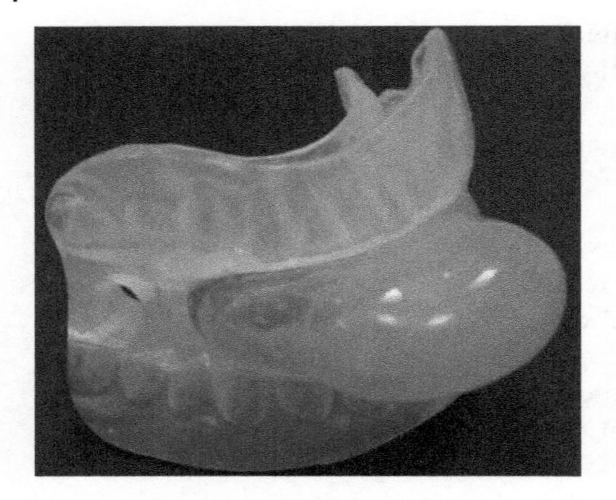

Figure 6.6 Tongue retaining device. *Source:* Courtesy of Lazard et al. (2009).

the extended lower jaw. This treatment is complicated with side effects of losing teeth, joint pain, muscle aches, and tissue sores. Recent studies (Leite et al. 2014) suggested that the MRD reduces the AHI in individuals with an enlarged base of the tongue and mild and moderate OSAS without damaging the function of the masseter and temporal muscles as determined by EMG.

Figure 6.6 shows TRD that involves a mouthpiece to cover the entire dental arches with a mandibular protrusion (Cartwright and Samelson 1982). It pulls the tongue slightly forward as the negative pressure created by the displacement of air from the lingual compartment of the device. The initial mandibular protrusion is 50–75% of maximal protrusion (O'Sullivan et al. 1995; Clark et al. 1996). The protrusion distance is reduced if the patient complains of pain, and it is increased if snoring remains unchanged after a three-week trial. TRD has been recommended to people with large tongues, no teeth, or chronic joint pain. A stomatologist takes a systematic dental examination for all patients. Severe periodontal diseases were considered to be a criterion for exclusion. TRD generates similar results as the MRD (Lazard et al. 2009). Therefore, it has been proposed as an alternative tool to CPAP, considering nasal obstruction as a contraindication.

4) **Invasive Therapy:** The patient's physical examination, including the patient's weight, presence or absence of tonsils, size of tonsils, length and thickness of the uvula, and palate and webbing of the palate (principally the posterior pillar) influence the choice of procedure. It is essential to note the vertical banding in the pharynx and lateral narrowing in the pharynx. Some information plays an essential role in determining the site of obstruction, including the size of the tongue, the possible obstruction from the lingual tonsils, and the bony structure of the face. This will be explained in more detail in the following section.

6.4 Surgical Treatments

In general, the main disadvantages of the surgical treatment are its invasive nature and complexity. It has also been reported that the success rates for the surgical interventions are relatively low, especially for morbidly obese patients and patients diagnosed with severe OSA.

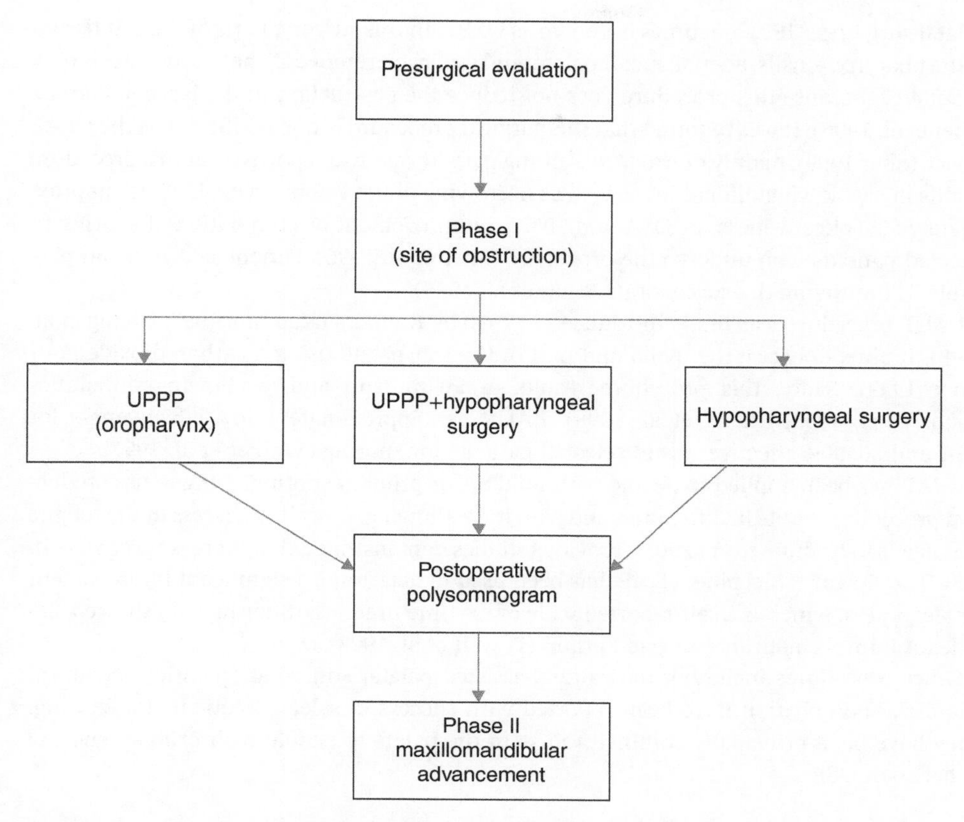

Figure 6.7 Surgical treatment algorithm for patients with sleep apnea.

Before surgical treatment, the severity of OSA (usually described by AHI or RDI) should be obtained; see Figure 6.7. To make sure patients fully understand the treatment plan, a frank discussion (includes the types of surgery available, the pain, complication rates, expected consequences, morbidity, and costs) is required. Various options should be discussed, including the desire of the patient to cure snoring, apnea, and symptoms of tiredness. Furthermore, the need for the evaluation of the patient after the surgical procedure and possible need for subsequent surgeries also should be discussed as these are commonly important parts of the treatment plan.

Figure 6.7 shows a surgical treatment algorithm for patients with sleep apnea. Phase I surgery, including surgery to the oropharynx, oropharynx plus hypopharynx or hypopharynx alone, is undertaken to address major areas of obstruction after the presurgical evaluation. Nasal surgeries may be used adjunctively during phase I or between phase I and phase II (Goldberg, AN, 2002).

6.4.1 Palatal Surgeries

Uvulopalatopharyngoplasty (UP3), laser-assisted uvulopalatoplasty (LAUP) and radiofrequency ablation of the palate (RFAP) are the most common procedures performed among palatal surgeries. UP3 sometimes is known as UPPP. In this surgery, parts of the soft tissues of the pharynx, tonsils, adenoids, soft palate, and uvula are removed. The success rate is only about 40%, because this procedure does not reduce the obstruction in the laryngopharynx (Sher et al. 1996). It was reported that this surgical procedure is not as efficient as the CPAP device (Sher 1999), mainly due to the resulting pain. It was also reported that this procedure results in swallowing difficulties as well as narrowing of the velopharynx. UP3 has approximately 50% effectiveness for OSA and 90% for improvement or elimination of snoring in selected patients who undergo the procedure (Sher et al. 1996). Patient selection can play a role in improving the success rate (Sher et al. 1985).

LAUP procedure was firstly introduced in 1990 by Kamami (Kamami 1994; Cheng et al. 1998). It aims to excise the uvula and part of the soft palate using a carbon dioxide laser (Sher 1999). Sadly, this procedure results in awful pain and swallowing difficulties (Astor et al. 1998; Cheng et al. 1998). LAUP has approximately 40% effectiveness for OSA and 80–90% effectiveness in selected patients for snoring (Walker et al. 1997).

RFAP has been applied to people with mild SA or primary snoring. Tongue base reduction procedure results in difficulties and pain in swallowing, as well as abscess in the tongue that needs to be drained (Li 2005). Previous studies demonstrated that there is no change in RDI. The Epworth Sleepiness Scale has been used to determine a significant improvement for sleepiness, which is a self-reported scale of daytime tiredness. Snoring data showed significant improvement in a selected group (Powell et al. 1998).

Other procedures including the cautery-assisted palatal stiffening operation, chemical sclerosis, and collation have been reported with success in selected patients for snoring. They have been principally confined to use in the palate to people with primary snoring rather than OSA.

6.4.2 Hypopharyngeal Procedures

Hypopharyngeal procedures are performed to treat the breathing blockage. These have several features such as genioglossus advancement (GA), partial glossectomy (PG), hyoid suspension (HS), tongue radiofrequency (TR), and tongue stabilization (TS). These can be performed alone or through a combination of tissue repositioning, tightening, removal, and shrinkage.

GA procedure is to cut the portion of the lower jaw, pulls the small piece of bone forward, and create more room for the tongue to relax during sleep without blocking breathing (Riley et al. 1993a). This procedure cannot be done alone as a treatment for OSA, and it may be used together with UPPP or hyoid advancement (HA). Risks of GA are injury to the teeth, teeth numbness, chin, and lower lip, and weakening of the lower jawbone (mandible).

PG procedure (Hendler et al. 2001) requires removal of a portion of the tongue (midline glossectomy, submucosal lingualplasty, submucosal minimally invasive lingual excision [SMILE], and lingual tonsillectomy). A surgical robot is used to perform this procedure. The procedure is based on the observation that people with SA appear to have enlarged tongues, likely due to having more muscle mass and deposition of fat within the tongue. Compared to other hypopharyngeal procedures, PG has higher risks which include troubles with speech or swallowing, bleeding, tongue numbness, tongue weakness, and change in the sense of taste.

The HS (Riley et al. 1994) procedure is based on expanding the airway by moving the hyoid bone forward. The hyoid bone is a U-shaped bone in the neck located above the thyroid cartilage (Adam's apple), which has attachments to muscles of the tongue and other muscles and soft tissues around the throat. This bone is mobile in humans, perhaps allowing the attached structures to collapse and cause airway blockage during sleep more easily. Through a neck incision, this procedure secures the hyoid bone either slightly downward to the thyroid cartilage or upward to the lower jaw. Risks of HS are difficulty swallowing, infection, and change in the appearance of the neck.

The TR involves a controlled cauterization of the tongue muscle that creates damage in the tongue, leading to the formation of a scar that shrinks the tongue slightly and makes the tongue less likely to fall back to block breathing during sleep. People undergoing TR should expect a total of 2–4 treatment sessions. Risks of TR are bleeding, infection, tongue weakness, and tongue numbness (Powell et al. 1999).

The TS holds the tongue in a different way. Through a skin incision on the neck, a screw is placed on the inside of the central part of the lower jaw. A stitch is attached to this screw and is then passed through the tongue on one side and is brought back through the other side. As the stitch is tied down, it reduces the ability of the tongue to move around, especially to fall back and block breathing during sleep. Risks of TR include tongue weakness, tongue numbness, change in the sense of taste, troubles with speech or swallowing, and tooth injury (Waite et al. 1989; Riley et al. 1993b; Hochban et al. 1997).

6.4.3 Other Procedures

There are other procedures such as mandible advancement (MA), maxilla–mandibular expansion (MME), tracheostomy surgery (TS) and bariatric surgery (BS). MA aims to reposition the tongue anteriorly to enlarge the retropalatal airway. Although the MME procedure is less invasive when compared to MA, in this operation, mouth distractors are used for long time to help stabilizing the expansion in both the mandible and maxilla. After that, the patient will need orthodontic therapy, which is hard to be accepted by the patient.

TS is commonly used for morbidly obese (BMI > 35) and patients with cardiovascular disease. It has been reported that the symptoms of hypersomnolence and arousals have disappeared within only 48 hours. The symptoms of arrhythmia were also remarkably decreased after the TS (Mihaescu et al. 2008a). Sadly, there are many disadvantages of TS, such as secretions, partial obstruction, and dyspnea due to the head movement. Some tissues can granulate at the TS site and esthetically unappealing contribute to the depression (Li 2005). BS is a commonly used treatment to reduce the body weight and attenuate OSA in morbidly obese patients diagnosed with moderate-to-severe OSA.

6.5 Continuous Positive Airway Pressure

Lung supportive devices, such as asthma nebulizers, positive airway pressure devices, or respirators, are used to restore or provide a proper respiration cycle in some patients. CPAP is the most effective long-term treatment for OSA. The benefits of the treatment

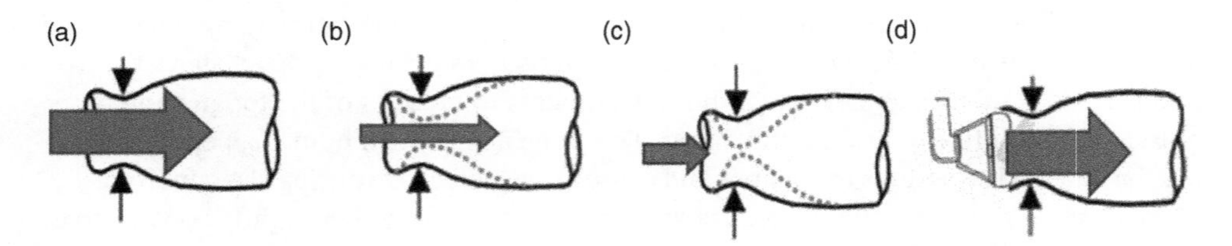

Figure 6.8 UA patency during inspiration in a person: (a) who is normal; (b) experiencing an obstructive hypopnea with snoring; (c) experiencing an obstructive apnea; and (d) subjected to nasal CPAP.

of sleep-disordered breathing include physiological, behavioral, and social. The physiological benefit is to cure sleep-disordered breathing and improve the patient's health. The main requirement of any lung supportive device to treat OSA is to reduce airway obstruction and improve breathing conditions. One of the main devices available in the market worldwide is the CPAP device.

6.5.1 CPAP Principle

Over the last four decades, the CPAP therapy has been the gold standard for the OSA treatment. It is used as an effective way for moderate-to-severe OSA conditions (Sullivan et al. 1981; Olson et al. 2003; Victor 2004; Basner 2007). CPAP provides air at an elevated pressure through a nasal or facial mask, creating a pneumatic splint that keeps the pharyngeal airway open during inspiration and expiration. Figure 6.8a illustrates the case of a normal subject during sleep in the supine position. During inspiration, there is a negative (lower than atmospheric) pressure in the lumen of the UA, and consequently, its soft wall would tend to collapse. However, in a normal UA, the surrounding muscles can exert sufficient force to maintain the airway open, regardless of negative intraluminal pressure during inspiration, allowing normal ventilation during sleep.

In contrast, in an OSA patient, the UA muscles are unable to withstand the collapsing force due to negative intraluminal pressure, so the UA tends to collapse. Depending on the degree of abnormal increase in UA collapsibility, the OSA patient can experience partial UA obstruction, as shown in Figure 6.8b, or total collapse, as shown in Figure 6.8c. In the former case, a hypopnea appears because the reduction in airway lumen results in an increased resistance high enough to reduce ventilation, even though the inspiratory effort is increased. When the UA is completely collapsed, as shown in Figure 6.8c, the patient is no longer able to inspire and experiences an obstructive apnea. In the most severe cases of OSA, the collapsibility of the UA during sleep increases considerably and collapse is induced even in cases where the intraluminal pressure is zero (atmospheric level) or slightly positive. Therefore, in these patients, the UA is collapsed not only during inspiration but also during expiration. Figure 6.8d shows a patient wearing a nasal CPAP during sleep. CPAP opens the UA and prevents its partial or total obstruction. The two closing arrows in Figures 6.8b and c indicate net collapsing force on the UA wall. The two opening arrows in Figure 6.8d indicate that the application of nasal CPAP results in a net force opening the UA.

6.5.2 CPAP Main Components

Figure 6.9 shows the main components of a CPAP system. It consists of an air delivery unit (ADU), a humidifier, a heated air delivery tube, a flexible tube, and a nasal or facial mask. There is also a filter at the air inlet to the system, which is not shown in the figure. A typical CPAP system consists of a fan that sucks air through the inlet filter and blows it at a defined pressure, typically between 4 and 20 cmH_2O, into the patient's mouth and/or nose through a flexible hose and a mask. The mask is an interface between the machine and the patient. There should be bias vent holes on the mask base or near the mask for expelling the exhaled air to make sure that the air received by the patient is fresh. The exhalation port can be an orifice in the mask wall or a particular device connecting the tubing outlet and the nasal mask. In the latter case, the air volume in the nasal mask is an additional small dead space for breathing.

During treatment, the device normally operates at a predetermined titration pressure, which is the minimum pressure at which obstructive events are prevented to maintain normal oxygenation levels and to decrease the frequency of arousals from sleep (Loube et al. 1999; Rakotonanahary et al. 2001; Kushida et al. 2005). To combat nasal dryness and irritation, sometimes cool or heated humidification is used between the fan and the mask to increase patient compliance by preventing drying of the mucous membranes of the UA (Horváth et al. 2008; Pressure CPAP machine, online, http://imgarcade.com/1/how-cpap-works).

Various CPAP devices use different components depending on the design philosophy adopted by the manufacturer. In the following sections, we will cover the general purpose of each element without going into details of the specifics.

1) **Air Handling Unit**

The primary function of the ADU unit is to provide the desired air pressure and airflow rates to maintain constant positive airway pressure at the mask. Most units use a centrifugal compressor-type unit, which is designed to deliver constant pressurized air within the pressure range of 4–20 cmH_2O. Their principal function is to introduce the air from the atmosphere at specific pressure into the respiratory system to maintain the airway open and avoid blockage during sleep. To achieve the effectiveness of the treatment, the pressure inside the mask has to remain constant, which is solved by an automatic feed-forward compensation

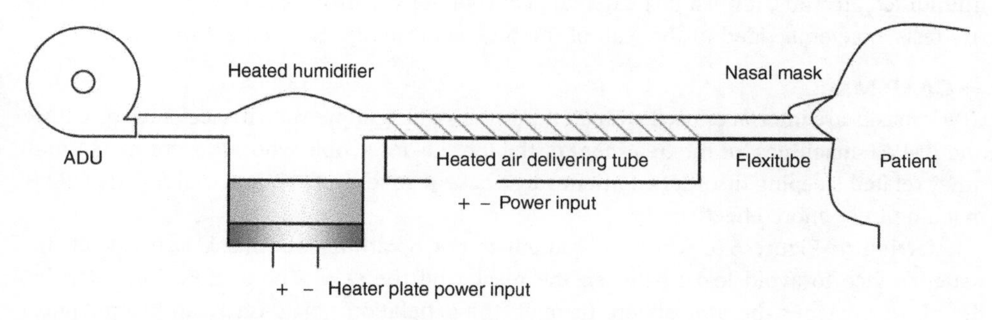

Figure 6.9 Schematic diagram of a CPAP system in use.

control. There are two basic types of pumps used on CPAP, and the oldest one is characterized by a specific fixed pressure used in a ramp function, so the system increases the pressure until it achieves the required pressure. Recent CPAP pumps track the airflow and adjust the pressure required to maintain the airway open at each stage. They can cope with changes in pressure requirements such as temporary nasal congestion or weight change. Power for the ADU is regulated by a pulsed width modulation (PWM) power regulator, controlled by a proportional integral and derivative (PID) controller. The air passes through a duct or a flexible tube to the humidifier.

The CPAP equipment, as shown in Figure 6.9, poses two main technical problems. First, given that the effective resistance of the exhalation port is considerable, the patient's breathing flow mainly circulates through the tubing and blower. Consequently, the blower should be designed to generate high pressure while presenting a low load (resistance) to breathing. Secondly, to keep the nasal pressure constant, the blower should be able to automatically adjust the generated airflow to keep the airflow constant through the exhalation port, regardless of the patient's breathing.

2) **CPAP Humidifier**

The humidifier has a heating source which could be a heated plate underneath the water reservoir to heat the water to certain temperatures. As the air passes over the warm water surface, evaporation delivers water vapor particles into the air. The CPAP heated tube conveys the humidified air from the humidifier to the mask. It is essential to have the latter tube heated to avoid any condensation, which may obstruct the air going to the patient lung and possibly produce difficult breathing or even choking of the patient.

Humidification is a comfort feature that is used in conjunction with a CPAP or a Bi-PAP machine. The humidity delivered from a humidifier helps the airways from drying out and makes the users feel more comfortable when they use CPAP or Bi-PAP machines. Although humidifiers are helpful for some patients, there is no clear evidence to recommend their systematic use for the CPAP therapy in OSAHS patients.

There are several different commercial humidifiers. Some of them can automatically detect the humidity in the room and automatically adjust the right amount of moisture that needs to be delivered.

3) **Heated Tube**

This is the main air delivery tube from the humidifier to the mask. This normally conveys humidified air; thus, it must be heated to avoid condensation as explained earlier. An electric resistance embedded in the wall of the tube is normally used to heat the tube.

4) **CPAP Mask**

CPAP masks are interfaces used with all noninvasive lung supportive devices such as CPAP and Bi-PAP machines for the treatment or the therapy for people who suffer from OSA and other related sleeping disorders. Patient must have a good fit mask and seal for the CPAP machine to be more effective.

Referring to Figure 6.9, when the patient is not breathing, the mask is fitted on the patient's face to avoid leaks between the mask and the skin. The airflow generated by the blower reaches the atmosphere through the exhalation port (American Sleep Apnea Association 1999; Piccirillo et al. 2000). When the patient inspires, an air fraction from the blower flow enters the lungs; hence, the airflow through the exhalation port is reduced. Therefore, for a given exhalation port, the pressure of CPAP can be adjusted by modulating the magnitude of the airflow generated by the blower. In addition to being the nasal pressure source, the airflow plays a vital role in avoiding rebreathing.

The currently available commercial CPAP masks can be classified into four groups: nasal, nasal pillow, full face, and specialty masks:

- The nasal-style CPAP mask is the most commonly used mask among CPAP users during a CPAP titration. It covers the nose only and uses headgear straps to secure the mask to patient's face.
- The nasal pillow CPAP mask is lightweight and has a pillow-style mask cushion that cradles around the nostril to provide a seal. These masks are slightly larger than oxygen tubing and are less invasive than the traditional nasal CPAP masks or full face CPAP masks.
- The full face mask (FFM) is designed to allow the CPAP or Bi-PAP users to breathe through their nose and mouth, and it offers a wide range of benefits. It helps reduce the overwhelming feeling of high CPAP pressures or assist with the inspiratory/expiratory cycle of using a Bi-PAP.
- OSA is a common disease in children, but the standard CPAP masks do not work well on children due to smaller faces and less facial definition. Some companies design specialty masks such as CPAP masks for children.

5) **Filter**

A filter is used in all types of CPAP machines and is usually located at the back of the CPAP machine at the air intake port. The two most widely used filter types include foam material and washable filter and disposable pollen filter. CPAP machines are sucking room air and filtering the air of dust, debris, and particles before the air is delivered to the user that is essential. Some CPAP machines offer more refined filters and hypoallergenic filters for people with allergies.

6.5.3 Titration Pressure

CPAP has been widely accepted in emergency medical services (EMS) with increasing frequency for treating various respiratory conditions since 1970 (Lin et al. 1995; Kosowsky et al. 2001; Kallio et al. 2003; Sullivan 2005). The application of nasal CPAP (Australia Innovates 2008) is currently the most commonly available treatment for OSA patients due to its reliability and effectiveness (The National Heart, Lung, and Blood Institute 2008). CPAP provides a specified amount of positive pressure to the airways to act as a pneumatic splint to maintain airway patency and to increase functional residual capacity (FRC). The value of nasal pressure that normalizes breathing during sleep does not depend on the severity of OSA of a particular patient, as measured by the number of nocturnal respiratory events, but depends on the degree of collapsibility of the patient's UA. Accordingly, to determine the optimal nasal pressure for treatment, each patient should be subjected to an individual CPAP titration procedure during sleep. This pressure is normally determined during a sleep study and by a medical doctor. The doctor will prescribe a CPAP machine at the pressure used to eliminate the patient's sleep-disordered breathing events. It is important to note that it is the air pressure (not the movement of the air) that prevents the apnea events. When the machine is turned on, but before the mask being placed on the head, a flow of air comes through the mask. After the mask is placed on the head, it is sealed to the face, and the air stops flowing. At this point, it is only the air pressure that accomplishes the desired result. This titration pressure can achieve several objectives:

- eliminate SA events;
- eliminate residual snoring;
- keep oxygen levels above 90%, allowing patients to get a restful night's sleep;
- maintain airway patency and recruit alveoli that may be full of fluid;
- decrease work of breathing and respiratory rate;
- reduce systolic blood pressure (Kallio et al. 2003).

6.6 Other Forms of CPAP

Several attempts have been made over the years to help in improving the CPAP device. There are many drawbacks to the use of CPAP, which will be explained in detail later. One of these is that an ordinary CPAP device produces almost a constant pressure that works against the breathing cycle during expiration. This may cause some serious discomfort to patients. This with many other problems has generated new ideas for improved devices. In this section, we summarize some of these ideas.

6.6.1 Bi-Level Positive Airway Pressure

Bi-PAP was first introduced in the 1990s. One limitation of CPAP is that the patient has to exhale against the extra pressure. This problem can be solved by using Bi-PAP. A control system is designed to adjust the pressure to be set at a higher level for inhalation and lower for exhalation, which is helpful for the patient's exhalation and lung compliance. The pressure settings are separated by a pressure support threshold of 3–6 cmH_2O. Bi-PAP devices are more suitable for two types of patients:

- patients with severe apnea who require high fixed CPAP pressure (i.e. 15 cmH_2O). Many sleep labs have a protocol that states when patients get to a certain pressure threshold to switch them over to bi-level;
- patients who are noncompliant may benefit from Bi-PAP devices that provide the comfort of exhaling against a lower pressure than the pressure they are breathing in.

Various companies have developed different Bi-PAP devices with different shapes and costs. However, these commercial Bi-PAP devices are higher costs compared to CPAP (Crystal 2008). This device makes up approximately 15% of the sleep therapy market.

6.6.2 Automatic Continuous Positive Airway Pressure

Auto-CPAP machines are similar to fixed pressure CPAP machines in which a single pressure setting is delivered; however, this pressure setting can automatically adjust overnight when the machine detects that the patient requires more pressure to stabilize the airway.

Auto-CPAP machines usually operate at a range of 4–20 cmH_2O and adjust operation pressures during the course of the night. An auto-CPAP machine can also be programmed to a fixed pressure CPAP device, and this is beneficial because a treating physician can determine the patient's compliance and AHI between these two therapy treatments.

Auto-CPAP and fixed CPAP machines differ significantly in therapy delivery, and ramp mode is simply a comfort feature that allows the user to fall asleep at lower pressure while the machine "ramps up" to the fixed pressure setting usually over a 20-minute period. While fixed pressure CPAP is the most prescribed device, many users are gravitating to auto-CPAP machines. Auto-CPAP machine is excellent for patients who do not like to go back to the sleep lab every one to two years for pressure adjustment because the machine detects the need for more or less pressure as needed. Various commercial auto-CPAP machines include ResMed S9 Autoset and Escape Auto, Philips Respironics System One REMstar Auto DS560HS, Fisher and Paykel ICON Auto, Devilbiss IntelliPAP Auto, and RESmart Auto.

6.6.3 Auto Bi-Level Machines

Auto bi-level machines combine the bi-level technology and auto-CPAP technology. Instead of having one fixed inspiratory pressure and one fixed expiratory pressure, these two pressure settings auto-adjust based on the therapy need. Similar to auto-CPAP machines, the auto bi-level devices can operate in two modes which are standard bi-level and auto-adjust modes.

A pressure support number is established to instruct the auto bi-level device on the differences in pressure between the inspiratory and expiratory pressure. This pressure setting is typically 3–6 cmH_2O. The inspiratory pressure is used to eliminate hypopnea, flow limitation, and any residual snoring. In contrast, the expiratory pressure is used to eliminate obstructive apnea events and stabilize the airway, keeping the airway patent and open. Various commercial auto bi-level devices are available with a cost relatively much higher than a standard CPAP.

Bi-level spontaneous timed (Bi-Level ST) devices operate with the same inspiratory pressure and expiratory pressure concept of traditional bi-level devices. However, they carry the fancy ST acronym. These devices are noninvasive ventilators. Unlike conventional Bi-PAP device where the user determines the inspiratory and expiratory pressure, the Bi-Level ST mode has a timed eupnoeic rate that establishes when patients breathe in and breathe out.

Bi-level ST devices usually are better choice for patients with chronic obstructive pulmonary disease, neuromuscular diseases, and who suffer from obesity hypoventilation. Device settings are specific and should be determined and adjusted upon the recommendation of the treating physician. Commercial examples are available, and in most cases, the cost is the main factor to decide.

6.6.4 Adaptive Pressure Support Servo-Ventilators

The adaptive pressure support servo-ventilators are not commonly used devices and require a sleep specialist to adjust device settings that is recommended for the treatment of complex SA. Available commercial adaptive pressure support servo-ventilators include Philips Respironics BiPAP AutoSV and the ResMed VPAP Adapt. There is no significant difference between these two devices. The algorithm for the Philips Respironics device targets peak flow, while the ResMed device targets minute ventilation, essentially the same common goal with slight variations in the delivery method. These devices do not have an auto-titrating algorithm. The component used to establish airway patency is a fixed pressure. The auto-adjusting component is the pressure support used to treat periodic breathing or complex central events.

6.7 Clinical Studies

With the several variations of CPAP technology, various devices have been utilized in clinical-based studies extensively, as given in the following list. To confirm its impact on OSA reported health consequences and improvements in the quality of life of patients with OSA are listed:

1) Applying CPAP therapy has reduced the risk of cardiovascular diseases (Badia et al. 2001);
2) A randomized controlled clinical trial has confirmed that applying CPAP has reduced the risk of stroke on patients with OSA (Chouly et al. 2006);
3) CPAP treatment in OSA is estimated to be associated with a 5.2% improvement in the arrhythmia risk associated with OSA (Farre et al. 1997);
4) Reported results of the clinically based study have confirmed that applying CPAP therapy has reduced morbidity and mortality (Navajas et al. 1998);
5) CPAP treatment may have a positive impact on reducing blood pressure, hypoxemia, rapid intrathoracic pressure changes, and secondary hemodynamic disturbances (Oostveen et al. 2003);
6) Applying CPAP regularly on patients with OSA has drastically reduced the desaturation of blood oxygen associated with OSA patients of OSA (Oostveen et al. 2003);
7) CPAP treatment could also play an important role in improving systolic function, hemodynamics, and subendocardial ischemia (Pham et al. 1995);
8) Applying CPAP has significantly improved vascular endothelial dysfunction, thereby reducing the inflammatory response and atherosclerosis associated with OSA patients (Vanderveken et al. 2005);
9) Applying CPAP has improved symptoms of excessive daytime sleepiness remarkably (Randerath et al. 2001a);
10) CPAP therapy reduced glucose levels observed in patients with OSA (Ficker et al. 2003);
11) Another study has reported that applying CPAP is considered an effective treatment for reducing nocturia associated with OSA (Reinikainen and Jaakkola 2003);
12) Using CPAP has improved the quality of life of patients with OSA (Keck et al. 2000a);
13) OSA therapy has improved some of the cognitive functions (Çengel 2007);
14) Another study finding is that applying the CPAP has improved the sexual quality of life for women, but not men (Randerath et al. 2002).

Previous clinical studies (Bersten et al. 1991; Cross et al. 2003; Mosesso et al. 2003; Australia Innovates 2008; The National Heart, Lung, and Blood Institute 2008) suggest that CPAP is the most effective treatment for OSA, which acts as a pneumatic splint, keeping the airway open during sleep. Its effect on local and systemic inflammation is debatable.

The following subsections present a summary of some clinical observations on CPAP applications.

6.7.1 CPAP

Skoczyński et al. found that CPAP treatment caused a statistically significant rise of total inflammatory cell count in nasal lavage of OSAS patients when compared with initial values (Skoczyński et al. 2008). The compression applied by CPAP on the nasal wall can cause a mechanical stimulus that triggers inflammation. Kohler et al. observed that four weeks of CPAP treatment has no beneficial effect on blood markers of inflammation and adiponectin in patients with moderate-to-severe OSA (Kohler et al. 2009). Fortuna et al. showed decreased airway nitric oxide levels, reflecting the correction of UA inflammation after CPAP treatment (Fortuna et al. 2011). Similarly, Tamaki et al. indicated that production of TNF-α by monocytes was significantly decreased after the therapy. While improvements have been demonstrated, they are not always certain, and the effect of CPAP on airway and systemic inflammation remains unclear (Tamaki et al. 2007).

Recently, Karamanh et al. evaluated the influence of three months of CPAP treatment on inflammation and oxidative stress markers in the airway and serum. Their research findings suggest that the CPAP therapy has a relevant effect on airways, and nitrotyrosine levels correlated well with the severity of OSA. The CPAP treatment decreases both inflammation and oxidative stress levels in airways in OSA patients, and it also helps reduce systemic oxidative stress levels in serum (Karamanh et al. 2014).

6.7.2 Auto-CPAP

Several clinical studies (Ayas et al. 2004; Masa et al. 2004; Nolan et al. 2006; Meurice et al. 2007) have been conducted to evaluate the auto-CPAP for the treatment of OSAHS. Measurement results observed that the auto-CPAP applies a lower mean nasal pressure than fixed pressure (standard) CPAP. Still, the effectiveness of auto-CPAP devices in reducing the number of sleep breathing events is similar to that of fixed pressure CPAP devices. However, currently available results cannot recommend the systematic application of auto-CPAP to the general spectrum of OSAHS patients, particularly considering its significant cost compared to conventional CPAP.

Simplified titration with auto-CPAP has been proven useful when applied to selected subpopulations of patients, for example, patients who exhibited a number of respiratory events when changing body posture, or patients who were treated with a high level of CPAP. However, the generalized use of auto-CPAP should be considered cautiously, as some patients require full polysomnographic CPAP titration in the sleep laboratory (Mulgrew et al. 2007; Sanders et al. 2008).

Previous studies have shown that the auto-CPAP devices can carry out simplified CPAP titration both at the sleep laboratory and at home. Auto devices can automatically determine the optimal CPAP instead of manually adjusting nasal pressure to determine the optimal pressure, thereby reducing the workload in sleep laboratories. CPAP titration at home has the advantage as the patients are sleeping in their actual environment, and the titration process can be extended to several nights at an affordable cost (Konermann et al. 1998; Rodenstein 2008).

6.7.3 Clinical Comparison Studies of Auto-CPAP and CPAP

Many clinical trials have compared auto-CPAP with the CPAP therapy and have suggested that the auto-CPAP therapy is associated with a decrease in mean pressure compared to CPAP therapy. Aside from this difference, auto-CPAP and CPAP are similar in several outcomes, including objective compliance, the ability to eliminate respiratory events, and the ability to improve subjective daytime sleepiness as measured by the Epworth Sleepiness Scale (d'Ortho et al. 2000; Hudgel and Fung 2000; Teschler et al. 2000; Randerath et al. 2001a; Planes et al. 2003; Massie et al. 2003; Noseda et al. 2004; Hukins 2004; Pevernagie et al. 2004; Cross et al. 2006; Nolan et al. 2007).

Cross et al. conducted a comparison study of CPAP titration at home and sleep laboratory to treat OSA. In this study, two patient groups (100 patients each group) were randomly selected for a standard one-night sleep laboratory CPAP titration and a three-night home-based CPAP titration (Cross et al. 2006). All titrations were performed using the automated CPAP machine (ResMed Spirit) and an individually fitted mask. After titration, all patients were issued with a fixed pressure CPAP machine (ResMed Sullivan 6) set at the fixed pressure determined from the titration study. Data were analyzed on an intention-to-treat basis. Their measurement results showed that there is no baseline difference between the two groups. The CPAP pressures were defined at titration (mean ± SEM: 10.6 ± 0.2 vs $10.4 \pm 0.2\,cmH_2O$, $p = 0.19$), and several mask leaks and initial acceptance rates were similar in the sleep laboratory and home-titrated groups. At a three-month follow-up, there was no significant difference in CPAP use (mean ± SEM: 4.39 ± 0.25 vs 4.38 ± 0.25 h/night; $p > 0.9$), Epworth Sleepiness Scale score (9.5 ± 0.5 vs 8.5 ± 0.5, $p = 0.14$), OSLER, and functional outcomes of sleep questionnaire between the sleep laboratory and home-titrated groups. The research findings suggest that automatic home-based CPAP titration is as effective as automatic in-laboratory titrations in initiating the treatment for OSA. The automatic home-based CPAP titration offers lower cost and potentially speeding up throughput and protecting sleep laboratory space for patients with complex problems.

6.8 Side Effects with CPAP Applications

Although CPAP has been an effective OSA treatment, a number of drawbacks and negative impacts have been reported. Over 45% of CPAP patients report negative side effects, including discomfort, crusting or dry nose, and nasal congestion (itching or rhinorrhea). The influence of CPAP on cerebral blood flow has been of interest to many researchers, and it has been speculated that the relatively high titration pressure generated by the CPAP has a significant side effect on CBF and may lead to conditions damaging to patients resulting in the possibility of stroke (Wiest et al. 2000; Hollandt and Mahlerwein 2003; Fischer et al. 2008). Furthermore, a considerable proportion of OSA patients who use the CPAP therapy report symptoms related to elevated air pressure and, despite its effectiveness, a low level of treatment compliance is reported (Sahin-Yilmaz et al. 2008). Long-term use of CPAP has also shown significant changes in nasal mucosa morphology (Constantinidis et al. 2000; Bossi et al. 2004; Malik and Kenyon 2004). Therefore, any means of reducing the titration pressure and those side effects are essential for safer and more compliant treatment.

In this section, we will summarize most of these drawbacks:

1) Buildup in-ear pressure (Wiest et al. 2001);
2) Poorly fitting mask (Martins de Araujo et al. 2000);
3) Air leaking from the mask leads to sleep disturbance (Martins de Araujo et al. 2000);
4) Chest discomfort due to exhaling against the air stream (Carter et al. 2002);
5) Intraocular buildup pressure (Todd et al. 2001);
6) Barotrauma including pneumocephalus and pneumothorax (British Standards Institution 1970);
7) Facial abrasions made by the CPAP mask (American National Standards Institute 1979);
8) Dryness in mouth and nose (Wiest et al. 2002);
9) Increasing nasal congestion events and rhinorrhea (Morrell et al. 1998);
10) The requirement to repeatedly titrate the CPAP due to changes in the applied pressure;
11) The exhaling process may not be as comfortable as required (Leite et al. 2014);

Collectively, these drawbacks may alter patient compliance, which may hinder the benefits of consistent CPAP therapy. Some research shows patients' poor adherence to the CPAP between 40 and 80 % (Kuna and Sant'Ambrogio 1991).

6.9 Significance of Pressure Oscillation

This section will start with summarizing the severe issues related to CPAP treatment, particularly the effect of CPAP on the UA dryness and the uncomfortable high titration pressure. These are the main driving rationales to make us look at a solution that will help overcome these two problems. The solution seems to be highlighted by the superimposed PO, which is the subject of the rest of this chapter.

6.9.1 Rationales

This section summarizes the main reasons for introducing superimposed pressure oscillation (SIPO) as an improved therapy technology.

1) **Humidity Requirement**

Patients receiving CPAP treatment report several nasal symptoms, such as dryness, sneezing, rhinorrhea, postnasal drip, nasal congestion, or epistaxis. They describe the symptoms as similar to an upper respiratory tract infection, i.e. having a cold (Hudgel 1992).

Drying of the airways produced as a side effect of CPAP might result from troubles in breathing air conditions. This is a transport process that controls the temperature and the humidity of the air during respiration. During inspiration, the air is in contact with the warm and moist nasal mucosa and is rapidly warmed and humidified. This heat and mass transfer process is produced because the difference in temperature creates a driving force and water concentration between the inspired air, which is at room temperature and humidity (Rubesin et al. 1987), and the airway surface liquid (ASL) layer, which is at body temperature and composed by 95% water and 5% of carbohydrate, protein, lipid, and inorganic material (America 1993). During expiration, the air loses heat and water to the outside environment. A healthy individual consumes 350 kcal of heat and 400 mL of water in one day to condition the inspired air considering moderate environmental conditions at about 25 °C and 50% of relative humidity, and only one-third is recovered during expiration (Richardson et al. 1980). However, under CPAP, there is an increase in turbulent effect during inspiration, and the air introduced in the respiratory system has higher pressure than in normal breathing. In normal breathing, the inspiratory linear velocity is between 6 and 18 m/s in the nasal valve, 2 m/s in the main passage, and 3 m/s in the nasopharynx during inspiration, and between 1 and 2 m/s, and 3 and 6 m/s at the nasal valve during expiration in order to maintain the airway open (Fernbach et al. 1983; Barney et al. 2010; Lydiatt and Bucher 2010; Matsuo et al. 2005; Fehrenbach and Herring 2015). During the application of these high pressures, the respiratory tissue is temporarily deformed, so secreting cells may be closed, blocking mucus secretion and hindering natural lubrication and normal air-conditioning of the airway. Toremalm (1961) and Roessleret al. (1998) explain in their work that the increase in turbulent effect causes drying and trauma of the mucosa, which means squamous metaplasia appears and ciliated cells are inactivated and reduced. Nevertheless, there are no further studies or experiments which analyze the response of the respiratory tissue or mucociliary activity under the CPAP therapy.

2) Pressure Requirements
In addition to the earlier noted side effects, it has been reported that the optimum operating pressure may change (reduce) after a period (for long-term CPAP therapy), and another titration may therefore be required to specify the new suitable operating pressure (Series et al. 1994). The same result as well as the cost were documented by Collard et al. (1997). Kribbs et al. reported that the high titration pressures resulted in increased patients' rejections (Kribbs et al. 1993). Berthon-Jones et al. reported rejection rates of more than 30% (Berthon-Jones et al. 1996), while others have reported 50–80% of rejection rates (Meurice et al. 1994 and Zozula et al. 2001), especially for patients with cardiovascular disease. Further, it was also reported that OSA patients using CPAP still have a poor quality of life and sleep, and daytime somnolence (Vgontzas et al. 1998; Barnes et al. 2004).

6.9.2 Pressure Oscillation

In this section, we will present some pieces of evidence on the significance of POs to help in reducing the UA dryness and the high titration pressure without jeopardizing the main therapy functionality of the CPAP.

POs have been proven to increase the activity of the UA muscles in both sleeping animals and humans. In 1968, Robin reported that the pressure waves resulting during human snoring have high frequencies ranging between 30 and 50 Hz (Robin 1968). In 1990, Plowman et al. conducted an experimental study on three tracheotomized sleeping dogs (weight range was 20–30 kg). They used PO waves similar to those occurring during human snoring (30 Hz) as previously described by Robin (Robin 1968), combined with an amplitude of ± 3 cmH_2O. They have reported that the activity of the genioglossus muscle was increased by the effect of the PO (Plowman et al. 1990). In 1992, Zhang et al. conducted an experimental study on 10 anesthetized supine dogs. They studied the effects of the PO at 10, 20, and 30 Hz, combined with an amplitude of $\pm 2.5\ cmH_2O$. They have concluded that the PO increased the activity of the pharyngeal receptors (Zhang and Mathew 1992).

In 1993, Henke et al. experimentally studied the effects of the PO on the UA muscles during the inspiratory breathing phase on 6 normal subjects and 10 patients with OSA. The PO was delivered to the UA via a nasal mask using a pressure wave ventilator. The normal subjects were allowed to sleep in any position, while OSA patients were asked to sleep in a supine position if they did not have apneic events while sleeping on their sides. They have reported the following (Henke and Sullivan 1993):

1) The activity of the UA muscles was increased as a result of the effect of the PO in both normal and OSA subjects, and
2) In some patients with OSA, the PO succeeded in keeping the airway open and prevented obstruction. This study had limitations: (i) the study was conducted during the inspiratory breathing phase only, while some of the respiratory problems and apneic events occur during the expiratory breathing phase, and (ii) the patients were allowed to sleep in different sleeping positions, which means that the reported sleeping data were not equivalent to one another as the compliance of the airway muscles depends not only on whether the subject is awake or asleep but also on the sleeping position itself.

In 1996, Brancatisano et al. reported that the PO resulted in increasing both the activity and stiffness of the UA muscles in anesthetized dogs, which in turn increased the UA potency and reduced its collapsibility (Brancatisano et al. 1996). In 1999, Eastwood et al. experimentally studied the effect of high-frequency-pressure oscillations (HFPOs) on the genioglossus muscles in three tracheotomized sleeping dogs. They used PO waves of 30 Hz, combined with amplitudes of ± 2 and $\pm 4\ cmH_2O$. They have concluded that the activity of the genioglossus muscles was increased due to the effect of HFPO during both expiratory and inspiratory breathing phases, but with more activity during the expiratory breathing phase (Eastwood et al. 1999).

In 2001, Badia et al. experimentally studied the effect of HFPO on the UA muscles in humans (seven male patients), diagnosed with moderate-to-severe OSA. These subjects had the following characteristics (Badia et al. 2001):

1) Ages ranged between 43 and 65 years;
2) AHI range was 22–64 events/h; and
3) BMI range was 28–32 kg/m^2.

They used PO with an amplitude of 1 cmH_2O at a frequency of 5 Hz (which is typically used in clinical studies) and a frequency of 30 Hz (which is typically produced during snoring in humans). They have reported that the use of PO did not restore the airflow during the apneic events or the length of the apneic event itself and did not affect the activity of the EMG nor the EEG. However, the limitations to this study are as follows: (i) the study was done only for a pressure amplitude of 1 cmH_2O, while the UA muscles have been proven to be activated (in animal models) when the amplitude of the PO ranges between 2 and 4 CmH_2O; (ii) all patients were allowed to sleep in their preferred position; (iii) the patients used the CPAP for the first time during the study; (iv) all patients were overweight or obese, and no normal-weight subjects were considered; and (v) the last two conditions have been proven to reduce both the activity and response of the UA muscles (Larsson et al. 1992; Deegan et al. 1995).

In 2005, Vanderveken et al. used the forced oscillation technique (FOT) in an experimental study to accurately specify the patency of the airway during sleeping disturbances. They used the PO with a frequency of 5 Hz, the smallest stimulating frequency for the UA muscles. The study was conducted on eight men diagnosed with OSA. Those patients were (i) between 42 and 58 years of age, (ii) 75–95 kg of weight, (iii) 24–30 kg/m^2 BMI, and (iv) 11–65 events/h RDI.

They have reported that the airway has an equal opportunity to get obstructed during the expiratory breathing phase as well as in the inspiratory breathing phase (Vanderveken et al. 2005). This result was following the development previously reported by Morrell et al. (1998), but at the same time contradicted most of the previous research, which reported that the UA is more likely to collapse during the inspiratory breathing phase rather than the expiratory breathing phase (Kuna and Sant'Ambrogio 1991).

6.9.3 Pressure Oscillations Superimposed on CPAP

As stated earlier, we are looking at the effect of superimposed PO on UA performance in terms of dynamic response and saliva secretion.

1) **UA Performance Due to PO**

To reduce tissue deformation, some research lines are focused on the possibility that many afferent neural receptors are stimulated by the vibration occurring during snoring. As stated earlier, Plowman et al. developed their study to examine the arousal responses that high-frequency oscillating pressures have on the UA muscles of awake and sleeping dogs focusing on the genioglossus, which is the muscle able to avoid the tongue to relax and close the UAs. They found that applying oscillating pressure waves at frequencies similar to those seen during snoring but at less magnitude order (30 Hz and ± 3 cmH_2O) produced reflex responses that help to maintain UA patency during sleep (Plowman et al. 1990). Posteriorly, Zhang et al., Brancatisano et al., and Eastwood et al. arrived at similar conclusions. Their experiments were also carried out with dogs at 30 Hz and ± 2.5 cmH_2O, finding that the vast majority of laryngeal mechanoreceptors are activated in these conditions and the reflex augmentation of soft palate muscle activity may serve to dilate the retropalatal airway (Zhang and Mathew 1992; Brancatisano et al. 1996 and Eastwood et al. 1999).

A further step was achieved by performing these experiments on humans. Henke et al. applied the HFPO (30 Hz) and low PO (4 cmH_2O) to the UA, via a nose mask, and genioglossus, sternomastoid, and diaphragm activity measured by EMG in sleeping humans (Henke and Sullivan 1993). An important finding was that for 46% of the trials in the patients with SA, the genioglossus activity was increased, producing a partial or complete reversal of the UA obstruction. However, it is seen that using lower oscillations (1 cmH_2O), there was not any change in the UA muscle activity (Pham et al. 1995; Farre et al. 1997; Navajas et al. 1998; Badia et al. 2001; Oostveen et al. 2003; Vanderveken et al. 2005; Chouly et al. 2006).

Following these scientific advances, an alternative to conventional CPAP was studied by applying superimposed oscillation pressures in order to maintain the airway open (Pham et al. 1995; Farre et al. 1997; Navajas et al. 1998; Oostveen et al. 2003; Vanderveken et al. 2005). The FOT is a noninvasive method that employs small-amplitude POs superimposed on the normal breathing, being a suitable tool for monitoring respiratory mechanics during mechanical ventilation and sleep. Randerath et al., Ficker et al., and Konermann et al. developed an auto-adjusting positive airway pressure device based on FOT (APAP-FOT) and were able to match more accurately the pressure required from patients based on the measurement of the UA impedance. These results show that APAP-FOT is as efficient as constant CPAP in the OSA treatment, but as the pressure applied can be highly reduced, patients prefer APAP-FOT for home treatment. However, these studies do not show the relation between the applied pressure and the dryness effect, if there is any (Randerath et al. 2001b; Ficker et al. 2003; Konermann et al. 1998).

2) **UA Humidification**

Intense research has been conducted to solve the dryness side effect focusing on humidification, because it is seen that dryness is alleviated with a rise in humidity on inspired air (Reinikainen and Jaakkola 2003). During quiet breathing at a room temperature, the nasal cavity heats the inspired air to 34 °C, and the air is humidified from 10 to 38 mgH_2O/L and reaches 37 °C and 44 mgH_2O/L in the main bronchi. During expiration, it is found 36 °C and 40 mgH_2O/L on the larynx, and 32–34 °C and 27–34 mgH_2O/L on the nares (Elad et al. 2008). So a hypothetic failure in air-conditioning under the CPAP therapy could be corrected by introducing the inspired air close to 37 °C and 44 mgH_2O/L (Mujica-Paz and Gontard 1997).

Regarding the ability to increase the absolute humidity in the inspired air during CPAP, three conditions were tested: heated humidifier, non-heated humidifier, and no-humidifier. Healthy individuals were tested at pressures of 5 and 10 mbar by measuring the relative humidity and temperature of the air at the junction between CPAP tube and the nose mask. The results, obtained under laboratory conditions, showed that heated humidifiers are superior to obtain an absolute humidity of 30 mgH_2O/L with 100% relative humidity at 30 °C (Wiest et al. 2000). These results can be compared with those obtained by Randerath et al., where the relative humidity of the inspired air and the water loss during respiration among cold passover, heated humidifiers, and without humidification under CPAP were compared (Randerath et al. 2002). They also concluded that the best results were obtained with heated humidification, where water loss was reduced by 38% compared to cold humidification, and absolute humidity during inspiration is 21.3 ± 5.1 mgH_2O/L and 33.6 ± 3.5 mgH_2O/L during expiration. Also, a comparison between two heated humidifiers adapted to CPAP was made between Somnowave®, Weinmann GmbH, Hamburg, Germany and Fischer & Paykel, Inc., Auckland, New Zealand, showing that there was no significant difference between them, so they were suitable for the treatment of the dry UAs under the CPAP therapy (Wiest et al. 2001).

One of the main elements which affects the humidity within CPAP therapy is the mouth leak. Another study focused on face mask instead of using the standard nasal CPAP to avoid mouth leaks was conducted (Martins de Araujo et al. 2000). Mouth leaks are significant because they cause unidirectional inspiratory nasal airflow and progressive drying of the nasal mucosa due to air from the lungs not passing over the nasal mucosa. This bypass promotes the release of inflammatory mediators, reduces the efficiency of CPAP treatment, and increases nasal airway resistance (NAR). An increase of NAR causes mouth leaks, which in turn aggravates the problem by increasing NAR setting up a cycle. Nasal inspiration is also essential for the defense against infiltrating particles and conditioning of the inspired air to alveolar conditions (saturated with water vapor at body temperature) to keep the lung's internal milieu by the time it reaches the pharynx (Wolf et al. 2004). Their results show that heated humidification reduces side effects such as NAR or dryness during the therapy but can be prevented by using a face mask. This becomes an alternative to cases when the CPAP therapy using heated humidification does not correct dryness caused by mouth leaks in OSA patients.

It is essential to consider that humidifier performance is susceptible to ambient conditions, so all humidifiers coupled with CPAP must be calibrated for each environmental situation (Todd et al. 2001; Carter et al. 2002). All calibration data must agree with the minimum standards set for breathing machines, in the UK 33 mg H_2O/L (British Standards Institution 1970) and in the USA 30 mg H_2O/L(American National Standards Institute 1979), but the minimum humidity required to avoid dryness during the CPAP therapy is still unknown. Finally, as discussed in this section, heated humidification during CPAP is the treatment of excellence for OSA because it can avoid dryness as a side effect in most of the patients. However, it fails to improve both the comfort and the acceptance patients (Wiest et al. 2002), so further research to reduce the size should be investigated.

6.10 Improvements on CPAP Therapy

This section focuses on validating the SIPO effect on improving CPAP efficacy by reducing the titration pressure and the possibility of removal of the humidification process. It will start by formulating comprehensive addition to the system. From the earlier mentioned literature and supportive pieces of evidence, we can develop the following hypotheses:

i) POs superimposed on CPAP pressure reduce the titration pressure and AHI by improving air access through the UA passage.
ii) POs superimposed on CPAP can stimulate the salivary glands and improve mouth dryness.

To validate these two hypotheses before running a clinical trial, two approaches were undertaken.

6.10.1 SIPO Modulate the Obstructed UA

The first hypothesis for SIPO requires two forms of signals: a mean pressure and an oscillating pressure. The former could be deduced from the fact that the OSA patient's UA tissue is a loose tissue at the back of the mouth. This tissue could be considered as a sagging elastic/viscoelastic diaphragm. Obviously, from vibration principles, it is challenging to modulate this element in its drooping shape and position. However, suppose a small pressure is applied to the diaphragm to make it stretched and under some tension. In this case, the diaphragm can be easily modulated by exciting it at various natural frequencies. With the differences between the diaphragm and the elastic/viscoelastic elements in the UA such as the tongue and uvula, the phenomena are identical. Applying a mean pressure to the tissue will put it under a stretched shape and make it easier to excite. Thus, a reduced titration pressure is an excellent choice to produce the necessary stretching in the tissue to make

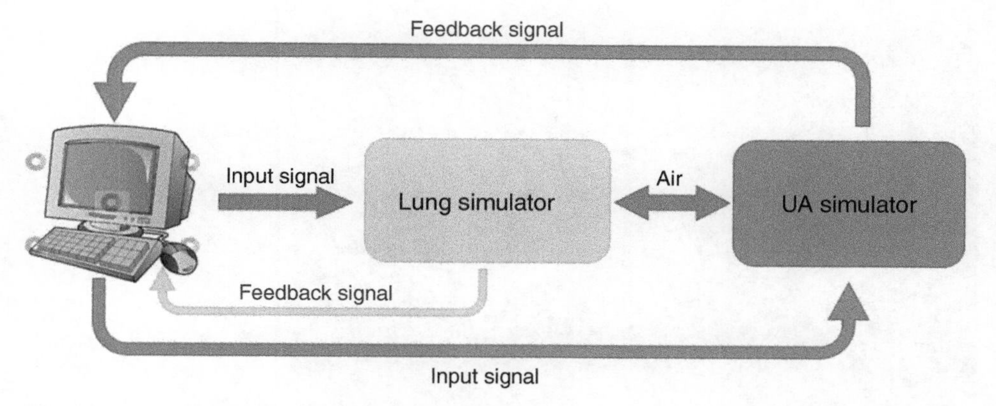

Figure 6.10 Experimental setup for testing the tissue.

it easier to excite. On the other hand, the required oscillation to modulate the UA is a small, superimposed pressure signal on the mean pressure.

Ashaat and Al-Jumaily conducted a detailed theoretical and experimental simulation to investigate the significance of SIPO on a CPAP therapy in reducing the titration pressure and expanding airway passages. Magnetic Resonance Imagining (MRI) scans for healthy and unhealthy human subjects were obtained and processed to be used in the simulation process. They investigated the dynamic characteristics of unhealthy UA models and presented the results of using the Finite Elements (FE), Computational Fluid Dynamics (CFD), and Fluid-Solid Interaction (FSI) methods in determining the dynamic response, UA pressure distributions, and the interaction between the airflow and the deformations of the soft tissues, respectively. The conditions of collapse and the effects of using PO superimposed on the CPAP during inspiratory and expiratory breathing phases were thoroughly investigated. An appropriate silicon rubber model for the tongue and uvula with physical properties close enough to human tissue was tested in a special experimental rig as shown in Figure 6.10 and validated as tested in a special experimental rig as shown in Figure 6.10 and validated that SIPO reduces the titration pressure and AHI. After investigating the results for the uvula and tongue as shown in Figure 6.11, and the combinations between them, they concluded the following (Ashaat and Al-Jumaily 2016):

1) The first few natural frequencies that stimulate both the uvula and tongue models are in the range of 5–40 Hz. For each patient specifics, the first and second natural frequencies for the unhealthy uvula and tongue models are close to one another; however, they are not as close as the results obtained from the healthy models (Robin 1968; Henke and Sullivan 1993; Brancatisano et al. 1996; Eastwood et al. 1999).
2) The natural frequencies for the unhealthy tongue models are slightly less than those for the healthy models. However, the natural frequencies for the unhealthy uvula models are slightly higher than those for the healthy models.
3) The third natural frequencies for the healthy and unhealthy uvula models are relatively different and higher than the reported frequency range. However, the healthy and unhealthy tongue models are very close to one another and fall within the reported frequency range (Robin 1968; Henke and Sullivan 1993; Brancatisano et al. 1996; Eastwood

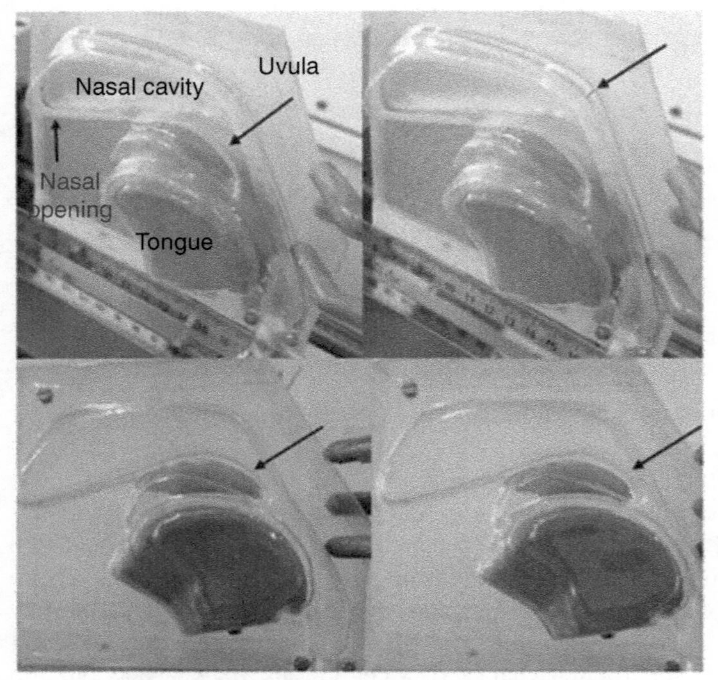

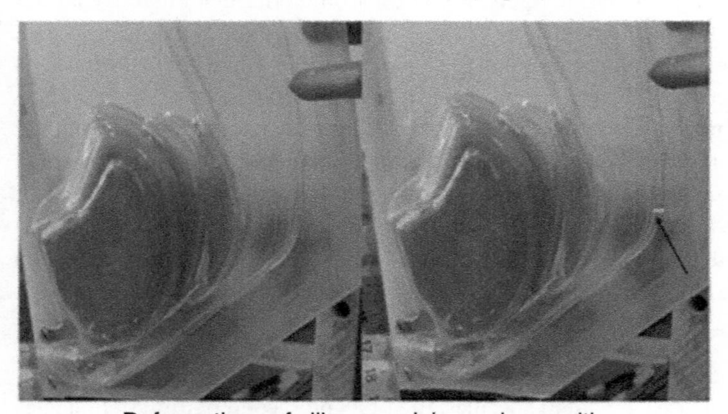

Figure 6.11 Tongue and Uvula experimental models stimulated by SIPO.

et al. 1999). The minimum air gaps at the rear of the uvula and tongue and the UA volume are more significant in the apneic group than in the healthy group of participants, which indicates that these two factors may not be responsible for the UA collapse.

4) The volumes of the uvula and tongue models in the apneic patients are much larger than those of the healthy subjects, contributing to the UA collapse.

The conducted simulations have succeeded in detecting the UA collapse during inspiration and expiration using the CFX Modeller on the ANSYS Workbench to determine the pressure distributions for the air model inside the UA, as well as using one-way FSI by using the static structural modeller on the ANSYS Workbench to determine the resulting UA deformations. The conditions for the collapse were determined by using two turbulent models, k-ε and Shear Stress Transport (SST). The results show that the SST turbulence model has resulted in better and more reasonable pressure distributions compared with the k-ε model, which recommends the use of the SST turbulence model to determine the dynamic characteristics of the UA in similar future studies. The results also show that the obtained conditions of collapse fall within the previously reported values of the esophageal pressures and suggest that the one-way FSI method is recommended in determining the dynamic characteristics of human UAs. After selecting the conditions of collapse, the boundary conditions at the nasal cavity were changed to simulate the use of the CPAP to prevent the UA collapse. The results showed that there was no sign for any obstruction or collapse with the use of the CPAP.

The effects of the POs on the performance of the UA were also investigated, and the results suggested that the use of the PO superimposed on the CPAP has succeeded in keeping the airway open with no sign of collapse and at lower pressure distributions compared to the obtained pressure distributions using the CPAP alone. From the obtained results, it was concluded that:

1) The obtained closing pressures fall within the reported pharyngeal pressures (Bijaoui et al. 2002; Tenhunen et al. 2011; Mihaescu et al. 2008b).
2) The use of the CPAP prevented the occurrence of UA collapse during inspiration and expiration.
3) The use of the PO superimposed on the CPAP prevented the occurrence of collapse at lower pressures when compared with the obtained values using the conventional CPAP alone, especially during inspiration.
4) The use of the SST turbulence model resulted in more reasonable pressure distributions and closer to reality compared to the obtained results from the k-ε turbulence model, and consequently, the SST turbulence model is recommended to be used in future similar studies.

The results obtained from conducting simulation experiments using silicon uvula and tongue (Figure 6.11) indicated:

1) The resulting deformations of the rig and the uvula using the CPAP are more significant than those without using the CPAP.
2) The resulting deformations in the upright position are more significant than the obtained values in the supine position, which suggests that the UA becomes less responsive in the supine position.
3) The resulting deformations increase with increasing the CPAP, which indicates that the CPAP can successfully stimulate the UA.
4) The use of the PO superimposed on the CPAP has resulted in more significant deformations compared with using the CPAP alone, which indicates that the UA can be stimulated more with the use of the PO.
5) The resulting deformations increase with increasing the frequency of the used PO, which indicates that the UA can be stimulated more when utilizing the PO with higher frequencies for the same amplitude.

6) The final computer simulation and experimental simulation have demonstrated that the use of the PO has resulted in larger deformations at lower UA pressure distributions, which indicates that the use of the PO can keep the UA open at much lower pressure distributions compared to the obtained results when using the CPAP alone, which in turn will reduce patient's rejection and consequently increase the CPAP marketability.

6.10.2 SIPO for Saliva Stimulation

To quantify the tracheal capacity of re-conditioning inspired air and investigate the effect of using CPAP and SIPO on the UAs' saliva generation, an experimental study was conducted at AUT-Institute of Biomedical Technologies (IBTec) (Adil 2019), Figure 6.12. The setup consisted of a glass chamber (Figure 6.12h) filled with fresh Krebs solution, fresh bovine trachea with a length of 28 cm, used as a model (c), glass connections designed to connect fresh bovine trachea (e), and a temperature/relative humidity evaluation kit (f). All parts were kept in a temperature-controlled chamber to maintain the temperature at 37 °C (b). In addition, this layout is connected to a lung simulator (a) placed outside the controlled chamber through a flexible hose to provide tidal breathing airflow.

All tracheas were tested before experimental data collection to ensure their viability. Different metrics have been reported to quantitatively characterize the mechanism of mucociliary, which can be used as an indication of tissue viability. Methods included measuring the ASL depth (Widdicombe 2002), the ciliary beat frequency (CBF) (Chilvers and O'Callaghan 2000), and the velocity of mucociliary transport (MCT) (Puchelle et al. 1982; Rubin et al. 1985). MCT was a preferred selection among the procedures for conducting the viability tissue test due to its simplicity and affordability of the required resources. A protocol was developed to perform MCT testing, and an observation chamber containing a modified trough made of sylgard (Dow Corning, USA) was assembled. The tracheal dorsal muscular portion was cut along its entire length and was incised to a set dimension (7.5 cm long and 2 cm wide).

Quantification of the value of the transepithelial potential difference (TEPD) was performed to understand and investigate the tracheal response toward the applied elevated air pressure provided by both the Lung Simulator (LS) and CPAP. In this study, the TEPD was quantified while conducting the experiment. Recording and determination of the behavior of TEPD were performed following a protocol based on previously reported research (Croxton 1993; Trindade et al. 2007). Briefly, the layout consisted of two electrodes made of thin copper wire, one of them connected and attached to the wet apical surface of the trachea and the other electrode was attached to the basolateral surface, overlaying the apical surface's electrode. The basolateral electrode was isolated securely to avoid any contact with the solution using parafilm. Both apical and basolateral electrodes were connected to an oscilloscope (purchased from Agilent, USA), via insulated alligator cables to record the TEPD. Before each experiment, the initial reading of TEPD was offset to maintain accuracy. The collected results were analyzed to construct a graph through plotting recorded TEPD values verse time.

The experimental setup was designed to record all temperature and relative humidity readings while performing and investigating different breathing patterns. To conduct such an investigation, four various temperature/humidity sensors were sourced from Sensirion

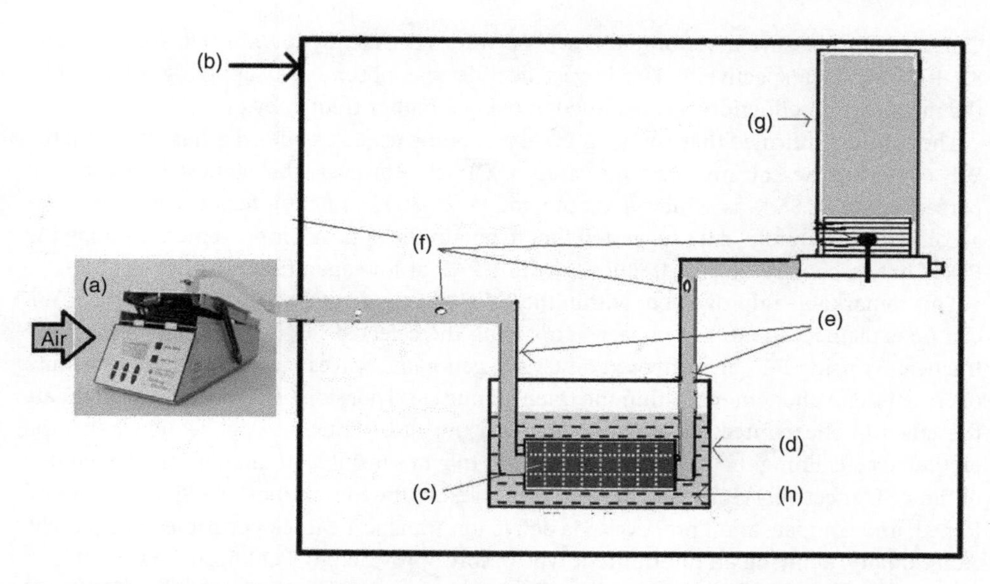

Figure 6.12 Ex vivo experimental setup to simulate normal breathing conditions; (a) lung simulator; (b) controled temp. chamber; (c) bovine trachea; (d) glass bath; (e) glass tube connection; (f) temperature and relative humidity sensors distributed within the ex vivo setup to record changes; (g) plastic below used to trap air inside the system; and (h) Krebs solution.

AG (Evaluation kit EK-H4, Switzerland) and were installed in specific locations within the respiratory setup (Figure 6.12f). These sensors were connected to the evaluation box through dedicated cables (cables with RJ45 plugs and pin-type connectors) to process data simultaneously. Obtained temperature and relative humidity data have been visualized using dedicated computer software EK-H4 Viewer (Sensirion, Switzerland).

Prior to installation, both temperature and RH sensors were calibrated following two methods. For temperature sensors, calibration was conducted against a standard thermometer using a calibrated oven. Data were collected and analyzed for accuracy. For relative humidity, sensors were calibrated against saturated salt solutions of lithium chloride (11.3%) and sodium chloride (75%) using a humidity calibrator (Keck et al. 2000b).

This work's assessment and evaluation parameters included parameters under normal breathing conditions, CPAP and combined CPAP with PO. Obtained data suggest that all these tracheas were viable during the experiment time. The result may also indicate the effectiveness of both the tracheal handling protocols and the developed protocol of MCT quantification.

After confirming the trachea viability, tracheas were connected to the experimental setup to determine the TWC of processed air and tracheal WF during normal breathing conditions. Both parameters were believed to serve as a reliable tool for the humidification status of the UA and to quantify tracheal thermal capacity. The initial investigation determined these parameters at normal breathing conditions.

The results indicated that applying PO waves at any of the investigated frequencies has decreased the TWC of the processed air when compared to those values obtained at applying CPAP (9.9 ± 0.4 mg/L). Using 5, 20, and 30 Hz frequency by the PO in addition to the

CPAP at this pressure deteriorated the TWC values drastically to 9.7 ± 0.9, 9.2 ± 1.3, and 8.7 ± 0.7 mg/L, respectively. The lowest decline was observed at applying 5 Hz, while the most severe reduction was recorded at using a higher frequency of 30 Hz.

The results confirmed that applying PO at any of the tested frequencies has improved the WF of the trachea at any pressure value of CPAP. However, the highest improvement percentage of 78.8% was achieved at applying PO at 30 Hz, in conjunction with the CPAP at 5 cmH_2O. Therefore, data suggested that to obtain the optimal improvement by applying PO, it has to be used at high frequency with CPAP at low operating pressure.

This remarkable improvement within the TWC and then within amounts of tracheal WF can be explained due to the impact of applying the selected PO waves to the ASL of the trachea. By using PO superimposed on CPAP, generated waves may disrupt the dynamics of cross-bridge phenomena within the tracheal muscle. Therefore, PO applications mitigate the effect of shear forces created by CPAP on the ASL. Thus, the application helps the epithelial cells lining the trachea perform their role of sensing and modulating the volume of the ASL effectively (Button et al. 2007). As a result, the importance of ASL can be maintained, and compensation processes via active ion transport can be performed adequately. Additionally, applying an intermittent type of force provided by PO improved the TWC of the passing air, while using a sustained flow rate pressure has deteriorated the TWC and caused fluid depletion to the tracheal ASL.

Recording the TEPD at the conditions described in this study indicated that applying high frequencies of PO (30 Hz) in combination with CPAP at 5 cmH_2O reduced the values of TEPD drastically. The results also showed that applied PO waves at frequencies of 30 Hz improved corresponding TEPD values by 44.7%. Using the lower frequency of PO has shown a subtle impact on the TEPD values, suggesting that complete muscle relaxation can only be achieved by applying higher frequencies at lower CPAP pressure.

6.11 Demonstrating SIPO Clinically

Based on the results from the earlier mentioned studies, clinical trials were essential to demonstrate the earlier stated two hypotheses. The trials were aimed to be fair trials of "usual CPAP" used at home by participants compared with the proposed technology. Usual CPAP refers primarily to the therapy mode and pressure prescribed for the patient due to their diagnosis of OSA. In contrast, comfort features are secondary settings, often adjusted by patients, and maybe turned off or added at different times during their therapy. They do not, however, change the efficacy of the "usual CPAP" therapy prescribed, and as such, removal of these comfort features was proposed, so study nights did not confer any additional risks to the participants in terms of patient care but also protected the accuracy of this study.

The use of the PO superimposed on the CPAP prevented collapse at lower pressures compared with the obtained values when using the conventional CPAP alone, especially during inspiration. The CPAP machine does produce not only positive air pressure but also has other features such as humidification and comfort. During trials, CPAP's comfort features were turned off because these would have affected the precision of the study.

All clinical trials on humans or animals need ethical approval to ensure participants' safety and comfort. Specific protocol steps should be followed depending on the project objectives to clarify the protocol, including the participant information sheet, consent form, invitation letter, and case report form. Each document helps to implement a smooth process for best results, following the ethical committee's guidelines. These trials were conducted with an appropriate ethical approval.

Breathing collapse is one of the most uncomfortable sleeping conditions that OSA patients suffer from. The CPAP therapy is not necessarily perfect. Thus, the main goal of the clinical trials is to enhance patients' sleep by reducing CPAP pressure and the AHI by adding superimposed POs. CPAP users also suffer from mouth and throat dryness after night long use of CPAP therapy. This uncomfortable dryness was the motivation to investigate the oscillation effect on salivary glands secretion.

The clinical trials included two procedures:

- applying a CPAP therapy at one night and the SIPO, to improve the UA performance on the second night;
- saliva secretion test to identify the significance of SIPO on UA dryness.

To start the trials, demographic data such as height and weight were collected. Other demographic data were also collected, such as age (years), neck circumference (cm), years of CPAP use (number of years), ethnicity, and CPAP's fixed therapeutic pressure (cmH_2O). After collecting demographic data, the saliva secretion test was started. A summary of the data collection procedures and their analysis is given in the following section.

6.11.1 Polysomnography setup

Polysomnography's Embla S4500 electrodes were attached to the participants' body parts using fitting kit shown in Figure 6.4 to measure the ECG, EEG, EMG, snoring pattern, oxygen saturation Pulse Oximeter (SPO2), EOG, and abdominal efforts signals using head (electrodes), face (eyes and chin electrodes), legs (electrodes), chest (belt), abdomen (belt), fingertip (oximetry), and neck (snoring sensor electrode). After attaching all the electrodes to the body, the other end of the electrode was attached to the matched port of the Polysomnogram (PSG) panel. The electrodes were checked via PSG device ports; each port showed a green light to indicate that the electrodes were attached correctly and sent a signal to the computer software. If a red light showed, the electrodes needed to be fitted and affixed in the correct position.

After the FFM was fitted to a participant's head and when the subject was preparing to sleep, the PSG signals were checked via verbal confirmation with the participants remotely through the monitoring room speaker by asking them to do the following: look to the left side, look to the right side, look up, look down, blink their eyes for a few seconds, make a snoring sound, grind their teeth, check mask for air leaks, and breathe normally. All these instructions gave a noticeable signal on the computer screen when followed.

For the first clinical trial, the devices that were used were either CPAP, SIPO and PSG, or CPAP with PSG. The CPAP trial was run after setting the CPAP machine to the new 30% reduced pressure, and the reduction value was estimated to be an appropriate value without significantly affecting the patient's conditions (Findley et al. 1985). This pressure was used on the first trial night after confirming the fixed CPAP pressure that had previously been calibrated by a medical doctor (physician). The CPAP trial night applied CPAP therapy and polysomnography.

In the second night trial, the following steps were assessed:

1) The saliva secretion test was applied.
2) PSG electrodes were attached using the procedure discussed earlier.
3) The oscillator hose was connected to the Y-shape adapter connected to the FFM.
4) Participants had to lie down and wear the mask.
5) The mask pressure was calibrated. The oscillator was run by operating the function generator and amplifier at the calibrated value.
6) The CPAP device was run at the reduced pressure value.
7) On the SIPO night, the same CPAP trial pressures were used with superimposed 1 cmH_2O pressure from the oscillator.
8) The mask air pressure was monitored, and any possible leaks were sealed.
9) The lights were turned off and the participant left to sleep.
10) Data were recorded from the monitoring room until participants woke up.

Participants were instructed to bring their CPAP device, and the clinical trial team checked their device setups to confirm the titrated CPAP pressure. To maintain the new CPAP pressure for both nights, the pressure was reduced by 30% from the original pressure. The trial's further pressure was applied on the CPAP nights. The same pressure was used for the subsequent trial along with oscillations. According to the literature (Haba-Rubio et al. 2015), the new CPAP pressure is considered appropriate.

Participants had CPAP/SIPO and PSG devices in their rooms, and the function generator and amplifier, and the monitoring computers were placed in the monitoring room. When the setup of the devices was complete, participants were ready to sleep. Participants took time to fall asleep, but the PSG counted their total sleep time, including lying in bed awake, to measure sleep efficiency. Participants were monitored during the night for around eight to ten hours in total.

During monitoring, assistance was required to maintain mask pressure. When the mask pressure dropped, the optimum pressure value was maintained by refitting and sealing the mask leak. Some participants used Vaseline in their regular routine to seal their mask leaks. Most participants did not prefer the FFM, and even using it during the trial made sleep uncomfortable for them, but was necessary for two reasons: (i) it is the only available mask that has an air pressure output port that can be connected to the PSG's pressure sensor by a cannula and (ii) modulating UA muscles (e.g. tongue, uvula, and salivary glands secretions) is the main goal of this investigation, and those organs are related to the mouth and nasal/pharyngeal and laryngeal airway.

After participants woke up, the trial ended, and the following procedures were followed:

1) Save their collected data in a specific coded folder;
2) Turn off the CPAP and SIPO;
3) Remove the mask and polysomnography electrodes;
4) Repeat the saliva collection test;
5) Sterilize the hoses and adapter pieces;
6) Clean and sterilize the polysomnography electrodes;
7) Change the bedsheets and pillowcases with the clean sterilized bed linen.

The PSG software can measure and record the selected parameters. However, in the case of OSA data measurements, the software's AHI parameter readings are inaccurate because they measure the mixed SA events per hour (obstructive and central SA). To solve the software issue, a manual scoring (Silicone tubing, online; Di-Tullio et al. 2018) was undertaken to calculate the SA events. According to the manually scored AHI, the software measured the other parameters that were included in the final individual report. While participants slept, the sleep laboratory technician wrote a sleep report. This report was designed to give the participant an indication of how the sleep study went on the trial night. The report information was collected using investigational equipment (PSG), such as CPAP pressure settings, sleep quality, treatment efficacy, and mask seal.

6.11.2 Saliva Collection Test

The saliva test was done on some of the recruited participants ($n = 15$). The saliva collection test was conducted before and after each trial. Two collection procedures were applied (Navazesh and Kumar 2008), stimulated and unstimulated. The unstimulated saliva collection test aimed to measure the saliva volume while UA muscles were at rest. The stimulated collection aimed to collect saliva while muscles were active. For saliva collection, the following materials were used: deionized water, disposable cylinders and funnels (salometer), masks and gloves, labels, sugar-free gum, and samples rack.

Tests procedure (unstimulated and stimulated) were as follows:

- before CPAP;
- after CPAP;
- before SIPO;
- after SIPO.

The duration of the unstimulated test was five minutes. Participants were instructed to follow these steps:

1) Rinse the mouth a few times and discard the water;
2) Swallow to void the mouth;
3) Wait for one minute and discard the saliva;
4) While sitting at rest without talking and with eyes open, record the time for five minutes;
5) Collect the saliva in the mouth and spit it into the funnel attached to the graduated cylinder (salometer).

However, the stimulation test took three minutes as follows:

1) Ask the participant to chew a piece of sugar-free gum;
2) Ask the participant to spit and discard saliva for the first two minutes;
3) Start recording time and ask the participant to continue chewing and spit each minute for three minutes into the funnel attached to the salometer.

After collecting the saliva samples, each sample was centrifuged to separate liquids and solids to collect the salivary gland flow rate after collecting the saliva samples. The labeled samples were weighed and subtracted from the tube weight (sample weight [gr] = total weight [gr] – tube weight [gr]). The centrifuge was set for 20 minutes at 4000 r.p.m. and set to run. After the sample had been centrifuged, solids and liquids were separated. The liquid was removed from the pipe and subtracted from the total weight of the sample to find the solid weight (solid content [gr] = total saliva sample weight [gr] – liquid weight [gr]). This procedure was followed for stimulated and unstimulated samples to calculate the flow rate. Flow rate (mL/min) = weight of stimulated sample/time of collection + weight of unstimulated sample/time of collection. Samples were analyzed before and after the CPAP therapy and after the SIPO therapy.

6.12 Concluding Remarks

Trials data of the sleep state of two groups were analyzed. The first trial was under the CPAP therapy, which is considered as the baseline or the term of reference, and the second was the SIPO trial. The results (Figures 6.13 and 6.14), which demonstrate the effectiveness of the SIPO therapy, were as follows:

1) **Respiratory Arousal Index**
The arousals that result from apnea, hypopneas, and periodic increases in respiratory effort per hour are expressed in an index called "respiratory arousal index" (RAI). Increased values of RAI indicate an increase of sleep events (Berry et al. 2017), whereas low values of RAI indicate that the sleep patterns are more likely standard, which leads to less fatigue and tiredness during wakefulness. Figure 6.13a shows the RAIC respiratory arousal index

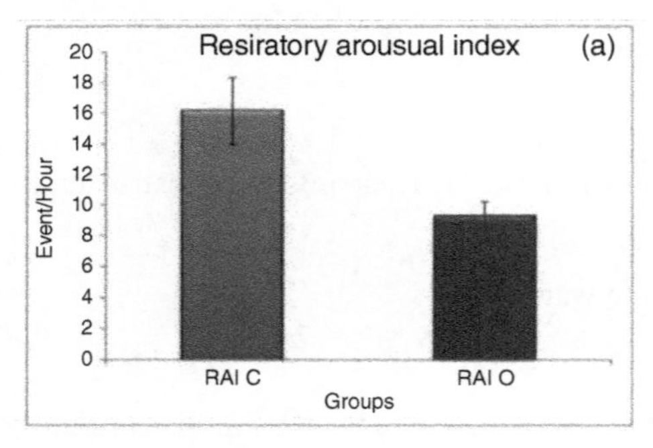

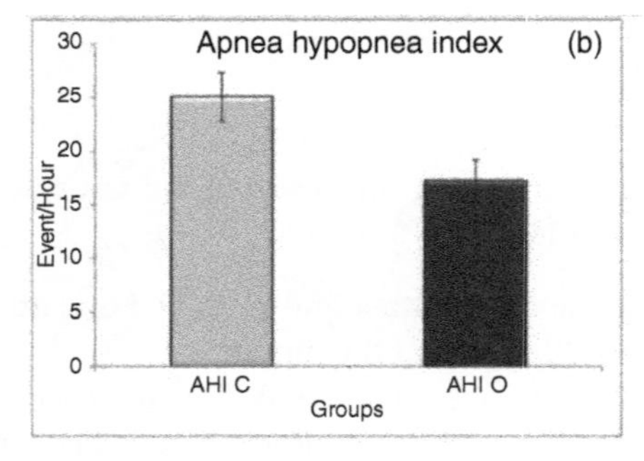

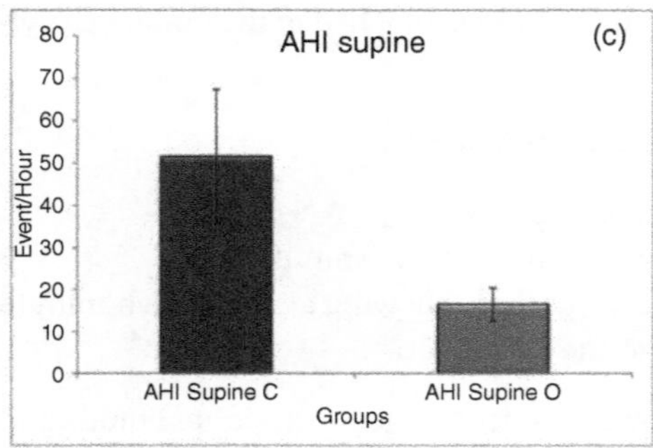

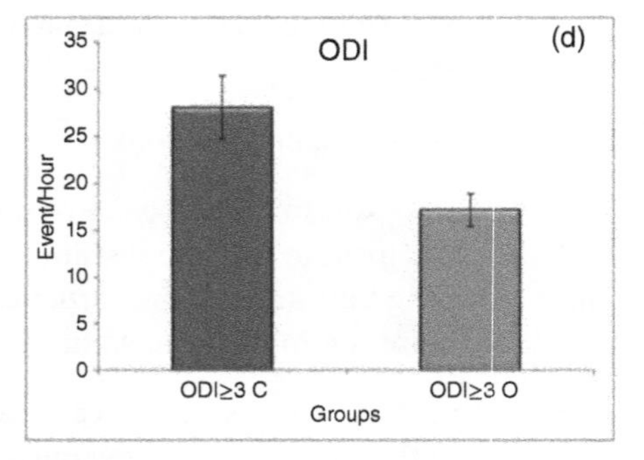

Figure 6.13 Comparison between (a) RAIC and RAIO; (b) AHIC and AHIO. *Source:* Based on Fino et al. (2020) and Troxel et al. (2020); (c) AHI supine CPAP and SIPO trials. *Source:* Based on Troxel et al. (2020); (d) ODIC and ODIO.

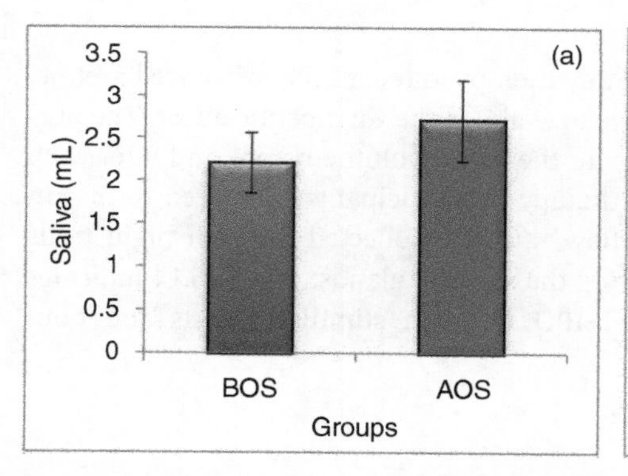

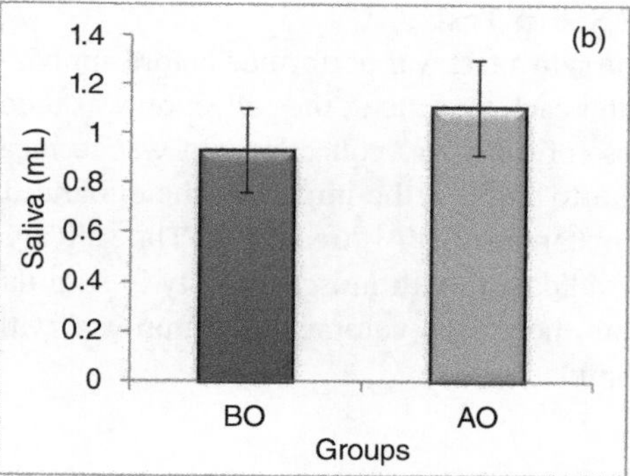

Figure 6.14 Saliva volumes before and after SIPO in (a) stimulation and (b) non-stimulation.

for the CPAP trial. RAIO shows the respiratory arousal index for the SIPO trial ($n = 33$). The RAIO was significantly decreased by 39.888% during the SIPO trial in comparison to the RAIC.

2) **Apnea–hypopnea index (AHI)**
To measure the severity of OSA, the AHI can indicate if the case is normal, moderate, or severe. This can be measured by polysomnography, which measures the oxygen drop in the blood associated with a breathing pause. There are three rules for analyzing the PSG data (Fino et al. 2020; Troxel et al. 2020). Figure 6.13b shows the AHIC indicating the AHI for the CPAP trial. The AHIO is the AHI for the SIPO trial. The AHIO was significantly decreased by 29.95% in comparison to the AHIC.

3) **Apnea–hypopnea index supine**
The polysomnography records their AHI on the existing sleep position when the participants slept by lying on their back with their face up. Decreasing AHI in this position means the airway was open during most of the sleeping time. High values indicate that collapse events per hour are increasing. A supine sleeping position is considered an exacerbating factor for OSA and causes more collapses (Troxel et al. 2020). Figure 6.13c shows AHI supine C recorded during the CPAP and SIPO trials for the participants. The orange bar indicates the AHI in the supine sleeping position in the CPAP trial ($n = 33$). The AHI supine O decreased significantly by 67.42% compared to that of the CPAP trial.

4) **Oxygen desaturation index (ODI)**
The precise oxygen saturation level in human blood should be around 95–100% (Peppard et al. 2009). When the oxygen level in the blood reduces, the blood pressure is lowered, and hypertension occurs. The oxygen desaturation index (ODI) is the number of times per hour of sleep that the blood oxygen level drops by a certain degree from the baseline (Mullins et al. 2021). The ODI is typically measured as a part of the formal sleep studies, such as diagnostic polysomnogram, home SA testing, or overnight oximetry. The ODI ≥ 3 O of the SIPO trial was significantly decreased by 38.036% compared to the CPAP trial; see Figure 6.13d.

5) **Saliva Tests**

The saliva test was performed before applying any therapy to record the reference baseline. After each treatment, the collection was redone to evaluate the therapeutic effect. The purpose of the saliva collection test was to measure the saliva volume before and after each trial to validate the impact of the enhanced therapy on participants compared to that in the standard CPAP treatment. The salivary flow rate was collected for both night trials to validate mouth muscle activity in stimulating the salivary glands. Figure 6.14 indicates that the saliva volume has improved with SIPO for both stimulated tests and non-stimulated tests.

Bibliography

Adil, A. (2019). An investigation into improving the CPAP and the electrical stimulation for the OSA treatment. PhD thesis. Auckland University of Technology.

Åkerstedt, T., Schwarz, J., Gruber, G. et al. (2019). Short sleep—poor sleep? A polysomnographic study in a large population-based sample of women. *Journal of Sleep Research* 28(4): e12812. https//doi.org/10.1111/jsr.12812.

Almeneessier, A.S., Alshahrani, M., Aleissi, S. et al. (2020). comparison between blood pressure during obstructive respiratory events in ReM and nReM sleep using pulse transit time. *Scientific Reports* 10(1): 1–10.

Alzoubaidi, M. and Mokhlesi, B. (2016). Obstructive sleep apnea during rapid eye movement sleep: clinical relevance and therapeutic implications. *Current Opinion in Pulmonary Medicine* 22(6): 545–554.

America, W.U. (1993). A National Sleep Alert. *Report of the National Commission on Sleep Disorders Research* 1.

American Academy of Sleep Medicine Task Force (1999). Sleep-related breathing disorders in adults: recommendations for syndrome definition and measurement techniques in clinical research. *Sleep Medicine* 22, 667–689.

American National Standards Institute (1979). *American National Standards for Humidifiers and Nebulizers for Medical Use. (ANSI Z79.9-179).* New York: The Institute.

American Sleep Apnea Association (1999). Sleep apnea information and resources. http://www.stanford.edu/~dement/apnea.html (accessed 8 June 2008).

Ashaat, S. and Al-Jumaily, A.M. (2016). Reducing upper airway collapse at lower continuous positive airway titration pressure. *Journal of Biomechanics* 49: 3915–3922.

Astor, F.C., Hanft, K.L., Benson, C. et al. (1998). Analysis of short-term outcome after office-based laser-assisted uvulopalatoplasty. *Otolaryngology – Head and Neck Surgery* 118(4): 478–480.

Australia Innovates (2008). CPAP sleep apnea control-machine to maintain breathing during sleep. http://www.powerhousemuseum.com/australia_innovates/?behaviour=view_article&Section_id=1030&article_id=10025 (accessed 8 June 2008).

Ayas, N.T., Patel, S.R., Malhotra, A. et al. (2004). Auto-titrating versus standard continuous positive airway pressure for the treatment of obstructive sleep apnea: results of a metaanalysis. *Sleep* 27: 249–253.

Badia, R., Farre, R., Rigau, J. et al. (2001). Forced oscillation measurements do not affect upper airway muscle tone or sleep in clinical studies. *The European Respiratory Journal* 18: 335–339.

Barnes, M., Mcevoy, R.D., Banks, S. et al. (2004). Efficacy of positive airway pressure and oral appliance in mild to moderate obstructive sleep apnea. *American Journal of Respiratory and Critical Care Medicine* 170(6): 656–664.

Barney, A., Bernabé, E.S., Nikolié, D., and Allen, R. (2010). Biomechanical modelling of palatal snoring. In: *2010 3rd International Symposium on Applied Sciences in Biomedical and Communication Technologies (ISABEL 2010)* (7 November 2010), 1–2. IEEE.

Basner, R.C. (2007). CPAP for obstructive sleep apnea. *The England Journal of Medicine* 356: 1751–1758.

Bassiri, A.G. and Guilleminault, C. (2000). Clinical features and evaluation of obstructive sleep apnea-hypopnea syndrome. *Principles and Practice of Sleep Medicine* 3: 869–878.

el Bayadi, S., Millman, R.P., Tishler, P.V. et al. (1990). A family study of sleep apnea: anatomic and physiologic interactions. *Chest* 98: 554–559.

Behbehani, K.K. (1995). Automatic control of airway pressure for treatment of obstructive sleep apnea. *IEEE Transactions on Biomedical Engineering* 42(10): 1007–1016.

Benumof, J.L. (2001). Obstructive sleep apnea in the adult obese patient: implications for airway management. *Journal of Clinical Anesthesia* 13(2): 144–156.

Berry, R.B., Rita, B., Charlene, G. et al. (2017). AASM scoring manual updates for 2017 (version 2.4). *Journal of Clinical Sleep Medicine: Official Publication of the American Academy of Sleep Medicine* 13(5): 665–666.

Bersten, A.D., Holt, A.W., Vedig, A.E. et al. (1991). Treatment of severe cardiogenic pulmonary edema with continuous positive airway pressure delivered by face mask. *The New England Journal of Medicine* 325: 1825–1830.

Berthon-Jones, M., Lawrence, S., Sullivan, C.E. et al. (1996). Nasal continuous positive airway pressure treatment: current realities and future. *Sleep* 19(9 Suppl): S131–S135.

Bijaoui, E.L. et al. (2002). Mechanical properties of the lung and upper airways in patients with sleep-disordered breathing. *American Journal of Respiratory and Critical Care Medicine* 165(8): 1055–1061.

Bossi, R., Piatti, G., Roma, E. et al. (2004). Effects of long-term nasal CPAP therapy on morphology, function, and mucociliary clearance of nasal epithelium in patients with obstructive sleep apnea syndrome. *Laryngoscope* 114(8): 1431–1434.

Brancatisano, A., Van der Touw, T., O'Neill, N., and Amis, T.C. (1996). Influence of upper airway pressure oscillations on soft palate muscle electromyographic activity. *Journal of Applied Physiology* 81(3): 1190–1196.

British Standards Institution (1970). *Specifications for Humidifiers for Use with Breathing Machines (BS 4494).* London: The Institution.

Button, B., Picher, M., and Boucher, R.C. (2007). Differential effects of cyclic and constant stress on ATP release and mucociliary transport by human airway epithelia. *The Journal of Physiology* 580(Pt. 2): 577–592.

Carter, B.G., Whittington, N., Hochmann, M., and Osborne, A. (2002). The effect of inlet gas temperatures on heated humidifier performance. *Journal of Aerosol Medicine* 15(1): 7–13.

Cartwright, R.D. and Samelson, C.F. (1982). The effects of a nonsurgical treatment for obstructive sleep apnea. The tongue-retaining device. *JAMA* 248: 705–709.

Çengel, Y.A. (2007). *Transferencia de calor y masa: un enfoque práctico*, 3ee. México, D.F.: McGraw-Hill.

Cheng, D.S.T., Weng, C.M.J., Yang, P.W. et al. (1998). Carbon dioxide laser surgery for snoring: results in 192 patients. *Otolaryngology – Head and Neck Surgery* 118(4): 486–489.

Chesson, A. et al. (1997). Practice parameters for the indications for polysomnography and related procedures. *Sleep* 20(6): 406–422.

Chilvers, M.A. and O'Callaghan, C. (2000). Analysis of ciliary beat pattern and beat frequency using digital high speed imaging: comparison with the photomultiplier and photodiode methods. *Thorax* 55(4): 314–317.

Chouly, F., Hirtum, A.V., Lagree, P.Y. et al. (2006). Simulation of the retroglossal fluid-structure interaction during obstructive sleep apnea. *Lecture Notesin Computer Science* 4072: 48–57.

Clark, G.T., Blumenfeld, I., Yoffe, N. et al. (1996). A crossover study comparing the efficacy of continuous positive airway pressure with anterior mandibular positioning devices on patients with obstructive sleep apnea. *Chest* 109: 1477–1483.

Collard, P., Pieters, T., Aubert, G. et al. (1997). Compliance with nasal CPAP in obstructive sleep apnea patients. *Sleep Medicine Reviews* 1(1): 33–44.

Constantinidis, J., Knöbber, D., Steinhart, H. et al. (2000). Fine-structural investigations of the effect of nCPAP-mask application on the nasal mucosa. *Acta Oto-Laryngologica* 120(3): 432–437.

Conway, W.A., Zorick, F., Piccione, P., and Roth, T. (1982). *Protriptylline in the treatment of sleep apnea. Thorax* 37: 49–53.

Cross, A., Cameron, P., Kierce, M. et al. (2003). Non-invasive ventilation inacute respiratory failure: a randomised comparison of continuous positive airway pressure and bi-level positive airway pressure. *Emergency Medicine Journal* 20: 531–534.

Cross, M.D., Vennelle, M., Engleman, H.M. et al. (2006). Comparison of CPAP titration at home or the sleep laboratory in the sleep apnea hypopnea syndrome. *Sleep* 29(11): 1451.

Croxton, T.L. (1993). Electrophysiological properties of guinea pig tracheal epithelium determined by cable analysis. *The American Journal of Physiology* 265(1 Pt 1): L38–L44.

Crystal G. (2008). What is BiPAP? http://www.wisegeek.com/what-is-bipap.htm (accessed 9 June 2008).

d'Ortho, M.P., Grillier-Lanoir, V., Levy, P. et al. (2000). Constant vs automatic continuous positive airway pressure therapy: home evaluation. *Chest* 118(4): 1010–1017.

Deane, S.A., Cistulli, P.A., Ng, A.T. et al. (2009). Comparison of mandibular advancement splint and tongue stabilizing device in obstructive sleep apnea: a randomized controlled trial. *Sleep* 32: 648–653.

Deegan, P.C., Mulloy, E., and McNicholas, W.T. (1995). Topical oropharyngeal anesthesia in patients with obstructive sleep apnea. *American Journal of Respiratory and Critical Care Medicine* 151(4): 1108–1112.

Di-Tullio, F., Glenda Ernst, G., Robaina, G. et al. (2018). Ambulatory positional obstructive sleep apnea syndrome. *Sleep Science (Sao Paulo, Brazil)* 11(1): 8–11.

Douglas, N.J., Thomas, S., and Jan, M.A. (1992). Clinical value of polysomnography. *The Lancet* 339(8789): 347–350.

Douglas, N.J., Luke, M., and Mathur, R. (1993). Is the sleep apnoea/hypopnoea syndrome inherited? *Thorax* 48: 719–721.

Eastwood, P.R., Satoh, M., Curran, A.K. et al. (1999). Inhibition of inspiratory motor output by high-frequency low-pressure oscillations in the upper airway of sleeping dogs. *Journal of Physiology* 517(1): 259–271.

Edwards, B.A., Eckert, D.J., Mcsharry, D.G. et al. (2014). Clinical predictors of the respiratory arousal threshold in patients with obstructive sleep apnea. *American Journal of Respiratory and Critical Care Medicine* 190(11): 1293–1300.

Elad, D., Wolf, M., and Keck, T. (2008). Air-conditioning in the human nasal cavity. *Respiratory Physiology & Neurobiology* 163(1–3): 121–127.

Emsellem, H.A. and Murtagh, K.E. (2005). Sleep apnea and sports performance. *Clinics in Sports Medicine* 24(2): 329–334.

Farre, R., Rotger, M., Monserrat, J.M., and Navajas, D. (1997). A system to generate simultaneous forced oscillation and continuous positive airway pressure. *The European Respiratory Journal* 10: 1349–1353.

Fehrenbach, M.J. and Herring, S.W. (2015). *Illustrated Anatomy of the Head and Neck-E-Book*. Elsevier Health Sciences.

Fernbach, S.K., Brouillette, R.T., Riggs, T.W. et al. (1983). Radiologic evaluation of adenoids and tonsils in children with obstructive sleep apnea: plain films and fluoroscopy. *Pediatric Radiology* 13(5): 258–265.

Ficker, J.H., Clarenbach, C.F., Neukirchner, C. et al. (2003). Auto-CPAP therapy based on the forced oscillation technique. *Biomedizinische Technik* 48: 68–72.

Findley, L.J., Wilhoit, S.C., and Suratt, P.M. (1985). Apnea duration and hypoxemia during REM sleep in patients with obstructive sleep apnea. *Chest* 87(4): 432–436.

Findley, L.J., Unverzagt, M.E. and Suratt, P.M et al. (1998). Automobile accidents involving patients with obstructive sleep apnea. *American Review of Respiratory Disease* 138(2): 337–340.

Fino, E., Plazzi, G., Filardi, M. et al. (2020). (Not so) Smart sleep tracking through the phone: findings from a polysomnography study testing the reliability of four sleep applications. *Journal of Sleep Research* 29(1): e12935.

Fischer, Y., Keck, T., Leiacker, R. et al. (2008). Effects of nasal mask leak and heated humidification on nasal mucosa in the therapy with nasal continuous positive airway pressure (nCPAP). *Sleep & Breathing* 12(4): 353–357.

Fleetham, J., West, P., Mezon, B. et al. (1982). Sleep, arousals, and oxygen desaturation in chronic obstructive pulmonary disease: the effect of oxygen therapy. *American Review of Respiratory Disease* 126(3): 429–433.

Fortuna, A.M., Miralda, R., Calaf, N. et al. (2011). Airway and alveolar nitric oxide measurements in obstructive sleep apnea syndrome. *Respiratory Medicine* 105: 630–636.

Goldberg, A.N. (2002). Obstructive sleep apnea: treatment algorithms. *Operative Techniques in Otolaryngology-Head and Neck Surgery* 13(3): 225–230.

Guilleminault, C., Partinen, M., Hollman, K. et al. (1995). Familial aggregates in obstructive sleep apnea syndrome. *Chest* 107: 1545–1551.

Haba-Rubio, J., Petitpierre, N.J., Cornette, F. et al. (2015). Oscillating positive airway pressure versus CPAP for the treatment of obstructive sleep apnea. *Frontiers in Medicine (Lausanne)* 2: 29.

Hans, M.G., Nelson, S., Luks, V.G. et al. (1997). Comparison of two dental devices for treatment of obstructive sleep apnea syndrome (OSAS). *American Journal of Orthodontics and Dentofacial Orthopedics* 111: 562–570.

Health, M.O. (2006). *Tatau Kahukura: Maori Health Chart Book*. Wellington: Minis. of Health.

Hendler, B., Costello, B., Silverstein, K. et al. (2001). A protocol for Uvulopalatopharyngoplasty, mortised genioplasty, and maxillomandibular advancement in patients with obstructive sleep apnea: an analysis of 40 cases. *Journal of Oral and Maxillofacial Surgery* 59: 892–897.

Henke, K.G. and Sullivan, C.E. (1993). Effects of high-frequency oscillating pressures on upper airway muscles in humans. *Journal of Applied Physiology* 75(2): 856–862.

Henke, K.G., Frantz, D.E., and Kuna, S.T. (2000). An oral elastic mandibular advancement device for obstructive sleep apnea. *American Journal of Respiratory and Critical Care Medicine* 161(2): 420–425.

Hiensch, R. and Rapoport, D.M. (2021). Upper Airway Resistance Syndrome. In: *Complex Sleep Breathing Disorders*, 103–115. Springer.

Hochban, W., Conradt, R., Brandenburg, U. et al. (1997). Surgical maxillofacial treatment of obstructive sleep apnea. *Plastic and Reconstructive Surgery* 99: 619–626; discussion 627–628.

Hollandt, J.H. and Mahlerwein, M. (2003). Nasal breathing and CPAP in patients with obstructive sleep apnea (OSA). *The Nose and Sleep-Disordered Breathing* 7(2): 87–93.

Horváth, G. and Sorscher, E.J. (2008). Luminal fluid tonicity regulates airway ciliary beating by altering membrane stretch and intracellular calcium. *Cell Motility and the Cytoskeleton* 65(6): 469–475.

https://rtsleepworld.com/2019/07/24/meet-somnotouch-psg-smallest-fully-aasm-compliant-psg-device/.

Huang, J.J. (2007). Modelling and optimization of air flow sensor within CPAP system. Doctoral dissertation. Auckland University of Technology.

Hudgel, D.W. (1992). Mechanisms of obstructive sleep apnea. *Chest* 101(2): 541–549.

Hudgel, D. and Fung, C. (2000). A long-term randomized crossover comparison of auto-titrating and standard nasal continuous positive airway pressure. *Sleep* 23: 1–4.

Hudgel, D.W. and Hendricks, C. (1988). Palate and hypopharynx – sites of inspiratory narrowing of the upper airway during sleep. *American Review of Respiratory Disease* 138(6): 1542–1547.

Huhtakangas, J.K., Huhtakangas, J., Bloigu, R. et al. (2020). Unattended sleep study in screening for sleep apnea in the acute phase of ischemic stroke. *Sleep Medicine* 65: 121–126.

Hukins, C.A. (2004). Comparative study of autotitrating and fixed-pressure CPAP in the home: a randomized, single-blind crossover trial. *Sleep* 27(8): 1512–1517.

Isono, S. et al. (1997). Anatomy of pharynx in patients with obstructive sleep apnea and in normal subjects. *Journal of Applied Physiology* 82(4): 1319–1326.

Kallio, T., Kuisma, M., Alaspaa, A., and Rosenberg, P.H. (2003). The use of prehospital continuous positive airway pressure treatment in presumed acute severe pulmonary edema. *Prehospital Emergency Care* 7(2): 209–213.

Kamami, Y.V. (1990). Laser CO2 for snoring. Preliminary results. *Acta Oto-Rhino-Laryngologica Belgica* 44(4): 451–456.

Kamami, Y.V. (1994). Outpatient treatment of sleep apnea syndrome with CO2 laser: laser-assisted UPPP. *Journal of Otolaryngology* 23(6): 395–398.

Karamanh, H., Özol, D., Ugur, K.S. et al. (2014). Influence of CPAP treatment on airway and systemic inflammation in OSAS patients. *Sleep & Breathing* 18(2): 251–256.

Keck, T., Leiacker, R., Heinrich, A. et al. (2000a). Humidity and temperature profile in the nasal cavity. *Rhinology* 38(4): 167–171.

Keck, T., Leiacker, R., Riechelmann, H., and Rettinger, G. (2000b). Temperature profile in the nasal cavity. *Laryngoscope* 110: 651–654.

Kiely, J.L., Murphy, M., and McNicholas, W.T. (1999). Subjective efficacy of nasal CPAP therapy in obstructive sleep apnoea syndrome: a prospective controlled study. *European Respiratory Journal* 13(5): 1086–1090.

Kohler, M., Ayers, L., Pepperell, J.C. et al. (2009). Effects of continuous positive airway pressure on systemic inflammation in patients with moderate to severe obstructive sleep apnoea: a randomised controlled trial. *Thorax* 64: 67–73.

Konermann, M., Sanner, B.M., Vyleta, M. et al. (1998). Use of conventional and self-adjusting nasal continuous positive airway pressure for treatment of severe obstructive sleep apnea syndrome. *Chest* 113(3): 714–718.

Koo, D.L., Kim, H.-R., and Nam, H. (2020). Moderate to severe obstructive sleep apnea during REM sleep as a predictor of metabolic syndrome in a Korean population. *Sleep & Breathing* 1–8.

Kosowsky, J.M., Storrow, A.B., and Carleton, S.C. (2000). Continuous and bilevel positive airway pressure in the treatment of acute cardiogenic pulmonary edema. *The American Journal of Emergency Medicine* 18: 91–95.

Kosowsky, J.M., Stephanides, S.L., Branson, R.D., and Sayre, M.R. (2001). Prehospital use of continuous positive airway pressure (CPAP) for presumed pulmonary edema: a preliminary case series. *Prehospital Emergency Care* 5(2): 190–196.

Kribbs, N.B., Pack, A.I., Kline, L.R. et al. (1993). Objective measurement of patterns of nasal CPAP use by patients with obstructive sleep apnea. *American Review of Respiratory Disease* 147 (4): 887–895.

Kuna, S.T. and Sant'Ambrogio, G. (1991). Pathophysiology of upper airway closure during sleep. *Journal of the American Medical Association* 266(10): 1384–1389.

Kushi Da, C.A., Littner, M.R., Morgenthaler, T. et al. (2005). Practice parameters for the indications for polysomnography and related procedures: an update for 2005. *Sleep* 28(4): 499.

Larsson, H., Carlsson-Nordlander, B. Lindblad, L.E. et al. (1992). Temperature thresholds in the oropharynx of patients with obstructive sleep apnea syndrome. *American Review of Respiratory Disease* 146(5): 1246–1249.

Lazard, D.S., Blumen, M., Lévy, P. et al. (2009). The tongue-retaining device: efficacy and side effects in obstructive sleep apnea syndrome. *Journal of Clinical Sleep Medicine: Official Publication of the American Academy of Sleep Medicine* 5(5): 431–438.

Lee, E.-M., Lee, T.H., Park, O.L. et al. (2020). Effective continuous positive airway pressure changes related to sleep stage and body position in obstructive sleep apnea during upward and downward titration: an experimental study. *Journal of Clinical Neurology* 16(1): 90–95.

Leite, F.G., Rodrigues, R.C., Ribeiro, R.F. et al. (2014). The use of a mandibular repositioning device for obstructive sleep apnea. *European Archives of Oto-Rhino-Laryngology* 271(5): 1023–1029.

Li, K.K. (2005). Surgical therapy for adult obstructive sleep apnea. *Sleep Medicine Reviews* 9(3): 201–209.

Li, W., Duan, Y., Yan, J. et al. (2020). Association between loss of sleep-specific waves and age, sleep efficiency, body mass index, and apnea-hypopnea index in human N3 sleep. *Aging and Disease* 11(1): 73–81.

Lin, M., Yang, Y.F., Chiang, H.T. et al. (1995). Reappraisal of continuous positive airway pressure therapy in acute cardiogenic pulmonary edema. Short-term results and long-term follow up. *Chest* 107: 1379–1386.

Ljunggren, M., Lindberg, E., Franklin K.A. et al. (2018). Obstructive sleep apnea during rapid eye movement sleep is associated with early signs of atherosclerosis in women. *Sleep* 41(7): zsy099.

Loube, D.I. (1999). Technologic advances in the treatment of obstructive sleep apnea syndrome. *Chest* 116(5): 1426–1433.

Loube, D.I., Gay, P.C., Strohl, K.P. et al. (1999). Indications for positive airway pressure treatment of adult obstructive sleep apnea patients: a consensus statement. *Chest* 115: 863–866.

Lydiatt, D.D. and Bucher, G.S. (2010). The historical Latin and etymology of selected anatomical terms of the larynx. *Clinical Anatomy* 23(2): 131–144.

Malik, N.W. and Kenyon, G.S. (2004). Changes in the nasal airway mucosa and in nasal symptoms following CPAP for obstructive sleep apnoea. *Australian Journal of Otolarynology* 7(1): 17.

Mannarino, M.R., Di Filippo, F., and Pirro, M. (2002). Obstructive sleep apnea syndrome. *European Journal of Internal Medicine* 23(7): 586–593.

Manon-Espaillat, R., Gothe, B., Adams, N. et al. (1988). Familial 'sleep apnea plus' syndrome: report of a family. *Neurology* 38: 190–193.

Marklund, M. (2021). Oral appliance therapy. In: *Management of Obstructive Sleep Apnea*, 185–211. Springer, Cham.

Marti, S., Sampol, G., Munoz, X. et al. (2002). Mortality in severe sleep apnoea/hypopnoea syndrome patients: impact of treatment. *European Respiratory Journal* 20(6): 1511–1518.

Martins de Araújo, M.T., Vieira, S.B., Vasquez, E.C., and Fleury, B. (2000). Heated humidification or face mask to prevent upper airway dryness during continuous positive airway pressure therapy. *Chest* 117(1): 142–147.

Masa, J.F., Jimenez, A., Duran, J. et al. (2004). Alternative methods of titrating continuous positive airway pressure: a large multicenter study. *American Journal of Respiratory and Critical Care Medicine* 170: 1218–1224.

Massie, C.A., McArdle, N., Hart, R.W. et al. (2003). Comparison between automatic and fixed positive airway pressure therapy in the home. *American Journal of Respiratory and Critical Care Medicine* 167(1): 20–23.

Mathur, R. and Douglas, N.J. (1995). Family studies in patients with the sleep apnea-hypopnea syndrome. *Annals of Internal Medicine* 122: 174–178.

Matsuo, K., Hiiemae, K.M., and Palmer, J.B. (2005). Cyclic Motion of the Soft Palate in Feeding. *Journal of Dental Research* 84(1): 39–42.

Menn, S.J., Loube, D.I., Morgan, T.D. et al. (1996). The mandibular repositioning device: role in the treatment of obstructive sleep apnea. *Sleep* 19: 794–800.

Menter, F.R. (1994). Two-equation eddy-viscosity turbulence models for engineering applications. *AIAA Journal* 32(8): 1598–1605.

Meurice, J.C., Dore, P., Paquereau, J. et al. (1994). Predictive factors of long-term compliance with nasal continuous positive airway pressure treatment in sleep apnea syndrome. *Chest* 105(2): 429–433.

Meurice, J.C., Cornette, A., Philip-Joet, F. et al. (2007). Evaluation of autoCPAP devices in home treatment of sleep apnea/hypopnea syndrome. *Sleep Medicine* 8: 695–703.

Michaelson, P.G., Allan, P., Chaney, J. et al. (2006). Validations of a portable home sleep study with twelve-lead polysomnography: comparisons and insights into a variable gold standard. *Annals of Otology, Rhinology, and Laryngology* 115(11): 802–809.

Mihaescu, M., Murugappan, S., Gutmark, E. et al. (2008a). Computational modeling of upper airway before and after adenotonsillectomy for obstructive sleep apnea. *Laryngoscope* 118(2): 360–362.

Mihaescu, M., Murugappan, S., Kalra, M. et al. (2008b). Large Eddy simulation and Reynolds-Averaged Navier-Stokes modeling of flow in a realistic pharyngeal airway model: an investigation of obstructive sleep apnea. *Journal of Biomechanics* 41(10): 2279–2288.

Morrell, M.J., Arabi, Y., Zahn, B. et al. (1998). Progressive retropalatal narrowing preceding obstructive apnea. *American Journal of Respiratory and Critical Care Medicine* 158(6): 1974–1981.

Morrison, D.L., Launois, S.H., Isono, S. et al. (1993). Pharyngeal narrowing and closing pressures in patients with obstructive sleep apnea. *American Review of Respiratory Disease* 148(3): 606–611.

Mosesso, V., Dunford, J., Blackwell, T., and Griswell, J. (2003). Prehospital therapy for acute congestive heart failure: state of the art. *Prehospital Emergency Care* 7(1): 13–23.

Mujica-Paz, H. and Gontard, N. (1997). Oxygen and carbon dioxide permeability of wheat gluten film: effect of relative humidity and temperature. *Journal of Agricultural and Food Chemistry* 45: 4101–4105.

Mulgrew, A.T., Fox, N., Ayas, N.T., and Ryan, C.F. (2007). Diagnosis and initial management of obstructive sleep apnea without polysomnography. *Annals of Internal Medicine* 146: 157–166.

Mullins, A.E., Williams, M.K., Kam, K. et al. (2021). Effects of obstructive sleep apnea on human spatial navigational memory processing in cognitively normal older individuals. *Journal of Clinical Sleep Medicine: Official Publication of the American Academy of Sleep Medicine*. 9080.

Mylavarapu, G., Mihaescu, M., Murugappan, S., and Gutmark E. (2010). Fluid structure interaction analysis in human upper airways to understand sleep apnea. *48th AIAA Aerospace Sciences Meeting Including the New Horizons Forum and Aerospace Exposition 2010*, p. 1264.

Navajas, D., Farre, R., Rotger, M. et al. (1998). Assessment of airflow obstruction during CPAP by means of forced oscillation in patients with sleep apnea. *American Journal of Respiratory and Critical Care Medicine* 157(5): 1526–1530.

Navazesh, M. and Kumar, S.K.S. (2008). *Measuring salivary flow: challenges and opportunities. The Journal of the American Dental Association* 139: 35S–40S.

Nieto, F.J., Young, T.B., Lind, B.K. et al. (2000). Association of sleepdisordered breathing, sleep apnea, and hypertension in a large community-based study. *JAMA* 283: 1829–1836.

Nolan, G.M., Ryan, S., O'Connor, T.M., and McNicholas, W.T. (2006). Comparison of three auto-adjusting positive pressure devices in patients with sleep apnoea. *The European Respiratory Journal* 2006(28): 159–164.

Nolan, G., Doherty, L., and McNicholas, W. (2007). Auto-adjusting versus fixed positive pressure therapy in mild to moderate obstructive sleep apnoea. *Sleep* 30(2): 189–194.

Noseda, A., Kempenaers, C., Kerkhofs, M. et al. (2004). Constant vs auto-continuous positive airway pressure in patients with sleep apnea hypopnea syndrome and a high variability in pressure requirement. *Chest* 126(1): 31–37.

O'Sullivan, R.A., Hillman, D.R., Mateljan, R. et al. (1995). Mandibular advancement splint: an appliance to treat snoring and obstructive sleep apnea. *American Journal of Respiratory and Critical Care Medicine* 151: 194–198.

Obstructive Sleep Apnea: Blocked Upper Airway. 2015. Mayo Clinic Website. http://www.emedicinehealth.com/script/main/art.asp?articlekey=135675&ref=129801.

Oksenberg, A., Arons, E., Nasser, K. et al. (2010). REM-related obstructive sleep apnea: the effect of body position. *Journal of Clinical Sleep Medicine: Official Publication of the American Academy of Sleep Medicine* 6(4): 343–348.

Olson, E.J., Moore, W.R., Morgenthaler, T.I. et al. (2003). Obstructive sleep apnea-hypopnea syndrome. *Mayo Clinic Proceedings* 78(12): 1545–1552.

Oostveen, E., MacLeod, D., Lorino, H. et al. (2003). The forced oscillation technique in clinical practice: methodology, recommendations and future developments. *European Respiratory Journal* 22(6): 1026–1041.

Palmer, L.J., Buxbaum, S.G., Larkin, E. et al. (2003). A whole-genome scan for obstructive sleep apnea and obesity. *American Journal of Human Genetics* 72: 340–350.

Palmer, L.J., Buxbaum, S.G., Larkin, E.K. et al. (2004). Whole genome scan for obstructive sleep apnea and obesity in African-American families. *American Journal of Respiratory and Critical Care Medicine* 169: 1314–1321.

Peppard, P.E., Ward, N.R., and Morrell, M.J. (2009). The impact of obesity on oxygen desaturation during sleep-disordered breathing. *American Journal of Respiratory and Critical Care Medicine* 180(8): 788–793.

Pevernagie, D., Proot, P., Hertegonne, K. et al. (2004). Efficacy of flow- vs impedance-guided autoadjustable continuous positive airway pressure: a randomized cross-over trial. *Chest* 126(1): 25–30.

Pevernagie, D., Masa, J.F., Meurice, J.C. et al. (2007). Treatment of obstructive sleep-disordered breathing with positive airway pressure systems. *European Respiratory Review* 16(106): 125–131.

Pham, Q.T., Bourgkard, E., Chau, N. et al. (1995). Forced oscillation technique (FOT): a new tool for epidemiology of occupational lung diseases? *European Respiratory Journal* 8(8): 1307–1313.

Piccirillo, J.F., Duntley, S., and Schotland, H. (2000). Obstructive sleep apnea. *JAMA* 284(12): 1492–1494.

Planes, C., d'Ortho, M., Foucher, A. et al. (2003). Efficacy and cost of home-initiated auto-nCPAP versus conventional nCPAP. *Sleep* 26(2): 156–160.

Plowman, L., Lauff, D.C., Berthon-Jones, M., and Sullivan, C.E. (1990). Waking and genioglossus muscle responses to upper airway pressure oscillation in sleeping dogs. *Journal of Applied Physiology* 68(6): 2564–2573.

Powell, N.B., Riley, R.W., Troell, R.J. et al. (1998). Radiofrequency volumetric tissue reduction of the palate in subjects with sleep-disordered breathing [see comments]. *Chest* 113: 1163–1174.

Powell, N.B., Riley, R.W., and Guilleminault, C. (1999). Radiofrequency tongue base reduction in sleep-disordered breathing: A pilot study. *Otolaryngology and Head and Neck Surgery* 120: 656–664.

Puchelle, E., Zahm, J.M., and Sadoul, P. (1982). Mucociliary frequency of frog palate epithelium. *The American Journal of Physiology* 242(1): C31–C35.

Qureshi, A., Ballard, R.D., and Nelson, H.S. (2003). Obstructive sleep apnea. *Journal of Allergy and Clinical Immunology* 112(4): 643–651.

Rakotonanahary, D., Pelletier-Fleury, N., Gagnadoux, F. et al. (2001). Predictive factors for the need for additional humidification during nasal CPAP therapy. *Chest* 119(2): 460–465.

Randerath, W., Galetke, W., David, M. et al. (2001a). Prospective randomized comparison of impedancecontrolled auto-continuous positive airway pressure (APAPfot) with constant CPAP. *Sleep Medicine* 2: 115–124.

Randerath, W.J., Schraeder, O., Galetke, W. et al. (2001b). Autoadjusting CPAP therapy based on impedance efficacy, compliance and acceptance. *American Journal of Respiratory and Critical Care Medicine* 163: 652–657.

Randerath, W.J., Meier, J., Genger, H. et al. (2002). Efficiency of cold passover and heated humidification under continuous positive airway pressure. *European Respiratory Journal* 20(1): 183–186.

Redline, S., Tosteson, T., Tishler, P.V. et al. (1992). Studies in the genetics of obstructive sleep apnea: familial aggregation of symptoms associated with sleep-related breathing disturbances. *The American Review of Respiratory Disease* 145: 440–444.

Redline, S., Tishler, P.V., Tosteson, T.D. et al. (1995). The familial aggregation of obstructive sleep apnea. *American Journal of Respiratory and Critical Care Medicine* 151: 682–687.

Reinikainen, L.M. and Jaakkola, J.J.K. (2003). Significance of humidity and temperature on skin and upper airway symptoms. *Indoor Air* 13: 334–352.

Reiter, J., Gileles-Hillel, A., Cohen-Cymberknoh, M. et al. (2020). Sleep disorders in cystic fibrosis: a systematic review and meta-analysis. *Sleep Medicine Reviews* 101279.

Richardson, M.A., Seid, A.B., Cotton, R.T. et al. (1980). Evaluation of tonsils and adenoids in sleep apnea syndrome. *The Laryngoscope* 90(7): 1106–1110.

Riley, R.W., Powell, N.B., and Guilleminault, C. (1993a). Obstructive sleep apnea syndrome: a surgical protocol for dynamic upper airway reconstruction. *Journal of Oral and Maxillofacial Surgery* 51: 742–747; discussion 748–749.

Riley, R.W., Powell, N.B., and Guilleminault, C. (1993b). Obstructive sleep apnea syndrome: a review of 306 consecutively treated surgical patients. *Otolaryngology and Head and Neck Surgery* 108: 117–125.

Riley, R.W., Powell, N.B., and Guilleminault, C. (1994). Obstructive sleep apnea and the hyoid: a revised surgical procedure. *Otolaryngology and Head and Neck Surgery* 111: 717–121.

Robin, I.G. (1968). *Snoring. Proceedings of the Royal Society of Medicine* 61(6): 575–582.

Rodenstein, D. (2008). Determination of therapeutic continuous positive airway pressure for obstructive sleep apnea using automatic titration: promises not fulfilled. *Chest* 133: 595–597.

Rodrigues, M.M., Gabrielli, M., Junior, O.G. et al. (2017). Nasal airway evaluation in obstructive sleep apnoea patients: volumetric tomography and endoscopic findings. *International Journal of Oral and Maxillofacial Surgery* 46(10): 1284–1290.

Roessler, F., Grossenbacher, R., and Walt, H. (1998). Effects of tracheotomy on human tracheobronchial mucosa: a scanning electron microscopic study. *Laryngoscope* 98: 1261–1267.

Rubesin, S.E., Jessurun, J., Robertson, D. et al. (1987). Lines of the pharynx. *Radiographics* 7(2): 217–237.

Rubin, B.K., Ramirez, O., and King, M. (1985). Mucus-depleted frog palate as a model for the study of mucociliary clearance. *Journal of Applied Physiology* 69(2): 424–429.

Sahin-Yilmaz, A., Baroody, F.M., Detineo, M. et al. (2008). Effect of changing airway pressure on the ability of the human nose to warm and humidify air. *The Annals of Otology, Rhinology, and Laryngology* 117(7): 501–507.

Sanders, M.H., Montserrat, J.M., Farré, R., and Givelber, R.J. (2008). Positive pressure therapy: a perspective on evidencebased outcomes and methods of application. *Proceedings of the American Thoracic Society* 5: 161–172.

Schmidt-Nowara, W.W., Meade, T.E., and Hays, M.B. (1991). Treatment of snoring and obstructive sleep apnea with a dental orthosis. *Chest* 99: 1378–1385.

Series, F., Cormier, I.M.Y., and La Forge, J. (1994). Required levels of nasal continuous positive airway pressure during treatment of obstructive sleep apnoea. *European Respiratory Journal* 7(10): 1776–1781.

Shahar, E., Whitney, C.W., Redline, S. et al. (2001). Sleep-disordered breathing and cardiovascular disease: cross-sectional results of the Sleep Heart Health Study. *American Journal of Respiratory and Critical Care Medicine* 163: 19–25.

Shamsuzzaman, A.S.M., Gersh, B.J., and Somers, V.K. (2003). Obstructive sleep apnea: implications for cardiac and vascular disease. *JAMA* 290: 1906–1914.

Sharma, S., Wali, S., Pouliot, Z. et al. (1996). Treatment of obstructive sleep apnea with a self-titrating continuous positive airway pressure (CPAP) system. *Sleep* 19(6): 497–501.

Sher, A.E. (1999). Surgical management of obstructive sleep apnea. *Progress in Cardiovascular Diseases* 41(5): 387–396.

Sher, A.E., Thorpy, M.J., Shprintzen, R.J. et al. (1985). Predictive value of Muller maneuver in selection of patients for uvulopalatopharyngoplasty. *Laryngoscope* 95: 1483–1487.

Sher, A.E., Schechtman, K.B., and Piccirillo, I.F. (1996). The efficacy of surgical modifications of the upper airway in adults with obstructive sleep apnea syndrome. *Sleep* 19: 156–177.

Siddiqui, F., Walters, A.S., Goldstein, D. et al. (2006). Half of patients with obstructive sleep apnea have a higher NREM AHI than REM AHI. *Sleep Medicine* 7(3): 281–285.

Skoczyński, S., Ograbek-Król, M., Tazbirek, M. et al. (2008). Short-term CPAP treatment induces a mild increase in inflammatory cells in patients with sleep apnoea syndrome. *Rhinology* 46: 144–150.

Stierer, T. and Punjabi, N.M. (2005). Demographics and diagnosis of obstructive sleep apnea. *Anesthesiology Clinics of North America* 23(3): 405–420.

Strohl, K.P., Saunders, N.A., Feldman, N.T. et al. (1978). Obstructive sleep apnea in family members. *The New England Journal of Medicine* 299: 969–973.

Sullivan, C.E., Berthon-Jones, M., Issa, F.G. et al. (1981). Reversal of obstructive sleep apnea by CPAP applied through the nares. *Lancet* 1: 862–865.

Sullivan, R. (2005). Prehospital use of CPAP: positive pressure = positive patient outcomes. *Emergency Medical Services* 34(8): 120, 122–4, 126.

Sun Y-C. (2005). CPAP system modelling. Master's degree. Auckland University of Technology

Sunnetcioglu, A., Sertogullarından, B., Ozbay, B. et al. (2016). Obstructive sleep apnea related to rapid-eye-movement or non-rapid-eye-movement sleep: comparison of demographic, anthropometric, and polysomnographic features. *Jornal Brasileiro de Pneumologia* 42: 48–54.

Susarla, S.M., Thomas, R.J., Abramson, Z.R. et al. (2010). Biomechanics of the upper airway: changing concepts in the pathogenesis of obstructive sleep apnea. *International Journal of Oral and Maxillofacial Surgery* 39(12): 1149–1159.

Tamaki, S., Yamauchi, M., Fukuoka, A. et al. (2007). Production of inflammatory mediators by monocytes in patients with obstructive sleep apnea syndrome. *Internal Medicine* 48: 1255–1262.

Tenhunen, M., Rauhala, E., Virkkala, J. et al. (2011). Increased respiratory effort during sleep is non-invasively detected with movement sensor. *Sleep & Breathing* 15(4): 737–746.

Teschler, H., Wessendorf, T., Farhat, A. et al. (2000). Two months auto-adjusting versus conventional nCPAP for obstructive sleep apnoea syndrome. *The European Respiratory Journal* 15(6): 990–995.

The National Heart, Lung, and Blood Institute (2008). How Is Sleep Apnea Treated? http://www.nhlbi.nih.gov/health/dci/Diseases/SleepApnea/SleepApnea_Treatments.html (accessed 8 June 2008).

Todd, D.A., Boyd, J., Lloyd, J., and John, E. (2001). Inspired gas humidity during mechanical ventilation: effects of humidification chamber, airway temperature prove and environmental conditions. *Journal of Peadiatrics: Child Health* 37: 489–494.

Toremalm, N.G. (1961). Air flow patterns and ciliary activity in the trachea after tracheostomy. *Oto-laryng Stockh* 53: 442–454.

Tortora, G.J., Derrickson, B.H., and Burkett, B. (2018). Principles of Anatomy and Physiology, 2nd Asia-Pacifice, Wiley. ISBN: 9780730354987.

Trindade, S.H.K., Júnior, J.F.D.M., Mion, O. et al. (2007). Métodos de estudo do transporte mucociliar. *Revista Brasileira de Otorrinolaringologia* 73: 704–712.

Troxel, W.M., Ann, H., Bonnie, G.D. et al. (2020). Broken windows, broken zzs: poor housing and neighborhood conditions are associated with objective measures of sleep health. *Journal of Urban Health: bulletin of the New York Academy of Medicine* 97(2): 230–238.

Vanderveken, O.M., Oostveen, E., Boudewyns, A.N. et al. (2005). Quantification of pharyngeal patency in patients with sleep-disordered breathing. *ORL: Journal for Otorhinolaryngology and Its Related Specialties* 67(3): 168–179.

Vgontzas, A.N., Tan, T.L., Bixler, E.O. et al. (1994). Sleep apnea and sleep disruption in obese patients. *Archives of Internal Medicine* 154(15): 1705–1711.

Vgontzas, A.N., Bixler, E.O., Tan, T.L. et al. (1998). Obesity without sleep apnea is associated with daytime sleepiness. *Archives of Internal Medicine* 158(12): 1333–1337.

Victor, L.D. (2004). Treatment of obstructive sleep apnea in primary care. *American Family Physician* 69(3): 561–568.

Waite, P.D., Wooten, V., Lachner, J. et al. (1989). Maxillomandibular advancement surgery in 23 patients with obstructive sleep apnea syndrome. *Journal of Oral and Maxillofacial Surgery* 47: 1256–1261; discussion 1262.

Walker, R.P., Grigg-Damberger, M.M., and Gopalsami, C. (1997). Uvulopalatopharyngoplasty versus laser-assisted uvulopalatoplasty for the treatment of obstructive sleep apnea. *Laryngoscope* 107: 76–82.

Widdicombe, J.H. (2002). Regulation of the depth and composition of airway surface liquid. *Journal of Anatomy* 201(4): 313–318.

Wiest, G.H., Fuchs, F.S., Brueckl, W.M. et al. (2000). In vivo efficacy of heated and non-heated humidifiers during nasal continuous positive airway pressure (nCPAP)-therapy for obstructive sleep apnoea. *Respiratory Medicine* 94: 364–368.

Wiest, G.H., Foerst, J., Fuchs, F.S. et al. (2001). In vivo efficacy of two heated humidifiers used during CPAP-therapy for obstructive sleep apnea under various environmental conditions. *Sleep* 24(4): 435–440.

Wiest, G.H., Harsch, I.A., Fuchs, F.S. et al. (2002). Initiation of CPAP therapy for OSA: does prophylactic humidification during CPAP pressure titration improve initial patient acceptance and comfort? *Respiration* 69: 406–412.

Wolf, M., Naftali, S., Schroter, R.C., and Elad, D. (2004). Air-conditioning characteristics of the human nose. *The Journal of Laryngology & Otology* 118: 87–92.

Wu, M.-F., Huang, W.C., Chang, K.M. et al. (2020). Detection performance regarding sleep apnea-hypopnea episodes with fuzzy logic fusion on single-channel airflow indexes. *Applied Sciences* 10(5): 1868.

Young, T., Palta, M., Dempsey, J. et al. (1993). The occurrence of sleep disordered breathing among middle aged adults. *The New England Journal of Medicine* 328: 1230–1235.

Zhang, S. and Mathew, O.P. (1992). Response of laryngeal mechanoreceptors to high-frequency pressure oscillation. *Journal of Applied Physiology* 73(1): 219–223.

Zhu, D.-M., Zhang, C., Yang, Y. et al. (2020). The relationship between sleep efficiency and clinical symptoms is mediated by brain function in major depressive disorder. *Journal of Affective Disorders* 266: 327–337.

Zozula, R., Rosen, R., and Phillips, B. (2001). Compliance with continuous positive airway pressure therapy: assessing and improving treatment outcomes. *Current Opinion in Pulmonary Medicine* 7(6): 391–398.

7

Pressure Oscillations in Asthma Treatment

7.1 Introduction

For more than 15 decades, airway smooth muscles (ASMs) contraction has been recognized as the primary driving mechanism for asthma attacks. Breathing has a strong relaxing and protection effect on ASM by inhibiting airway constriction. It is crucial to understand the behavior of ASM, which helps to understand the reversible airway obstruction associated with asthma. Many authors worldwide, including our group, have conducted intensive research on computational modeling of ASMs in an attempt to understand their behaviors and responses. However, after about 25 years of work in this field, nothing can replace the need for real-life experimental programs to understand physiological systems and their elements, in particular cells such as ASM. In any physiological system, including a small cell, it is not possible to have a complete model which can address the effects of all active and passive components in the system. Obviously, there are many physiological, physical, and environmental triggering mechanisms that make it impossible to duplicate all real-life activities in a mathematical or computational model. In brief, nothing will replace the experimental and laboratory studies on a cell such as a smooth muscle. Therefore, this chapter focuses on experimental investigations, which are summarized in the rest of the chapter.

This chapter briefly describes the basic concepts of asthma, ASM physiology, and dynamics through both experimental data and theoretical modeling results. To further illustrate the behavior of contracted ASM, a fading memory model (FMM) is presented in this chapter. A detailed in vivo investigation of the effect of superimposed pressure oscillations (SIPOs) on asthmatic subjects is presented. This consists of presenting the respiratory system of a mouse model, detailing the sensitization processes to produce an acute and a chronic asthmatic model, all necessary protocols for testing the model to demonstrate asthmatic status, all required methods to check the effect of SIPO on the lung performance of the asthmatic models when they are exposed to an asthmatic attack, and a comprehensive comparison between the acute model and the chronic model response. Finally, the preventive and protective potential of pressure waves superimposed on a mean pressure and breathing patterns before asthma occurs is discussed.

Pressure Oscillation in Biomedical Diagnostics and Therapy, First Edition. Ahmed Al-Jumaily and Lulu Wang.

Table 7.1 Assessment of asthma severity.

Case		Frequency of symptoms	
		Daytime: Any coughing, tight chest, and wheezing	**Nocturnal: Any coughing, tight chest, wheezing, and night awakening**
Intermittent or mild I		≤2 per week	≤1 per month
Chronic persistent	Mild II	3–4 per week	2–4 per month
	Moderate III	>4 per week	>4 per month
	Severe IV	Continuous	Frequent

7.2 Asthma

This section discusses the asthma disease, its impact, and the effect it has on respiratory performance. The ranges of medicinal treatments currently available are discussed with their efficacy, cost, and applications.

Asthma is one of the most common chronic (long-term) diseases that affect approximately 300 million people worldwide (Bateman et al. 2008). It causes the airways in the lungs to become irritated and inflamed (swollen). Consequently, this makes them very twitchy, sensitive, and reactive to various irritants resulting in narrowing and reduced airflow through the lungs.

Asthma has two key defining features: (i) a history of respiratory symptoms during the night and early morning, including wheezing, chest tightness, coughing, breathing difficulties, and (ii) variable expiratory airflow limitations. These asthma symptoms vary from person to person, but some problems such as infection and exposure to allergens will cause asthma to worsen. Asthma severity is based on the presenting symptoms (Table 7.1) (Lalloo et al. 2007), and it is a variable that does not predict the response to treatment. In general, it is of little value for patients who are already on treatment.

The reasons why people get asthma are not fully understood. Some genetic and environmental factors interact to cause asthma are proposed by many researchers, including atopy (an inherited tendency to develop allergies), asthmatic parents, respiratory infections during childhood, and contact with some airborne allergen or exposure to some viral infections in infancy or in early childhood when the immune system is developing. The asthma triggers vary from individual to individual. Some most common asthma triggers are allergens, infections, pollutants, occupational agents, foods, psychological factors, cold air, drugs, exercise, hyperventilation, and others such as perfumes or chemical agents.

7.2.1 Types of Asthma

Asthma can be divided into six categories: nonallergic asthma, allergic asthma, aspirin-exacerbated respiratory disease (AERD), exercise-induced asthma, cough variant, and occupational asthma (Barnes et al. 1992).

1) **Nonallergic asthma** is often developing when people are in their adulthood. Patients with nonallergic asthma normally have low allergy antibody levels which cannot be

identified by allergy skin tests or blood tests. Normally, an upper respiratory infection (cold, flu, and rhinovirus) sets off their asthma. As soon as cold or flu symptoms appear, patients are typically prescribed a short course of inhaled corticosteroids (ICSs) for 10–14 days.

2) **Allergic asthma (AA)**, also called atopic asthma, is one of the most common types of asthma, particularly common in kids. Some allergy antibodies (immunoglobulin E or IgE) are produced in some individuals because of atopy. These antibodies can react on subsequent exposures to the same allergen. The reaction that ensues in the body following these subsequent exposures may produce the symptoms of asthma. AA occurs when an allergy sets off an asthma flare-up. Some common allergies include mold, roaches, pollens, and pet dander, more details can be found in Barnes et al. (1992).
3) **Exercise-induced asthma (EIA)** occurs more often in younger people. Many asthmatics get asthma due to physical exertion. Any type of sports or physical exertion leads asthmatics to cough, hard breathing, and chest tightness, and these symptoms improve when they stop the sports. Symptoms for EIA usually occur 5–10 minutes after completing the exercise and, in most cases, resolve within 30–45 minutes. Taking inhaled bronchodilator about 15 minutes before exercise is the typical treatment for EIA.
4) **Cough variant asthma** is characterized by a dry hacking cough. It often occurs when asleep or awake and affects both adults and children, and it happens more often in children. The asthma symptoms of chest tightness, wheezing, or difficulty breathing are not presented in these patients who often have normal lung function. They may have hypersensitivity of the airways because people with typical asthma may react to triggers such as exercise, exposure, or allergens and irritants. Coughing is the only symptom in patients with cough variant asthma.
5) **Occupational asthma** is caused by exposure to an agent encountered in the work environment. It generally affects adults of working age as they become exposed to sensitizing chemicals at their workplaces. Over 300 substances have been associated with occupational asthma, including small molecules, irritants, platinum salts, animal biological products, and plants (Global Initiative for Asthma Report 2014). These substances are all capable of provoking the immune system to produce allergic antibody (IgE).
6) **Nocturnal asthma** has similar signs and symptoms to classical asthma. Almost 75% of nocturnal asthma symptoms occur between midnight and 8 a.m., with peaking at 4 a.m. The lung functions are usually at the lowest levels during this period because normal body functions facilitate changes according to the time of day or night.

Narrowing of airways can be caused by asthma, which interferes with the normal movement of air in and out of the lungs. Three main factors cause the narrowing of airways: inflammation, bronchospasm, and hyperreactivity. Inflammation is the most important factor driving the narrowing of the bronchial tubes. It increases the thickness of the wall of the bronchial tubes and results in a smaller passageway for air to flow through. The inflammation occurs in response to an allergen or irritant and results from the action of chemical mediators. The inflamed tissues produce an excess amount of "sticky" mucus in the tubes. The mucus can clump together and form "plugs" that can clog the smaller airways. Specialized allergy and inflammation cells (eosinophils and white blood cells) accumulate at the site and cause tissue damage. These damaged cells are shed into the airways, thereby contributing to the narrowing. Bronchospasm can occur in all people and can be brought on by inhaling dry or cold air, which causes the airway to narrow further. On the other hand, hyperreactivity in asthmatics, the chronically inflamed and constricted airways become highly sensitive or reactive to triggers such as allergens, irritants, and infections. Exposure to these triggers may cause more inflammation and narrowing. These three factors make exhaling difficult, which results in a typical wheezing sound since forceful exhalation is required to overcome the narrowing. Asthmatics also frequently cough to expel the thick mucus plugs. Reducing the airflow may cause less oxygen to pass into the bloodstream, and in severe cases, carbon dioxide may dangerously accumulate in the blood.

7.2.2 Asthma Diagnostics

Common asthma diagnostic methods include physical examination, personal medical history, lung function, and other tests. To properly diagnose asthma, patients are always recommended to discuss their medical history and condition with a physician.

Personal medical history plays an important role in identifying asthma. Several causes of similar respiratory symptoms and history are less than fully repeatable over time. Research has indicated that 7–13% of children have been identified with changes in three months after a symptom history check, although the prevalence rate remained stable (Peat et al. 1992).

During the **physical examination**, doctors generally pay special attention to the patient's eyes, ears, throat, chest, skin, and lungs. A pulmonary function test is included in physician examination to detect how well people exhale air from their lungs. Asthma is often recognized by using a pattern of one or more characteristics during the examination. Research showed that it is best to confirm asthma by evidence of variable or reversible airflow obstruction accompanying symptoms (Venables et al. 1984). Previous studies found that bronchitis or infection in children could be misdiagnosed as nocturnal or post-exercise coughing as asthma, (Koning 1981; Clifford et al. 1989).

There are several asthma diagnostics tests. These include but are not limited to:

1) **Lung function tests**, normally taken before and after inhaling a bronchodilator medication, which helps open the patient's airways. One likely has asthma if one's lung function improves significantly with the use of a bronchodilator. A trial with asthma medication can improve lung's function. Spirometry and peak expiratory flow (PEF) tests are the two main tests used to diagnose asthma. Spirometry is recommended for confirming the diagnosis of asthma. During the spirometry test, a human subject takes a deep breath, then exhales forcefully into a mouthpiece that is connected to a spirometer. The amount of air that the subject can breathe in and out is measured by the spirometer. PEF test is the most straightforward lung function test, which requires one to breathe in as deeply as they can and then blow into a peak flow meter as hard and as fast as possible. The peak flow meter measures the rate at which the patient can force air out of their lungs. Peak flow variability has been compared with methacholine challenge in community populations (Higgins et al. 1992). A clinical study showed that methacholine challenge is a more sensitive marker for diagnosing asthma than the peak flow or the exercise challenge (Siersted et al. 1996).
2) **Several specialized tests** such as airways responsiveness, inflammation, or allergy tests may also be required for some people. Airways responsiveness is absent in some people with other clear evidence of asthma but may be variably present in some kids and adults

without significant respiratory symptoms (Lee et al. 1983; Sears et al. 1986; Enarson et al. 1987; Salome et al. 1987; Pattemore et al. 1990). Subjects are requested to breathe in a medication that deliberately irritates or constricts their airways, causing a small decrease in the forced expiratory volume in 1 second (FEV_1) measured using spirometry, and possibly triggering mild asthma symptoms. Non-asthmatic does not respond to this stimulus. This test includes inhaling progressively and increasing the amount of the medication at intervals, with measurement of FEV_1 by using spirometry in between to check if it falls below a certain threshold.

3) **Airway inflammation testing** is helpful in some cases for checking inflammation in the airways. It can be done by taking a sample of mucus or measuring nitric oxide concentration by using a special machine. Airway inflammation is considered if a high level of nitric oxide is recorded.
4) **Allergy tests** are carried out to check if the subject is allergic or sensitive to certain substances known to cause occupational asthma. To confirm whether asthma is associated with specific allergies such as dust mites, pollen, or foods, a skin test or blood test is often required.

7.2.3 Asthma Treatment

Various clinical asthma treatment guidelines to control asthma are published (New Zealand Guidelines Group 2002; National Institutes of Health 2007). These focus on reducing impairment and risks. Reducing impairment is by reducing the frequency and intensity of symptoms and functional limitations currently or recently experienced by the patient, while reducing risk aims at reducing the likelihood of future asthma attacks and progressive decline in lung function or medication side effects. Fanta et al. reported several ways to manage asthma symptoms, including monitoring disease progression, avoiding environmental stimuli, using controller medications, and using quick-relief drugs (Fanta and Hockberger 2021).

Traditionally, the asthma symptoms have been treated using pharmaceutical bronchodilators and inflammation suppressors. Currently, other alternatives are being developed to reduce the intake of drugs. In the following text, the existing treatments for asthma are discussed. In general, treatment of asthma can be grouped as pharmacotherapy or non-pharmacological therapies. The first section discusses medicinal therapies, while the second focuses on alternative, non-medicinal therapies.

7.2.3.1 Pharmacotherapy Treatments

Asthma cannot be cured but can be treated with pharmaceutical medications which are potentially and often, a life-long treatment. Pharmacotherapy treatment of asthma is determined by severity on presentation, current asthma medication, patient profile, and a level of control (Lambert 2013). These medications are of two types, bronchodilators and anti-inflammatories. The first involves relievers for acute asthma symptoms with short-acting bronchodilators and rapid onset of action, and the other involves controllers for chronic asthma using drugs with an anti-inflammatory and/or sustained bronchodilator action (Lalloo et al. 2007).

The first is used to stimulate the relaxation of ASM tissue. Bronchodilators are used in this regard. β-2 adrenoreceptor agonists relaxes ASM via pathways responsible for stimulating the cellular messenger cyclic adenosine monophosphate, abbreviated as cAMP. This class of drugs is used for pulmonary diseases, including asthma (Wolthers 2009). The pathways affected by cAMP stimulation are utilized to relax ASM components used in contraction. These agonists have both short- and long-term action depending on the specific drug that is being used. Short-term acting agonists are helpful in treating mild-to-acute asthma attacks over few hours and include drugs such as isoproterenol (ISO), salbutamol, and terbutaline. Long-term acting agonists are not as quick to act against asthma symptoms, but to maintain therapeutic objectives over time. Salmeterol and formoterol are two such drugs. Targets of both short- and long-term agonists are different subtypes of adrenoreceptors along the surface of cells. Some agonists are nonspecific and stimulate multiple subtypes, with notably different outcomes. This is exemplified by ISO, which targets β-1 and β-2 receptors, though in ASM, only available are β-2 receptors (Dahlen et al. 2011).

The second is used to target the ASM pathways responsible for the narrowing of the airway lumen due to inflammatory responses. Corticosteroids are used to inhibit the phospholipase A2 pathway inside the cell, which is responsible for producing and regulating inflammatory molecules. Prostaglandins and leukotrienes are molecular participants in physiological pathways which (in this context) ultimately constrict ASM, increase goblet cell production of mucus, and upregulate the number of inflammatory cells in the matrices around the airways. However, there are always risks associated with the use of any of these medications. The inefficiency of drug delivery, tolerance, and the development of side effects over time are facts of the therapeutic approach to asthma treatments (Ducharme and di Salvio 2004).

Children asthma management is a highly challenging task. Prevention of asthma exacerbations is a critical component in ongoing asthma management. The use of anti-inflammatory medication is an essential tool in asthma management. Wolthers reported that ICSs are able to reduce inflammation and hyperresponsiveness of the bronchioles (Wolthers 2009), which have been considered as a first-line treatment for mild persistent asthma (Rachelefskyl 2009).

Montelukast (MON) was proposed as an alternative treatment option to ICS (Dahlen et al. 2011). MON has been used as either second-line monotherapy or combined therapy with ICS, but it can possibly be employed as a first-line controller medication for mild to moderate persistent asthma. MON blocks cysteinyl leukotrienes, which reduces eosinophil migration, bronchoconstriction, and mucous hypersecretion. Ducharme et al. conducted an investigation to compare the use of MON and ICS in children and adults with mild to moderate persistent asthma. ICS is a twice-daily inhaled medication and requires greater cooperation from the child. Compared with ICS, MON is easier for administration as it is a once-daily medication (Ducharme and di Salvio 2004). Knorr et al. found that MON has fewer safety concerns (Knorr et al. 2001). Dahlén reported that MON can be used directly in children with asthma that have concurrent rhinitis (Dahlén 2006). Children who take ICS on a long-term basis may have a delay in their bone growth. Pharmacologically MON has been considered a safe alternative (Pedersen et al. 2007; Kazani et al. 2010). Montella et al. found adverse events, including pharyngitis, headache, gastrointestinal disorders, and rash, although rates of these events are similar to those receiving an active medication or placebo (Montella et al. 2012).

7.2.3.2 Non-pharmacological Treatments

Due to the many side effects of pharmaceutical treatments, many attempts have been undertaken to find alternative nonpharmaceutical therapies. There were several attempts to reduce the risks of asthma irritation. Barrier methods to control exposure to the allergen can be beneficial for adults with allergic asthma. Environmental approaches such as matricidal agents, air filtration devices, special vacuuming, conventional or steam cleaning of carpet and household furniture, or domestic mechanical ventilation and heat recovery are found to be beneficial for an adult or a child with allergic asthma symptoms.

Allergic asthma symptoms can be reduced by immunotherapy and medications in adults; however, the efficacy relative to other therapies is unknown and some serious risks are associated with the use. Alternative treatments of asthma have been developed to decrease the amount of pharmaceuticals used as therapeutics. Buteyko breathing, continuous positive airways pressure (CPAP), and bronchial thermoplasty are three main areas of consideration for finding benefits away from total dependence on drug therapies. Buteyko is characterized by high-volume, low-frequency nasal breathing. The positive pressure of CPAP helps to offset observed negative pressure present in the airways of sleep apnea sufferers (Yim et al. 2007). The negative pressure reduces the airway lumen size and presents additional challenge to people with asthma who also suffer from sleep apnea. The use of CPAP in studies of a population of dual-sufferers indicated a reduction in critical biochemical markers for asthma (Kamm 1999). Bronchial thermoplasticity is a costly, invasive technique that removes ASM mass from the airway wall. The introduction of a flexible bronchoscope through either the nose or the mouth to the lungs allows for the delivery of radio frequency energy to the targeted tissue. Thermal destruction of the airway tissue aims at decreasing the total contraction force during an asthmatic attack (Kazani et al. 2010).

Pervious research on various breathing methods and CPAP showed that ASM contractility can be reduced by changing breathing volume, respiratory rate, and mean lung volume (Singh et al. 1990; Kamm 1999; Manocha et al. 2002; Cox et al. 2006; Yim et al. 2007). Current knowledge and research indicate that definitive physical treatment for asthma is still a future (necessary) accomplishment. Current therapies are largely focused on the chemical pathways involved in the asthmatic response, with a smaller percentage of efforts focusing on the alternative treatments. The application of mechanically generated physiological pressure oscillations is a potential area of research that looks to address the effective physical treatment of people with asthma.

7.3 Airway Smooth Muscles (ASMs)

ASMs play an important role in an asthmatic attack. In general, there are two different stages of asthma, acute and chronic. The former is not affected by tissue remodeling, while the latter, with its consistent inflammatory insult over time, is characterized by the resultant airway remodeling. Smooth muscles from the patients who died by acute exacerbation were increased much greater compared to people who died from other diseases (Kamm 1999).

Asthma is associated with ASM contraction and shortening of the airway walls. This results in constriction and occlusion of the airway lumen leading to breathing difficulties.

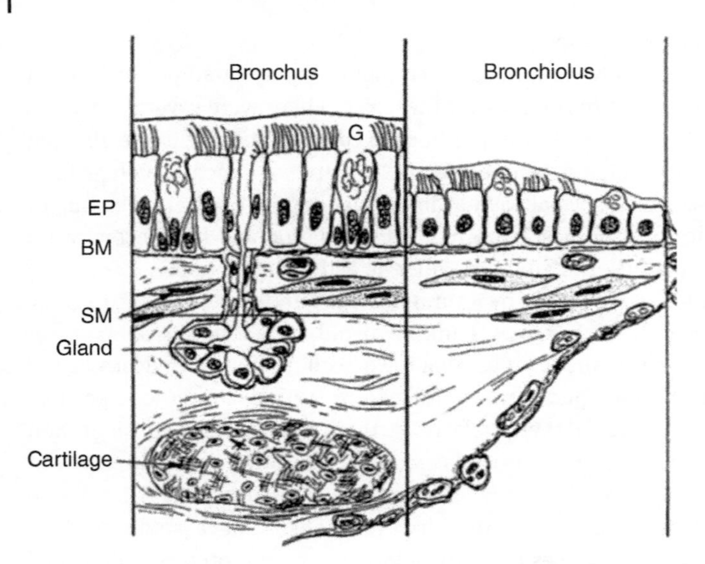

Figure 7.1 Schematic representation of the airway wall. *Source:* Shen et al. (1997a)/with permission of The American Physiological Society.

Previous studies suggest that airway hyperresponsiveness (AHR) may increase the cross-bridge cycling rates and shorten the ASM (Kamm 1999; Guyton and Hall 2000; Ganong 2001). The shortening of the ASM regulates the airway luminal diameter and narrows the airway. Asthma treatments could be greatly improved by better understanding the ASM and the airway dynamics. This section discusses the physiological properties of ASM and the contractile apparatus.

7.3.1 Structure of Airway Smooth Muscle

Figure 7.1 shows a typical airway wall structure. Airways are covered with the epithelium (EP), over the inner surface as its first defense from external intruders, like lumens of other human organs. This lining of pseudo-stratified ciliated columnar cells sits on top of a basement membrane (BM) that separates smooth muscles (SMs) of the lamina propria layer. A subepithelial collagen layer is located behind the basement membrane, but it is considerably thicker and more important than the basement membrane (Shen et al. 1997a). Contraction and relaxation times are slower than in skeletal and cardiac muscles, are rhythmical, and are the two main characteristics of smooth muscles (Salpeter et al. 2006).

7.3.2 ASM Function in Health and Disease

Research findings suggest that changes in the ASM phenotype play a fundamental role in pathogenesis of the lung diseases. It has been shown that the levels of contractile proteins such as smooth muscle-specific α-actin, desmin, myosin heavy chain (MHC), and myosin light chain (MLC) kinase are expressed at the higher levels in freshly isolated ASM compared to that found in cultured cells (Halayko and Solway 2001; Forsythe et al. 2002). This loss of a contractile phenotype in cultured ASM can be reversed when cells are cultured in serum-free medium. Most experimental results were obtained from canine and rat ASM cells, while direct relevance of the in vitro findings to the in vivo state remains unclear (Wong et al. 1998).

Seow and Fredberg reported that alterations in intrinsic properties of ASM are important for promoting nonspecific AHR (Seow and Fredberg 2001). Many researchers (Amrani and Panettieri 2002; Rizzo et al. 2002; Grunstein et al. 2002) showed that a variety of cytokines (IL-1β, IL-5, TNFα, and IL-13) could modulate the contractile phenotype of ASM by inducing a nonspecific hyperresponsiveness to G-protein-coupled receptor agonists (GPCRs).

ASM mass increases because hyperplasia and hypertrophy occur in subjects with severe chronic asthma or preterm infants with chronic lung disease (Sward-Comunelli et al. 1997; Panettieri Jr 1998). The increases in ASM mass are still not apparent and even ASM hyperplasia may represent an injury–repair response to the chronic inflammatory process. Several potential mitogens and critical signaling molecules have been identified using cultured ASM (Simon et al. 2002; Brar et al. 2002).

ASM may orchestrate and perpetuate airway inflammation by its ability to secrete cytokines (IL-1, IL-5, IL-6) and chemokines (IL-8, eotaxin, and Regulated on Activation, Normal T Expressed and Secreted (RANTES)) and by expressing a variety of cell adhesion molecules (ICAM-1, VCAM-1, CD44, and integrins) (Lazaar and Panettieri 2001). ASM also secretes Vascular Endothelial Growth Factor (VEGF) Inhibitor, a molecule important in angiogenesis (Knox et al. 2001). Interestingly, the cell signaling mechanisms by which cytokines modulate ASM synthetic responses appear disparate from those of hematopoiesis derived cells and, as such, may offer a novel therapeutic target to inhibit airway inflammation.

7.3.3 ASM and Airway Responsiveness

Schild et al. reported that asthmatic patients released more histamine and responded with contraction to challenges like house dust or pollen compared to non-asthmatics (Schild et al. 1951). Freedman and Benson demonstrated that airway wall thickening and/or baseline ASM tone could amplify the airway narrowing caused by a subsequent stimulus, supporting the concept that AHR is a manifestation of airway disease, not a root cause (Freedman 1972; Benson 1975). Woolcock et al. showed that the increase in maximal achievable airway narrowing in response to histamine is an important feature of AHR in asthma. Most of non-asthmatic subjects could inhale high concentrations without much airway narrowing (Woolcock et al. 1984).

Many studies have focused on additional properties of ASM that could be important in generating AHR and on additional explanations for AHR that did not involve a fundamental change in ASM phenotype (Cerrina et al. 1986, 1989; Whicker et al. 1988; Bai 1990, 1991; Bramley et al. 1994; Chin et al. 2012). Moreno et al. presented a theoretical analysis of the geometric factors that could link ASM activation and excessive airway narrowing (Moreno et al. 1986; James et al. 1989). James et al. and Wiggs et al. quantified the potential contribution of airway wall remodeling to increase airway narrowing (James et al. 1989; Wiggs et al. 1992). Lambert et al. concluded that the increase in smooth muscle mass is potentially the most important structural change to explain AHR (Lambert et al. 1993). Oliver et al. confirmed the importance of increased muscle mass and suggested that increased muscle could explain the failure of asthmatics to respond to deep inspirations (Oliver et al. 2007).

Other potential contributions to AHR include the amount of shortening, the velocity of shortening, reduced relaxation, and a reduced effect of strain on the reduction of force that occurs with breathing and deep inspiration. Ma et al. examined the maximal shortening and the shortening velocity of primary isolated ASM cells from asthmatic and normal subjects, and they found both, greater maximal shortening and faster shortening associated with an increased expression of MLC kinase (MLCK) (Ma et al. 2002). Leguillette et al. studied the relative expression of two isoforms of human myosin in the ASM of asthmatic and non-asthmatic subjects. Their research findings suggest that the changes in the relative proportion of two myosin isoforms could increase ASM shortening velocity and could increase AHR in asthmatics (Léguillette et al. 2009).

7.3.4 Mechanical Properties of Airway Smooth Muscle

At the molecular level, a major determinant of the mechanical properties of the smooth muscle cell is the extent of the actin–myosin cross-bridge cycling, where a high rate of cross-bridge formation is leading to muscle contraction. Huxley described the foundation knowledge for muscle contraction. He proposed that muscle contraction is caused by the cyclical attachment and detachment of thick filaments to thin filaments, with the net effect being the relative sliding of these filaments. The cycle of attachment and detachment of myosin heads to actin is ATP-dependent (Huxley 1957). The relative displacement of the filaments originates in the mechanical configuration change (power stroke) of the myosin heads, which extend from the thick filament (Craig and Megeman 1977; Xu et al. 1996).

Figure 7.2 shows the regulation of ASM contraction. Generally, calcium induces ASM contractions in two pathways: intracellular calcium concentration and MLC calcium sensitivity. For the first pathway, the calcium ions (Ca^{2+}) from intracellular sarcoplasmic reticulum stores ligands to GPCR, such as acetylcholine and methacholine, inducing the activation of phospholipase C (PLC), which in turn leads to the formation of the inositol

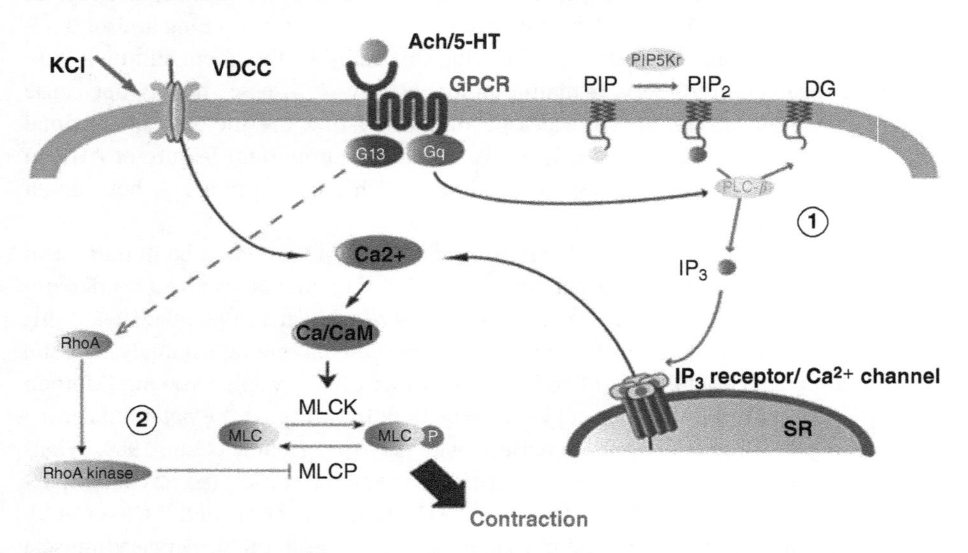

Figure 7.2 Regulation of ASM contraction, 1: intracellular Ca concentration, GPCR-IP3 pathway; 2: Ca sensitivity, MLCP phosphorylation-RhoA/ROCK pathway. *Source:* Kudo et al. (2013)/Frontiers Media SA/CC BY 3.0.

triphosphate (IP3) (Chen et al. 2002). IP3 occurs to release Ca^{2+} from sarcoplasmic reticulum (SR) stores, and then Ca^{2+} forms a calcium–calmodulin (CaM) complex, activates MLCK which phosphorylates regulatory MLCs (rMLCs) forming phosphorylated-MLC (p-MLC) (Berridge 2009). Finally, shortening and contraction occur by the activation of actin and myosin cross-bridges (Gunst and Tang 2000).

Kudo et al. proposed the second pathway. In this pathway, the p-MLC is regulated by MLC phosphatase (MLCP), which converts p-MLC back to inactive MLC. MLCP is negatively controlled by Ras's homolog gene family member A (RhoA) and its target Rho kinase such as Rho-associated, coiled-coil containing protein kinase (ROCK), which phosphorylates myosin phosphatase target subunit 1 (MYPT-1) (Kudo et al. 2002).

The primary constituents of the contractile apparatus in the ASM cell are thin (action) and thick (myosin) filaments. The filament consists of filamentous actin (F-actin) and several proteins such as tropomyosin and caldesmon (Gunst and Tang 2000). The F-actin molecules are arranged in a double-stranded helix with a continuous strand of tropomyosin along its length. Troponin, a regulatory protein associated with the actin filament in skeletal muscle, is absent in ASM. Caldesmon is thought to replace troponin in ASM, although its importance in regulating contractile activation is not fully understood (Vibert et al. 1993; Lecarpentier et al. 2002). Thin filaments are anchored to dense cytosolic bodies or membrane-associated dense plaques and form hexagonal arrays around the thick filaments. Each thick filament is surrounded by 15–30 thin filaments, although it can only interact with up to 6 thin filaments. The noncontractile components in ASM make up the cytoskeleton.

Many research groups have investigated the length adaptive effects on actin mobility and polymerization (Wang et al. 2000; Fredberg 2001; Fabry and Fredberg 2003). Inhibition of actin polymerization results in a notable reduction in ASM force generation (Gunst et al. 2003). Actin can detach from one dense body to another body within the cell (Ali et al. 2007). Previous studies demonstrated that myosin action might be more critical than actin polymerization concerning length adaptation in ASM. The myosin to actin filament ratio increased when the muscle contracted (Kuo et al. 2003). Myosin filament content increased up to 80% within seconds of activation (Smolensky et al. 2005). Thick filament density dropped approximately 30% after large length oscillations in flexible ASM strips, which confirms the ability of dephosphorylated myosin filaments (Kuo et al. 2001). Moreover, the presence of thin filaments facilitates myosin filament assembly (Seow 2005). This process could explain the development of optimal overlap of thin and thick filaments upon activation by restricting the development of thick filaments to locations along thin filaments.

7.4 Breathing Dynamics and ASM

This section focuses on the role of ASMs in the breathing process and development of the asthmatic attack.

7.4.1 ASM Dynamics

ASM dynamics involves passive and active (contractile) components. The passive part dominates in a relaxed, unstimulated muscle, while the active dominates in a contracted muscle.

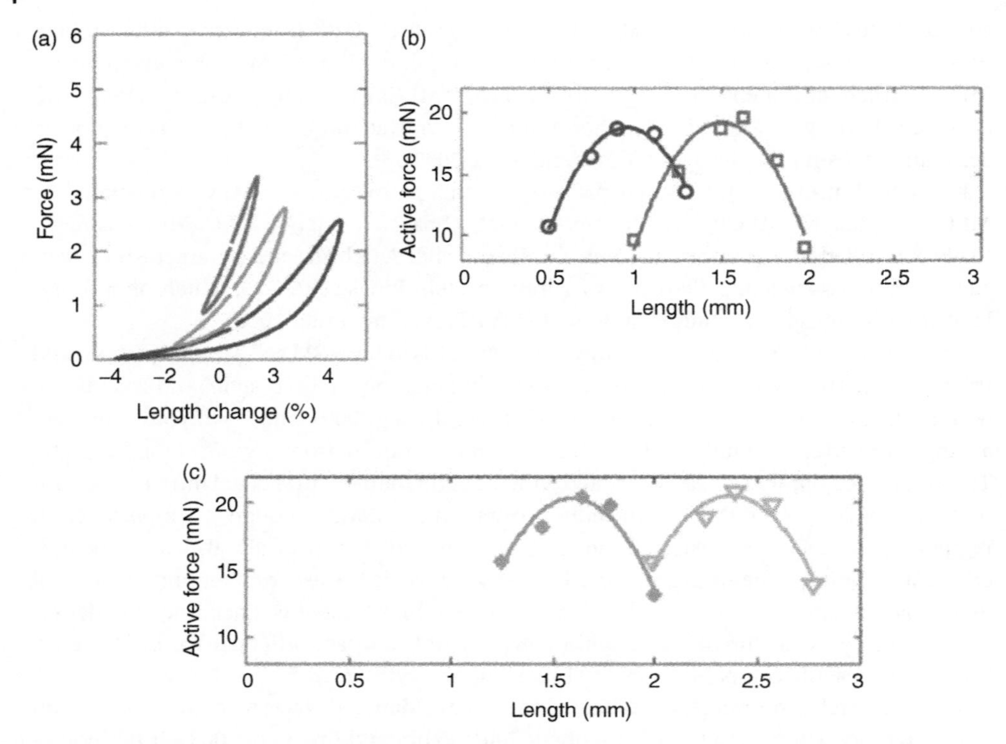

Figure 7.3 Characteristics of ASM behavior: (a) Force-length loops; (b) Force-length curves; (c) Length adaptation.

This section describes only characteristic behaviors of the active element of ASM (Donovan 2013).

- Force–length loops, Figure 7.3a: ASM exhibits a characteristic, nonlinear, hysteretic force–length "loop" where the degree of hysteresis and nonlinearity depends on the length's amplitude and frequency of oscillations (Mijailovich et al. 2000; Bates et al. 2009a). It has the shortest timescale with length oscillations measured in seconds.
- Force–length curves, Figure 7.3b and c: the peak of this force–length curve, where maximal force is exerted, is at what is called the adapted length. Increases or decreases away from this adapted length result in reduction in exerted force, and the typical shape might be roughly characterized by an inverted quadratic (Gunst and Stropp 1988; Wang et al. 2001). Changes in the length that ASM affects the ability of the muscle to exert force are measured in minutes.
- Length adaptation (Figure 7.3b and c): If force–length loops occur at a new length for a sufficiently long time, ASM will adapt to its new size, exerting peak force occurring at this new length (Wang et al. 2001; Bossé et al. 2008). ASM can exert maximal force at any length.

7.4.2 Modeling of Airway Smooth Muscle Dynamics

Huxley et al. described the sliding filament theory of striated muscle (Huxley 1957). Hai and Murphy developed a four-state cross-bridge model to explain the latch state in smooth muscle instead of modeling the myosin heads as either attached or detached (Hai and Murphy 1998a, 1998b). They subsequently released an expanded version model that merged their original model with the location-dependent attachment and detachment constants from Huxley's model (Koning 1981).

Yu et al. modified the four-state cross-bridge model to simulate the smooth muscle response to transient changes in muscle length (Yu et al. 1997). Similar to Zahalak, a Gaussian distribution was employed in skeletal muscle to calculate the cross-bridge length distribution (Zahalak 1981).

Mijailovich et al. used the four-state model integrated with Huxley's model to study the effects of tidal breathing on phosphorylation levels and force development in ASM tissue. Results showed that imposed length fluctuations decrease the mean number of attached bridges, depress muscle force and stiffness, and increase force–length hysteresis (Mijailovich et al. 2000).

Anafi and Wilson proposed an empirical model for the dynamic force–length behavior of ASM. The model predicted the experimental data for force–length loops of the muscle under length oscillations, but it was not defined outside the scope of periodic oscillation, (Anafi and Wilson 2002).

Lecarpentier et al. and Blanc et al. directly modified Huxley's model from skeletal muscle to ASM. They also compared experimental results from animals and humans (Lecarpentier et al. 2002; Blanc et al. 2003). Lambert et al. proposed a plasticity model for the steady-state length adaptation calculations. This model was concordant with morphometric observations that show an increase in myosin filament density when the muscle is adapted to a longer length. The model provided a framework for the design of experiments to quantitatively test various aspects of smooth muscle plasticity in terms of the geometric arrangement of contractile units and the muscle's mechanical properties (Lambert et al. 2004).

Hai et al. provided a better fit to the experimental data of Hai and Murphy (Hai and Murphy 1998a, 1998b) by adding an ultra-slow latch-bridge cycle to the four-state model (Hai and Kim 2005). Silveira et al. developed a 2-D cytoskeletal network model to account for cytoskeleton remodeling and the length adaptation in ASM. This is the first model to explain the process of length adaptation in smooth muscle as a function of time. The model showed good agreement with data on the length dependence of force production, shortening velocity, and compliance for fully adapted ASM at different lengths (Silveira et al. 2005). However, the model was contradicted by the recent publication.

Fredberg et al. investigated bovine tracheal ASM at a breathing frequency of 0.33 Hz, and oscillation amplitudes from 0.25 to 8% of the reference length (Fredberg et al. 1997). Shen et al. conducted similar tests on ASM from dogs with frequencies up to 20 Hz and amplitudes ranging from 1 to 10% of the reference length (Shen et al.1997b).

Bates and Lauzon improved Hill's equation by adding another nonlinear elastic term to account for tissue rheology. The model was fitted to experimental data for force–length loops of ASM under length oscillations (Bates and Lauzon 2005).

Wang et al. developed a mathematical model to study how calcium, myosin light chain kinase (MLCK), and myosin light chain phosphatase (MLCP) interact to regulate ASM cell contraction. This model has an ability to predict that oscillations in calcium concentration cause a significantly greater contraction than an elevated steady calcium concentration (Wang et al. 2008).

7.5 Length Oscillation Bronchodilation

This section discusses the potential of length oscillation superimposed on breathing patterns as a means for ASM relaxation. It contains the development of the filament sliding model, simulation and experimental data, and a discussion on the status of the research in this field.

7.5.1 Filament Sliding Model

Al-Jumaily et al. proposed a FMM for ASM dynamic response. The parameters were adapted from the FMM developed by Hunter et al. (Hunter et al. 1998) for cardiac muscles to ASM. Compared to the existing models described in Section 7.3, the FMM model provides simplicity and good ASM response representation. In this work, the effect of longitudinal oscillation simulated by the FMM on ASM response was presented in terms of static stiffness and hysteresivity (Al-Jumaily and Du 2011). Initially, the model establishes an unspecified function response to a transient change in muscle length based on Hill's equation:

$$\frac{\left(\frac{T}{T_0}\right) - 1}{\left(\frac{T}{T_0}\right) + a} = -\frac{V}{aV_0} \tag{7.1}$$

where T denotes a responding force, T_0 is the isometric force prior to any length changes, and V_0 ($V_0 = a_1/A_1 a$) is the maximum shortening velocity at zero load. A_1 and a_1 are determined from the quick isometric release by considering a transient step change in the length of the muscle in isometric contraction. The published experimental data (Sieck and Prakash 1997; Takahashi et al. 2000; Stephens et al. 2003) was used to represent the parameter a, which leads to a value for a between 0.32 and 0.38.

The Heaviside and Dirac's delta functions are often used for modeling phenomena associated with step response and impulse, respectively. In this work, data on transient step changes in length were obtained from Meiss (1982). The FMM model was developed to simulate the following two cases:

7.5.2 Finite Duration for Length Steps

Any realistic transient step change in real time takes a finite duration. The velocity of shortening is given by Al-Jumaily et al. (Al-Jumaily and Du 2011):

$$\dot{\lambda} = \begin{cases} 0 & t < 0 \\ \frac{\Delta\lambda}{\Delta t} & 0 \ll t \ll \Delta t \\ 0 & t > 0 \end{cases} \tag{7.2}$$

where $\Delta\lambda$ is a length change, t is time, and Δt is finite time duration. The initial response for different finite time durations is given by (Al-Jumaily and Du 2011):

$$\frac{\left(\frac{T}{T_0}\right) - 1}{\left(\frac{T}{T_0}\right) + a} = \Delta\lambda \sum_{i=1}^{3} A_i \frac{[1 - exp\,(-\alpha_i t)]}{\alpha_i} \tag{7.3}$$

where A_i and α_i are constant parameters. A contracted ASM tissue placed under longitudinal sinusoidal oscillations was considered. The tissue length λ can be obtained as:

$$\lambda = 1 + \Delta\lambda \exp(j\omega t) \tag{7.4}$$

where ω is frequency. The response of the muscle tissue to oscillations is given by:

$$\frac{T}{T_0} = \frac{1 + Q_a}{1 - Q} \tag{7.5}$$

where Q can be obtained by:

$$Q = \Delta\lambda j\omega \sum_{i=1}^{3} A_i \frac{\exp(j\omega t)}{\alpha_i + j\omega} \tag{7.6}$$

7.5.3 ASM Response

Figure 7.4 shows the initial drop in force (T_1/T_0) for a finite duration of length steps as a function of $\Delta\lambda$ at different durations of length change. It can be clearly seen that the force drop for smaller length change durations is larger at each $\Delta\lambda$. Perhaps the cross-bridge cycling cannot keep up with the shortening at these faster length changes, resulting in increased detachment of myosin heads from actin.

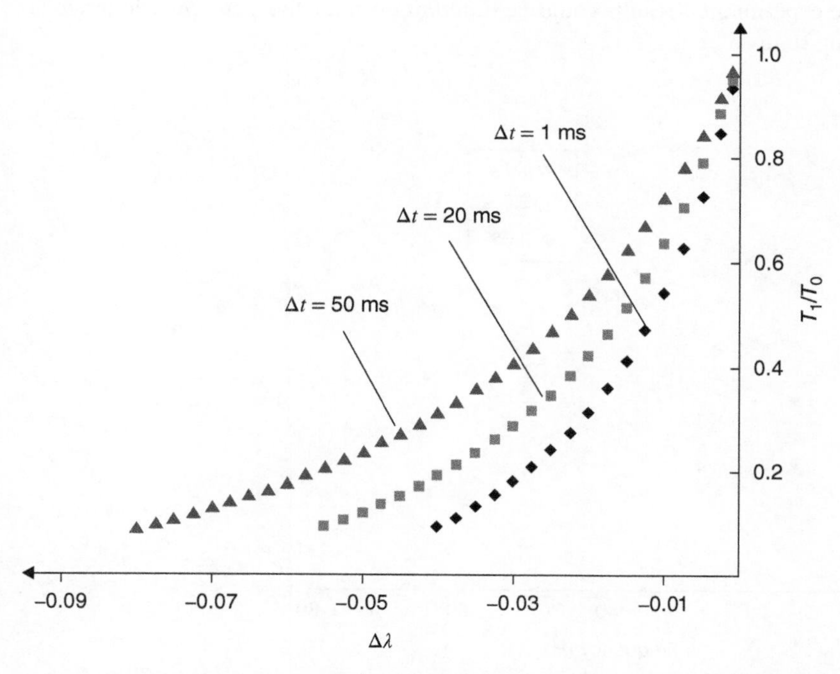

Figure 7.4 Effects of the initial force response to different shortening times and length shortening ratios.

Figure 7.5 displays the effect of oscillations on the static stiffness, a dimensionless quantity defined as the ratio of force prior to oscillation, to that directly after oscillation instability (Du et al. 2007). It can be seen that the static stiffness decreases in a seemingly exponential relation with the frequency of oscillation until about 25 Hz, where almost no variation after that is observed. The stiffness decreases as the amplitude of oscillation increases.

Figure 7.6 shows a comparison between simulation and available experimental results (Du et al. 2007). There is a complete agreement between the two (Donovan 2013). The observed correlation between the stiffness reduction and frequency can be partly explained using the cross-bridge dynamics. The cross-bridge cycling is likely disturbed by the oscillations, and the cross-bridge rates are relatively slow (Hai and Murphy 1998a, 1998b).

Figure 7.7 presents the phase–frequency relationship of the FMM model. The phase increases with an increase in the amplitude of oscillation. Therefore, the hysteresivity increases with an increase in the amplitudes of oscillation, which indicates more disturbed cross-bridge cycling (Du et al. 2007).

The accuracy of the FMM model highly depends on the parameters selected from the experimental data. The constant parameter (A_i) determines the initial fall in force (T_1/T_0), which is presumably mainly related to the number of broken cross-bridge attachments resulting from the stretch. The other constant parameter (a_i) has the units of s^{-1} to determine the muscle response time (recovery rate), which represents the time for reattachment of the disrupted cross-bridges. The second contributing factor to the inaccuracies in T_1 is the uncertainty involved with biological variations in smooth muscles (Seow 2000; Bai 2004). The experimental results could be different even for the same muscle tissue at different testing times.

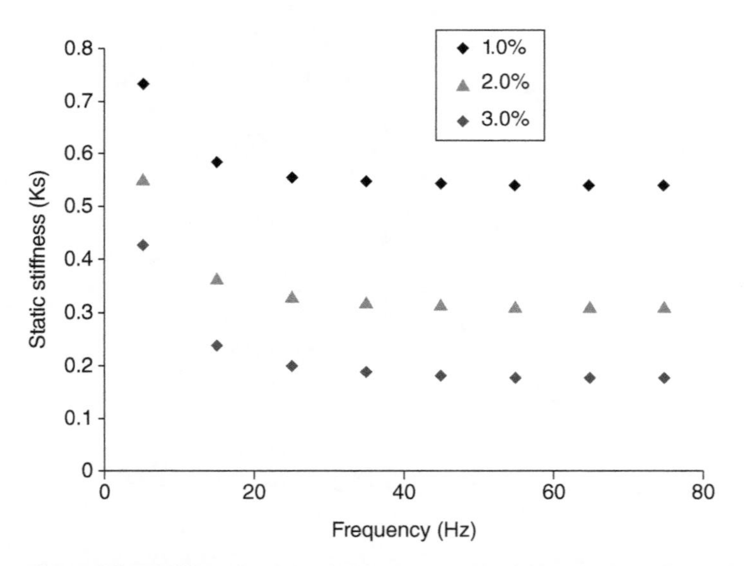

Figure 7.5 Stiffness from the fading memory model versus frequency with different amplitudes for the longitudinal oscillation.

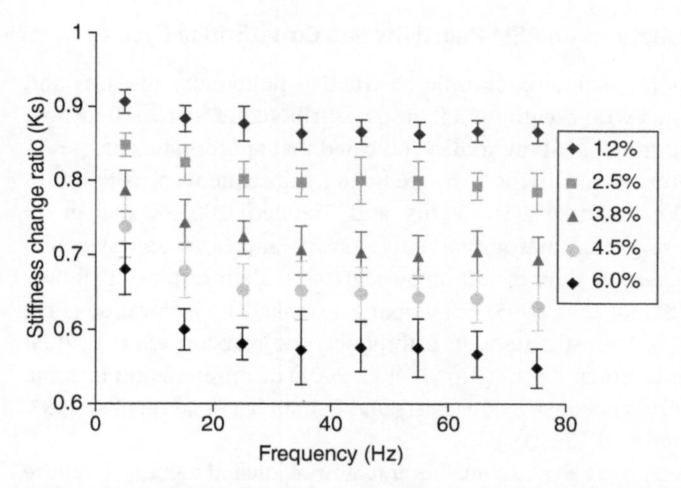

Figure 7.6 Stiffness change versus frequency for different oscillation amplitudes, mean values ±95% confidence interval (n = 11).

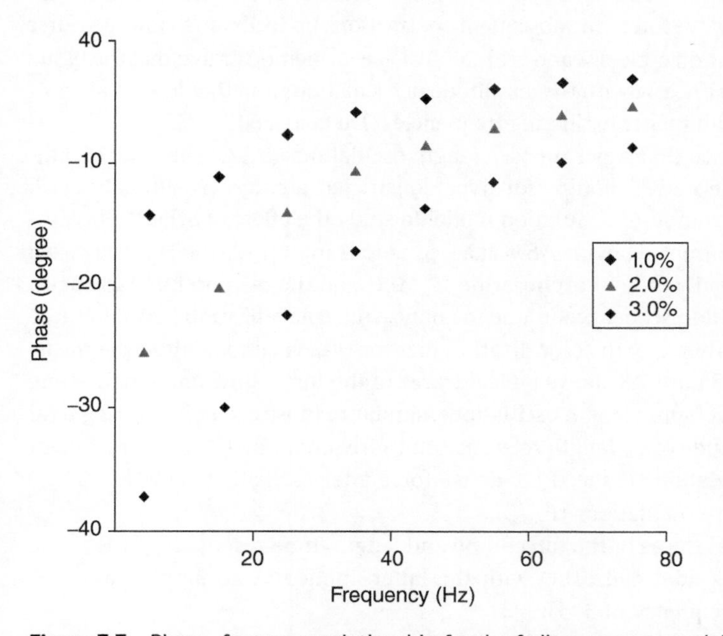

Figure 7.7 Phase–frequency relationship for the fading memory model.

7.6 Length Oscillation Bronchoprotection

This section discusses protective potential of the pressure waves superimposed on breathing patterns before the attack occurs. It discusses the concepts of ASM plasticity and glassy behavior of smooth muscle cells.

7.6.1 Effect of Length Oscillations on ASM Reactivity and Cross-Bridge Cycling

Several pulmonary dysfunctions, including chronic obstructive pulmonary diseases and asthma, are normally associated with breathing difficulties attributed to increased airflow resistance caused by airway narrowing. Many studies indicated that appropriate changes in breathing patterns limit bronchoconstriction by reducing subsequent sensitivity to chemicals (Wang et al. 2000; Fredberg 2001; Fabry and Fredberg 2003; Gunst et al. 2003). Previous publications suggested that normal subjects with increased airway resistance to inhaled methacholine reverse their condition by controlling their respiratory movement (Freedman et al. 1988; Skloot et al. 1995) or by taking a single deep inspiration (Fish et al. 1981; Brown et al. 2001). Unfortunately, in asthmatics, the relaxant effect of tidal breathing or deep inspiration is either absent (Brown et al. 2001) or minimal and in some patients, it may even worsen bronchoconstriction (Berry and Fairshter 1985; Lim et al. 1987, 1989; Gunst et al. 1990; Lakser et al. 2002).

Length oscillations have been proposed as a contributor to non-medicinal therapy to relieve ASM hyperconstriction and have been studied by many researchers (Gunst 1983; Fredberg et al. 1997) with a particular focus on simulating the effect of tidal breathing on ASM reactivity. Wang et al. found that length oscillations applied to ASM before cholinergic stimulation significantly reduced constrictive forces in subsequent contractions up to about 30 minutes after several contraction–relaxation cycles (Wang et al. 2000). Du et al. demonstrated that the tissue stiffness decreased almost linearly with the amplitude of oscillations and that lower values of stiffness were obtained with higher oscillating frequencies (Du et al. 2007).

Al-Jumaily et al. proposed that superimposed length oscillations (SILOs) on tidal breathing can contribute to alternative therapy for hyperconstricted airways (Al-Jumaily et al. 2012). They conducted a computer simulation model to study the effect of SILO on breathing oscillations on pre-contracted bovine ASM at a frequency range of 10–100 Hz. They used two specific antibodies against the phosphoserine 19 MLC and the α-smooth muscle actin and applied an immunofluorescence technique to analyze the colocalization between these two filaments. ImageJ software with colocalization method was used to analyze the measured experimental data. Figure 7.8 shows typical traces of the force–time and length–time response of the tissue undergoing length oscillations, similar to those occurring during tidal breathing, with an amplitude of 4% length reference and a frequency of 0.33 Hz. It is evident that the mechanical oscillations reduced the active force after oscillation cessation (F_{post}) compared with that before oscillations (F_{prior}).

Figure 7.9 shows typical traces of the force–time and length–time responses of the tissue undergoing simultaneous tidal and SILO with the latter applied at an amplitude of 1% length reference and a frequency of 50 Hz.

7.6.2 Concluding Remarks

Compared with pure tidal oscillations (Figure 7.8), Figure 7.9 shows that SILO further decreased the active force after oscillation cessation during the recovery period. Furthermore, the muscle does recover slowly; however, the active force during the recovery period (for up to at least 20 minutes after oscillation cessation) is still lower than that before the application of oscillation (i.e. at the plateau phase of the ACh-induced contraction).

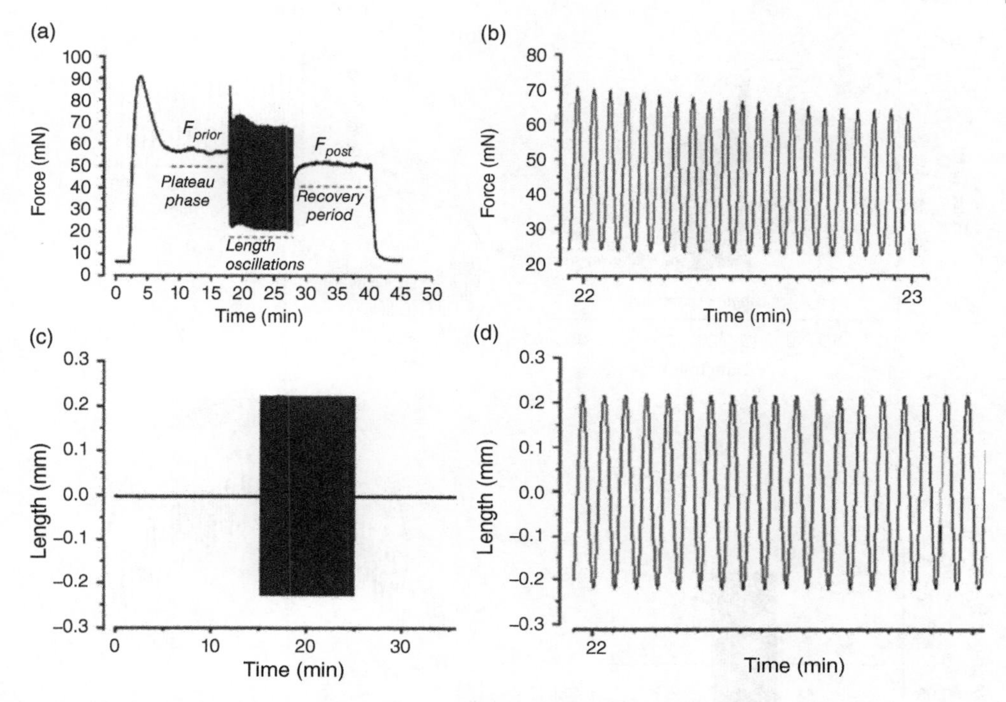

Figure 7.8 Typical traces of force–time and length–time response during tidal-length oscillations. (a) Complete force contractile response. (b) Expanded trace of the force over 1 minute. (c) Complete representative trace of the tissue length fluctuations. (d) Expanded trace of length over 1 minute. In all the experiments, the tidal strains were applied at an amplitude of 4% and frequency of 0.33 Hz.

The results demonstrate that both tidal and SILO reduce the active force in contracted ASM for a relatively long term and that the latter enhances the force reduction of the former. It is found that the reduction is frequency- and time-dependent. Additionally, the colocalization analysis method indicated that length oscillations caused the detachment of the actomyosin connections and that this condition is sustained even after the cessation of the length oscillations (Al-Jumaily et al. 2012).

The effect of frequency variation on the active force (expressed as % of the plateau force) over the period after oscillation cessation is demonstrated in Figure 7.10. Despite the fact that at 10 and 100 Hz, there is no significant force reduction with SILO, at 5, 10, and 20 minutes, the SILO of 20, 50, and 70 Hz do significantly increase force reduction due to tidal oscillation. The relationship between the active force and the SILO frequency at the end of all length oscillations is shown in Figure 7.11. Force values were recorded at 5, 10, 15, and 20 minutes of the recovery period and re-plotted in a dose–response fashion, where the dose was taken to be the frequency of SILO. The results indicate that the SILO frequencies reduce the active force in a manner that is both dose- and time-dependent. The dose–response effect is consistent at 5 and 10 minutes of the recovery period, whereas this effect seems to wear off with time. At 15 and 20 minutes after cessation of length oscillation, the dose–response effect is not observed as seen from the corresponding curves shape. There is no effect on the active force reduction at the frequency of 100 Hz (Al-Jumaily et al. 2012).

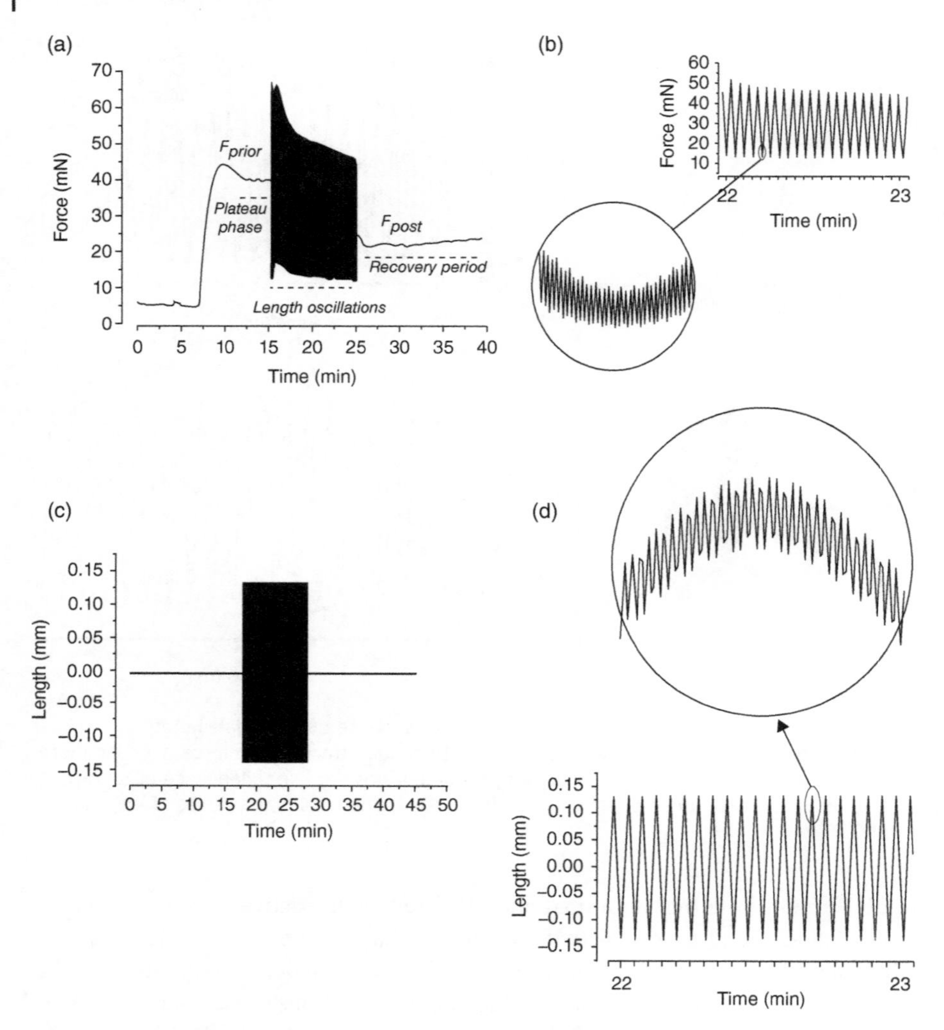

Figure 7.9 Typical traces of the force–time and length–time response with both tidal-length oscillations and SILO simultaneously applied: (a) complete force contractile response, (b) expanded trace of the force over 1 minute, (c) representative of the complete trace of the tissue-length fluctuations. (d) Expanded trace of the length over 1 minute. In these traces, the SILOs were set at an amplitude of 1% reference length and a frequency of 50 Hz.

7.7 Engineering Perspectives of Contraction–Relaxation Mechanism

Contraction of smooth muscles is initiated by the receptor or stretch-mediated activation of actin and myosin. The phosphorylation of the myosin 20 kDa light chain by the MLCK is a prerequisite for contraction (An et al. 2007; Marieb 2005). The intracellular Ca^{2+} concentration increases in response to stimuli, which binds with the protein CaM to form a complex, Figure 7.12. This complex then activates the MLCK to phosphorylate the myosin light

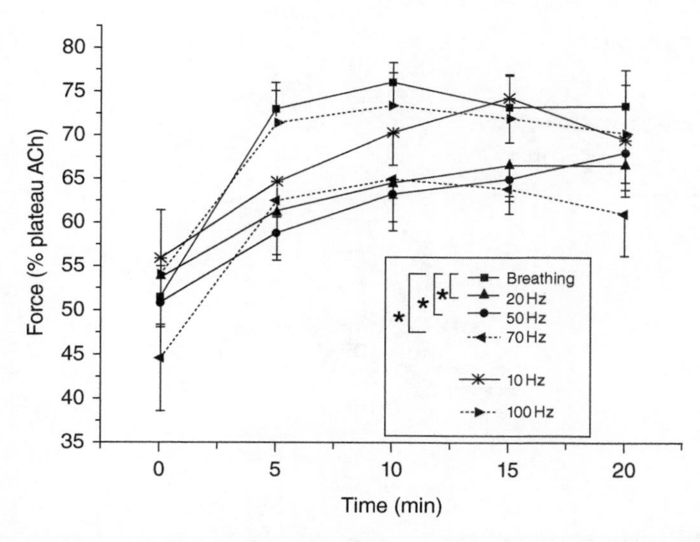

Figure 7.10 Force–time response during isometric contraction after applying either the tidal-length oscillations alone (square), or with SILO tested at a different frequency (stars, circles, and triangles).

chain 20 (MLC 20). Agonists bind to the serpentine receptors on the ASM cell membrane leading to an increased PLC activity. PLC acts as a specific enzyme in the formation of IP3 and DG (diacyl glycerol). IP3 triggers an increase in the intracellular Ca^{2+} from the sarcoplasmic reticulum (SR). DG, along with Ca2+, activates protein kinase C (PKC), which activates specific contraction-promoting proteins, leading to initiation of the actin–myosin cross-bridging (Fabry and Fredberg 2003).

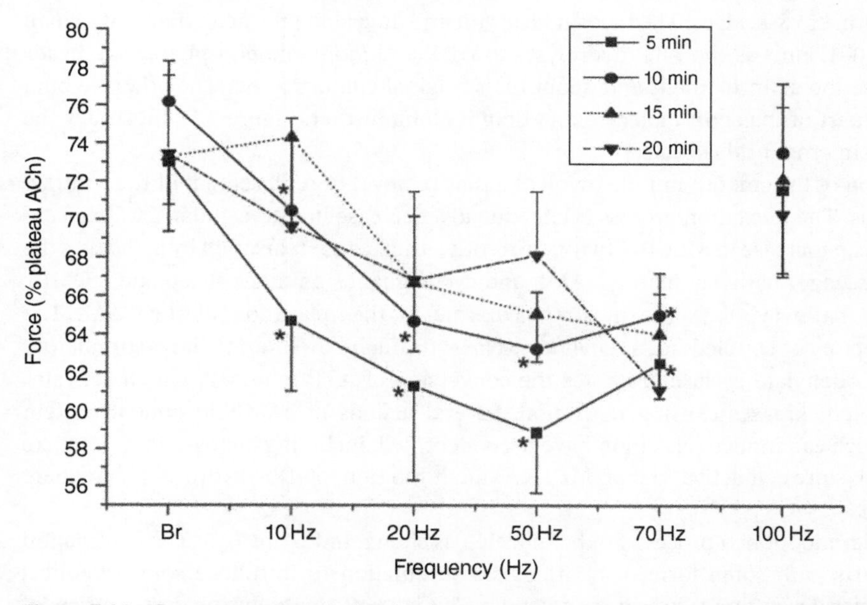

Figure 7.11 Dose–response relationship between the active force and the superimposed frequency at the recovery period.

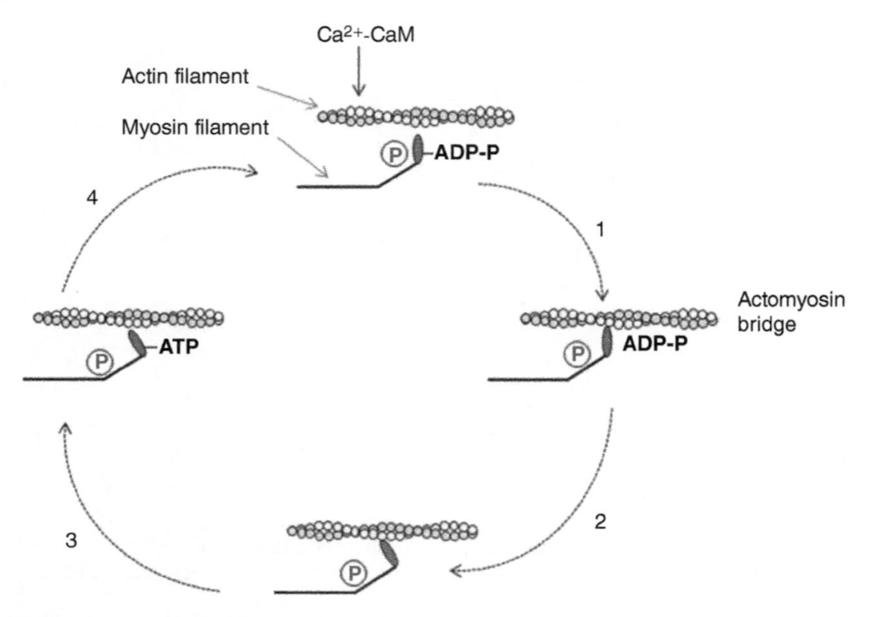

Figure 7.12 Actin–myosin cross-bridge cycle.

The cross-bridge cycle comprises the attachment and detachment of myosin heads to the actin filament. The energy for a cycle is provided by the hydrolysis of ATP. Myosin functions as an ATPase utilizing ATP to produce a molecular conformational change of part of the myosin and produce movement. Movement of the filaments over each other happens when the globular heads protruding from myosin filaments attach and interact with actin filaments to form cross-bridges. The myosin heads tilt and drag along the actin filament a small distance (10–12 nm) as the ATP hydrolyses to ADP and inorganic phosphate. The heads then release the actin filament and adopt their original conformation. They then re-bind to another part of the actin molecule and drag it along further. Figure 7.12 illustrates the actin–myosin cross-bridge cycle.

Relaxation of the smooth muscle involves either removal or replacement of the contractile stimulus. The relaxation process occurs due to a decrease in intracellular Ca^{2+} concentration and an increase in MLCP activity. Currently, drugs cause relaxation by activating the second messenger pathway mainly cAMP and cyclic guanosine monophosphate (cGMP). The cAMP pathway plays the most critical role in the relaxation of the ASM. The β-adrenoceptor is coupled to adenylate cyclase through an intermediary stimulatory G-protein. Adenylate cyclase promotes the conversion of ATP to cAMP, which activates certain protein kinases causing relaxation. Several actions of cAMP-dependent protein kinases which can induce relaxation have been identified, including increased Ca^{2+} uptake by internal stores, inactivation of MCLK, and inhibition of IP3 hydrolysis (Fredberg et al. 1999).

The earlier mentioned process can be modeled as shown in Figure 7.13. The contraction process starts with some form of neurological stimulation to introduce some chemical process which leads to a physical contraction. The current treatment process consists of using some beta-blockers to block the calcium channels and result in relaxation. However,

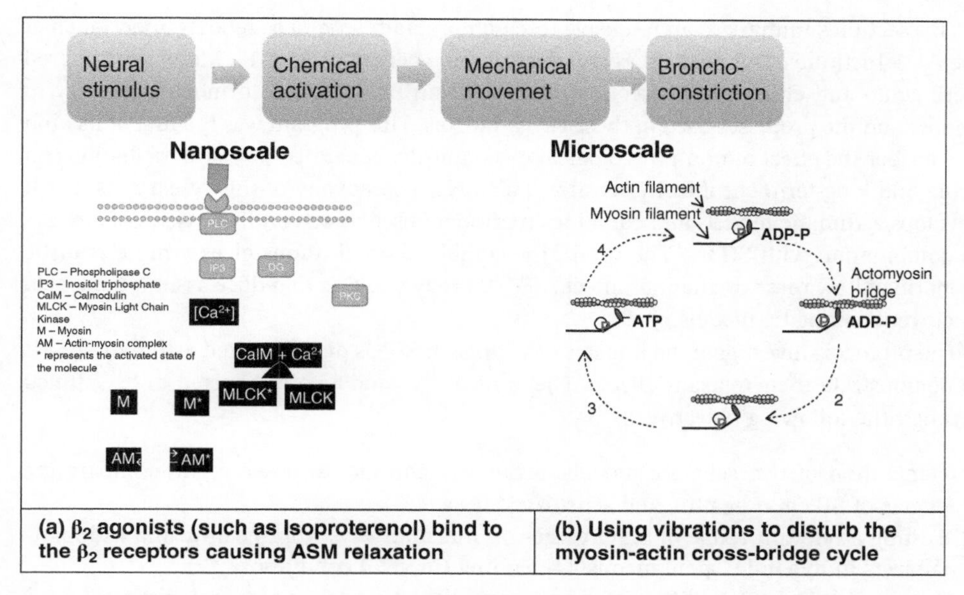

Figure 7.13 Engineering perspective of: (a) pharmaceutical and (b) mechanical relaxation.

the proposed mechanical approach is to introduce oscillation (vibration) to disturb the cross-bridge cyclic process and produce relaxation. The evidence presented in the previous section emphasizes the fact that disturbance of the cross-bridge cycling leads to relaxation. Although all the pieces of evidence presented earlier were concluded from tissue experiments, in the following section, we introduce the concept as it is applied to animal models.

7.8 Animal Models

Existing studies regarding the effect of oscillations on ASM have some limitations including but are not limited to:

1) It is assumed that the relaxation of contracted airways is due to the direct effect on the cross-bridge cycle when it could be more related to an adaptive response of the tissue (Hai and Murphy 1998a)
2) Most of the studies have been carried out on isolated smooth muscle, ignoring other structures that may play a role during asthma attacks such as cartilage, parenchyma, etc.
3) Most of the experiments have been done in tissues obtained from healthy animals, even though it has been proven that these oscillation patterns act differently in asthmatic smooth muscle (Kumar and Foster 2001).
4) Most studies have focused only on physiological oscillation patterns of deep inspiration and breathing (Whorwell et al. 1986; Bowler et al. 1998; Al-Jumaily and Alizad 2008; Jo-Avila et al. 2014)
5) Not many studies have been conducted in the presence of bronchodilator medication.

This section summarizes an intensive research program on mouse models undertaken at the AUT Institute of Biomedical Technologies in the period 2011–2019. Models considered were acute and chronic. This section starts with an introduction to mouse anatomy to understand the processes used in these two protocols. This program was the first of its kind to consider the effect of non-physiological mechanically generated pressure oscillations on acute and long-term chronically sensitized airways. The novelty of this research is that it develops asthmatic models and combines traditional oscillations on these types of airways in combination with SIPOs. The use of the combined applications gives a more realistic scenario of how these oscillations affect ASM, thereby working to reduce assumptions held in current asthmatic models.

The protocols investigate the impacts of a range of SIPOs on contracted asthmatic ASMs to demonstrate their relaxant effects. The aims of the study were expected to be fulfilled through the following objectives:

1) Generation of two separate models, acute and chronic, to observe and compare the effects of SIPO on healthy and sensitized airways.
2) Testing a relevant range of SIPO values on intact airways from healthy and asthmatic subjects in dynamic, spontaneously breathing (in vivo) conditions.
3) Assessing the effects of SIPO in the two established models, acute and chronic.

7.8.1 Mouse Anatomy

A brief discussion of pertinent murine airway components is given in this section, with Figure 7.14 illustrating the mouse upper airway anatomy as an initial reference point.

a) **Pharynx and Epiglottis**

Mice are nasal breathers. All inhaled air and substances pass initially through the nostrils to the nasal cavity and then to the nasopharynx prior to reaching the pharynx. The pharynx is

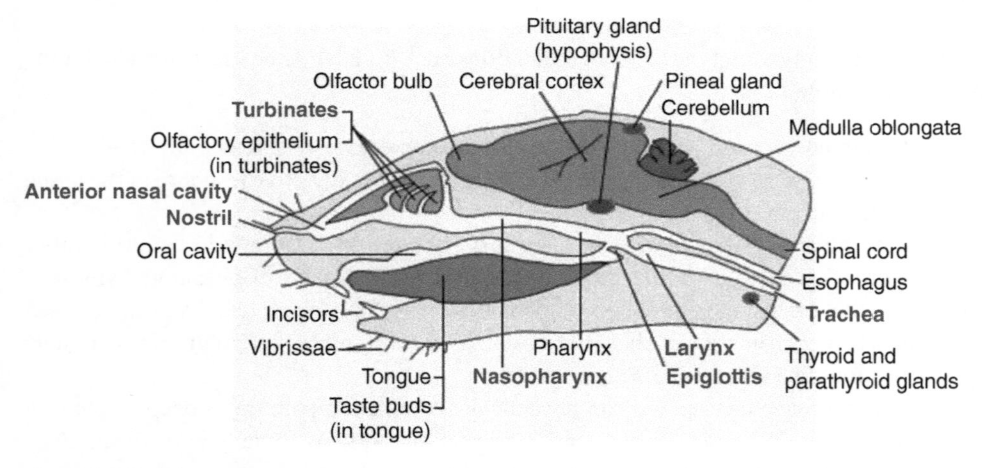

Figure 7.14 Mice upper airways anatomy. Of particular interest to this study are the nasal and oral cavities as well as the esophagus and trachea. *Source:* Pathology (2013)/Frontiers Media SA/CC BY-4.0.

continuous with both the nasal and oral cavities. The epiglottis covers the opening to the larynx when food is swallowed, preventing food from entering the trachea and respiratory tract.

b) **Larynx**
The larynx is separated from the pharynx by the epiglottis and extends to the trachea. The tissue character of the larynx consists of respiratory epithelium, with elastic fibers, glands, cartilage, and vocal cords.

c) **Trachea**
The trachea connects the larynx and primary bronchi. The trachea tissue contains multiple incomplete hyaline cartilage rings that support an open lumen for air movement and are spaced along its length. The ends of the cartilage are connected by ASM, which contracts and relaxes to influence the diameter of the trachea. A tracheotomy cuts a small hole in the connective tissue between adjacent cartilage rings, and a cannula is inserted into the lumen of the trachea. The cannula is secured with surgical silk, closing the tracheotomy's hole, allowing the mouse to maintain spontaneous breathing patterns.

d) **Lungs**
The mouse's right lung is divided into four lobes, and the left lung is a single lobe, as shown in Figure 7.15. The lungs receive air that is passed from the trachea through the bronchial tree to the site of gas exchange in the alveoli. The trachea divides into the primary bronchi, which still contain cartilage and respiratory epithelium. The primary bronchi then branch into bronchioli and the terminal bronchioli with their alveoli. (Unlike humans, mice do not have respiratory bronchioli between terminal bronchioli and alveoli.)

As the airways decrease in size, their walls become thinner, containing less connective tissue and smooth muscle. The associated epithelium also becomes more simplified, allowing for gas and drug exchange with pulmonary vasculature. Not all inhaled air participates in gas exchange, because it does not reach the alveoli, and the alveoli are unable to transport the molecules.

e) **Ventilation**
The normal breathing rate for a mouse is approximately 150 breaths/min, with a tidal volume of air between 0.15 and 0.2 mL (Jo-Avila et al. 2014). As mentioned, not all inhaled air

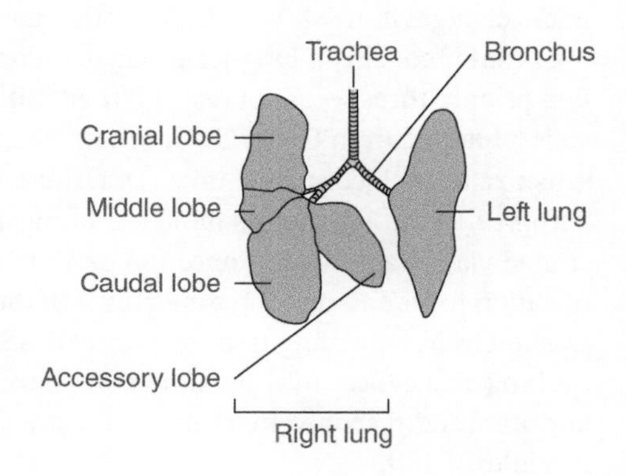

Figure 7.15 Mice lower airways. *Source:* Pathology (2013)/Frontiers Media SA/CC BY-4.0.

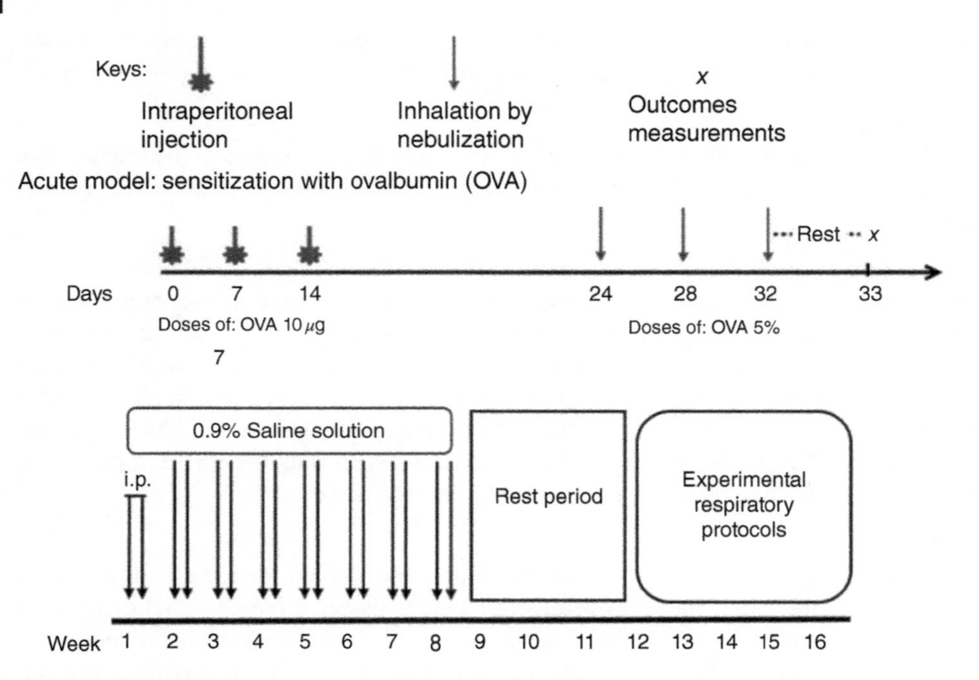

Figure 7.16 Sensitization protocols: Upper-Acute model; Lower Chronic Model.

perfuses the alveoli. This residual volume of air constitutes the dead space within the respiratory system. Using the percentage of human dead space (Berne and Levy 2008) as an approximation, 30% of the mouse tidal volume is equivalent to ~50 μL of dead space.

7.8.2 Acute and Chronical Asthmatic Models

The acute model is a short-term sensitization protocol close to one month in duration, using Ovalbumin (OVA) as an allergen (allergen). The allergen is prepared in the presence of alum which acts as an adjuvant, increasing the immunological response of the mouse. The allergen is administrated via intraperitoneal injection and nebulization. With this model, the objective is to observe the short-term response of ASM to the allergen, such as inflammation markers and initial changes in ASM. The details for the sensitization protocol are presented in Figure 7.16a.

Chronic model is a long-term sensitization protocol, requiring eight weeks of sensitization prior to three weeks of rest and then initiation of experiments. A mixture of three allergens, obtained from Greer Labs (Lenoir, NC, USA), was prepared in 0.9% saline solution to sensitize the subjects: Dust mite (D. farinae, 0.3 μg/μL), Ragweed pollen (A. artimissifolia, 3.3 μg/μL), and Aspergillus mold (A. fumigatus, 0.3 μg/μL). The allergen mix was administrated via intraperitoneal injection on the first two days of sensitization, followed by nebulization for the remaining time points of the protocol. The objective of the chronic model was to observe the long-term response of ASM to the allergen mix, as measured primarily by lung resistance and dynamic compliance, and with an evaluation of inflammatory and immune response markers. The details for the sensitization protocol are illustrated in Figure 7.16b.

7.8.3 Model Limitations

Many criteria exist for defining an acceptable animal model of asthma. Core criteria for consideration:

- Immunological and respiratory sensitivity mediated by IgE and/or IgG to the antigen that results in bronchoconstriction;
- An increase in airway resistance;
- Chronic inflammation of the airways, with an associated increase of eosinophils and cytokines;
- Nonspecific hyperresponsiveness;
- Excessive production of mucus, supplemented by goblet cell metaplasia and enlargement of the submucosal glands; and
- Airway tissue remodeling, including thickening of the collagen and smooth muscle layers (Karol 1994; Benayoun et al. 2003; Ulrich et al. 2008).

7.9 Model Sensitization

As mentioned earlier, no single animal model provides all of the earlier mentioned criteria. Fortunately, a variety of models have been developed to study the individual features of asthma. The design of a sensitization protocol to establish an animal model must take into account technical details such as the type of allergen, use of adjuvants, and the method of administration of the allergen. Common types of allergens used are OVA and House Dust Mite and custom-made allergen mixes. OVA is the most popular antigen as it is readily available, and the animal prior exposure can be easily prevented through the environment (Zosky and Sly 2007; Bates et al. 2009b). Adjuvants such as alum, heat-killed bordetella pertussis, and ricin are used to improve the immunological response against the allergen. Among the adjuvants, alum is the most commonly used as it promotes and enhances the reaction against the allergen but has the inconvenience of inducing an immunologic response by itself (Marieb 2008). The allergen can be administrated by different methods, among them are intraperitoneal (i.p.) injection, subcutaneous injection, and aerosolization. The choice of the sensitization protocol will depend on the features of asthma that need to be present in the model.

Asthmatic models are referred to as either acute or chronic in an effort to represent and replicate features of acute or chronic human asthma. In animal models, the terms are also indicative of the duration of allergen exposure (sensitization) and the maintenance of the disease state after sensitization has ended. Acute asthma in a murine model has multiple interpretations which can also overlap with chronic models' insights. While acute characteristics are interpreted as being less advanced physiologically and immunologically than chronic asthma characteristics, acute subjects still present (milder) asthmatic consequences of lung resistance and dynamic compliance that are also associated with longer exposures to allergens. Indeed, since the allergen exposure is for shorter durations in acute models, an acute model can even be described as a (short exposure) chronic asthmatic state. Throughout the discussions of this work, any reference to acute models of asthma is deemed equivalent to also being a short-term chronic model.

Figure 7.16 shows the sensitization protocols for the acute and chronic models. The acute model is divided in two stages: (i) first stage: induction of immune response (injection of allergen) from day 0 to 14 and (ii) second stage: induction of AHR (nebulization of allergen) from day 24 to 32.

However, for the chronic model days 1 and 2 of week 1 introduce the DRA allergen mix in 0.9% saline solution via i.p. injection, and all other days introduce nebulized DRA allergen mix. General weight, behavior, and health of the mice were checked on the days of protocol chronic sensitizations as well as on days without manipulations.

7.9.1 Sensitization Assessment

The asthmatic models were evaluated to determine if the sensitization process was successful. This was performed using different techniques, which evaluate AHR through bronchoconstriction using the mechanical respiratory resistance R_L and compliance C_{dyn} and the increased immunological response with enzyme-linked immunosorbent assay (ELISA) to determine IgE levels and bronchoalveolar lavage (BAL) to observe cellularity changes in the airways. The combination of the mechanical parameters and chemical indicators provides a more reliable control to confirm the success of the sensitization protocol.

7.9.2 AHR/Plethysmography

AHR is measured in tracheotomized mice inside a custom-built thermoregulated pneumotachograph/plethysmograph setup. This system was developed and calibrated in-house. Figure 7.17 shows a schematic of the modified plethysmograph (Kumar and Foster

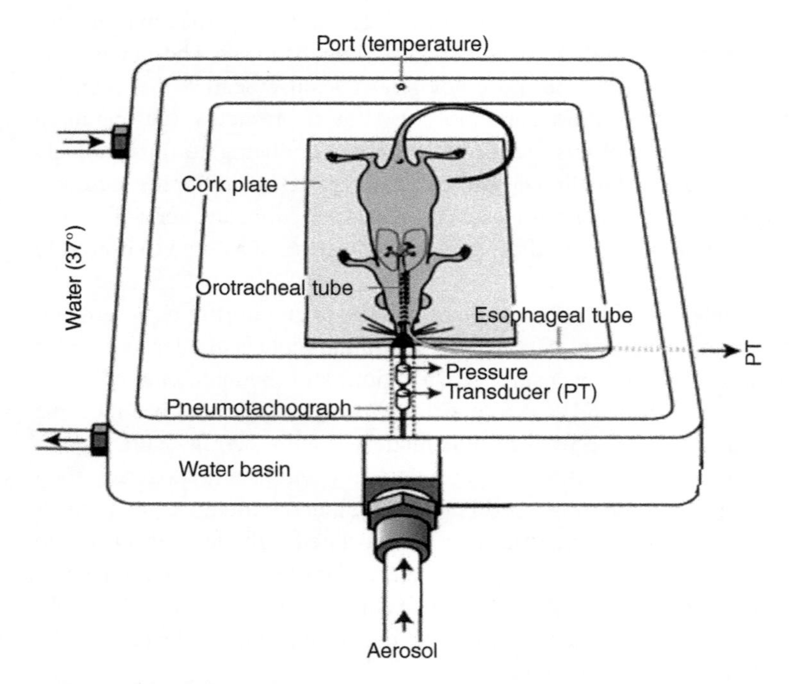

Figure 7.17 Sketch of a plethysmograph diagram. *Source:* Based on Kumar and Foster (2001).

2001). This technique evaluates the bronchoconstriction in the control and sensitized animal models by measuring the lung resistance and dynamic compliance. These two parameters are considered the most significant to evaluate bronchoconstriction in airways in this type of model. Through this technique, we also expect to assess the degree of AHR in both healthy and asthmatic mice.

AHR is assessed as an increase in RL after a challenge with aerosolized ACh in anesthetized and spontaneously breathing mice Balb/c. The invasive technology and methodologies selected for this study collect data for determining values of R_L and C_{dyn} as indicators of bronchoconstriction. To calculate these indicators, pulmonary parameters of tidal airflow V, tidal volume T_v, and transpulmonary pressure P_{tp} are required. The tidal airflow is measured in mice using a pneumotachometer. Flow restriction and pressure differences are evaluated across two detector ports. The analog signals generated from the differentials are proportional to the air velocity and are integrated by a low-pressure transducer for measurement of V in the raw data of the murine breathing cycle. The tidal volume is measured by combining the tidal flow over a known period of breathing cycles. Figure 7.18 depicts murine breathing cycles and the classification of volumes that occur during breathing. Tidal volume is determined by the difference in the amount of air present at the completion of each inspiration and expiration in the breathing cycle. It does not describe the total lung volume. Software is used in the study to monitor and record continuous values of T_v as changes in V are detected. The transpulmonary pressure is the differential pressure measured between the pleural and alveolar pressures. P_{tp} is the positive pressure which prevents the lungs from collapsing and is difficult to acquire as a measurement. It is determined by the pressures (physically monitored at the level of the lungs) before inspiration but also at the end of expiration. Measurement of P_{tp} is accomplished by using a fluid-filled catheter that is inserted into the esophagus of the mouse, with the terminal end of the tube positioned at the level of the mid-thorax (Welker et al. 2004). The catheter is connected to a pressure transducer which sends data to a computer for incorporation into parametric values. The fluid in the catheter is displaced as the lungs expand and contract, causing a

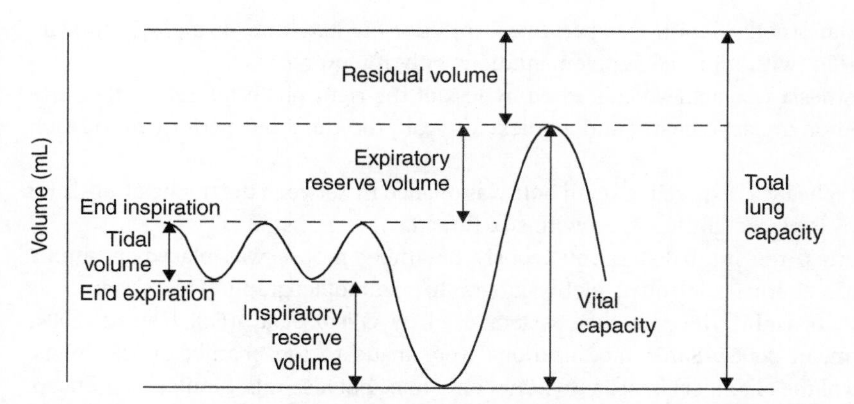

Figure 7.18 Lung volumes are occurring during breathing cycles. Total lung capacity is the sum of several volumetric classifications. Tidal volume is the smallest volumetric unit illustrated here and is the difference between end inspiration and end expiration volumes. Inspiratory and expiratory reserve volumes and residual lung volume also add to the total lung capacity.

volumetric change converted by the transducer into a transpulmonary pressure reading. Critical to this method is the catheter placement in the esophagus and maintaining proximity to the neighboring trachea and bronchi. Slight adjustments of the catheter's placement can yield maximum values of P_{tp} for use in subsequent analyses.

The pulmonary resistance is a measurement of the respiratory system's opposition to airflow. Pleural pressure values, measured with the esophageal catheter, are compared against the airflow measured by the pneumotachometer. In the airways, resistance is lower with a larger lumen diameter, and resistance increases with narrowing the airway lumen. Applied to an asthmatic model, the narrowing of the airway lumen results in larger resistance values or a greater opposition to airflow in the lungs. Using measured parameters of T_p and *flow*, the R_L can be determined by:

$$R_L = Tp/flow \tag{7.7}$$

which can be determined before and after aerosol challenges of control or bronchoconstriction agents. Pulmonary resistance is important to assess in murine asthma studies as it indicates the effectiveness of a treatment's ability to decrease the opposition to airflow, and thereby increase the flow of air available for respiratory functions.

The dynamic compliance is calculated from specific parameters measured during respiratory cycles. In this calculation of C_{dyn}, tidal volume and transpulmonary pressure values, measured from the end of a preceding cycle to the end of a current respiratory cycle, are defined as the dividend and the divisor, respectively (Salome et al. 1987). Computer software and programs can then determine C_{dyn} from these two known pulmonary parameters. Dynamic compliance is interpreted physiologically as the total (elastic and airway) pulmonary resistance. It is important in murine asthma studies to assess how medication affects the volume of air taken into the airways:

$$C_{dyn} = T_v/P_{tp} \tag{7.8}$$

The protocol used to determine the parameters was as follows:

- Mice were anesthetized with intraperitoneal injections of ketamine 40 mg/kg and xylazine 10 mg/kg (with minimal supplementations only if required).
- When anesthesia was achieved, assessed by loss of the right pinch toe reflex, the mice were placed on a plastic board, and a quick surgical procedure was performed to reach the trachea.
- Once the trachea was exposed, a small hole was opened in between the trachea rings, and a cannula was placed and fastened with silk thread.
- Following this, the intubated spontaneously breathing mouse was placed in supine position in a thermo-controlled plethysmograph/pneumotacograph. This system was designed to resemble the reference system used by Glaab et al. (Glaab et al. 2004, 2007; Hoymaan 2006). Some modifications were made to the chamber (size, inlets, and outlets of the chamber) in order to fit not only mice but also rats. Similar transducers were used.
- The tracheal tube was connected directly to the pneumotachograph to determine the tidal flow, which was connected to a differential pressure transducer.

- To measure transpulmonary pressure, a water-filled tube was inserted into the esophagus to the level of the mid-thorax and coupled to a pressure transducer. The analog signals obtained from this transducer were digitalized and recorded.
- Different doses of bronchoconstrictor and bronchodilator were administered through the nebulizer which was connected to the mouse lungs via the pneumotachometer. A 5 mL of test solution (normal saline solution, or a specific concentration of Ach or ISO) was placed in the nebulizer tank.
- Each mouse was initially challenged for two minutes with saline solution. After 2 minutes of nebulization, the nebulizer was switched off before recording the respiratory parameters for 10 minutes. The recording time corresponds also to a resting time for the animal, since it will not undergo any additional treatment or manipulation during this time. After this, dose responses for ACh and ISO were performed to determine the optimal concentration of the drugs to be used).
- AWR to ACh 10^{-2} M in all mice groups was assessed to observe the response of ASM (this concentration was determined during the previous step and showed the higher contraction on the sensitized animals).

This dose was tested several times and was administered before the addition of ISO (which was used to induce bronchodilation) as follows: nebulization of ACh 10^{-2} M for two minutes followed by two minutes of rest; then nebulization with ISO 10^{-6} M for two minutes followed by two minutes of rest; the previous step was repeated with ISO 10^{-5} M, ISO 10^{-4} M, ISO 10^{-3} M, ISO 10^{-2} M, ISO 10^{-1} M, and ISO 1 M, and finally two minutes of nebulization of saline and two minutes of recovery. As shown in Figure 7.19, using R_L ($cmH_2O/mL/s$) as indicator of bronchoconstriction in vivo, a dose–response test for ACh was performed. According to this, the best concentration to induce bronchoconstriction on our sensitized acute animals was of 10^{-2} and 10^{-4} M for the chronic model even though 1 M showed a similar response (Figure 7.19). As per literature, lowest concentration results in better practical results when combined with ISO (Hoymaan 2006), because higher concentrations tend to affect the response of the airways in the presence of ISO. Considering this finding, the concentration selected for future in vivo experiments will be Ach 10^{-2} M. This concentration differs from the one used on the experiments in vitro, which was Ach 10^{-4} M (data not shown). This concentration was also determined using dose–response tests before defining the experimental protocols.

When a similar dose–response test was performed for ISO as shown in Figure 7.20, it was found that several different concentrations provided a similar reduction on the bronchoconstriction induced by Ach 10^{-2} M. These concentrations were 10^{-6} M, 10^{-5} M, 10^{-4} M, and 10^{-1} M (Figure 7.19). Considering that all these concentrations reduced in a similar manner, the bronchoconstriction was induced, and we decided to use the lower dose for our experiments, which was ISO 10^{-6} M.

Using the concentrations mentioned and selected in the previous paragraphs, a quick comparison between healthy and sensitized subjects was performed to observe any changes. We found using *t*-test that sensitized airways were more responsive than healthy airways to Ach 10^{-2} M (p-value < 0.01 vs p-value 0.18, respectively, as observed in Figure 7.20) and that ISO 10^{-6} M was less effective to induce relaxation on pre-constricted airways from sensitized subjects when compared with healthy airways (p-value 0.18 vs p-value < 0.05, respectively, as observed in Figure 7.21 using the same concentrations for ACh and ISO. This behavior was also observed later on in the in vitro tests.

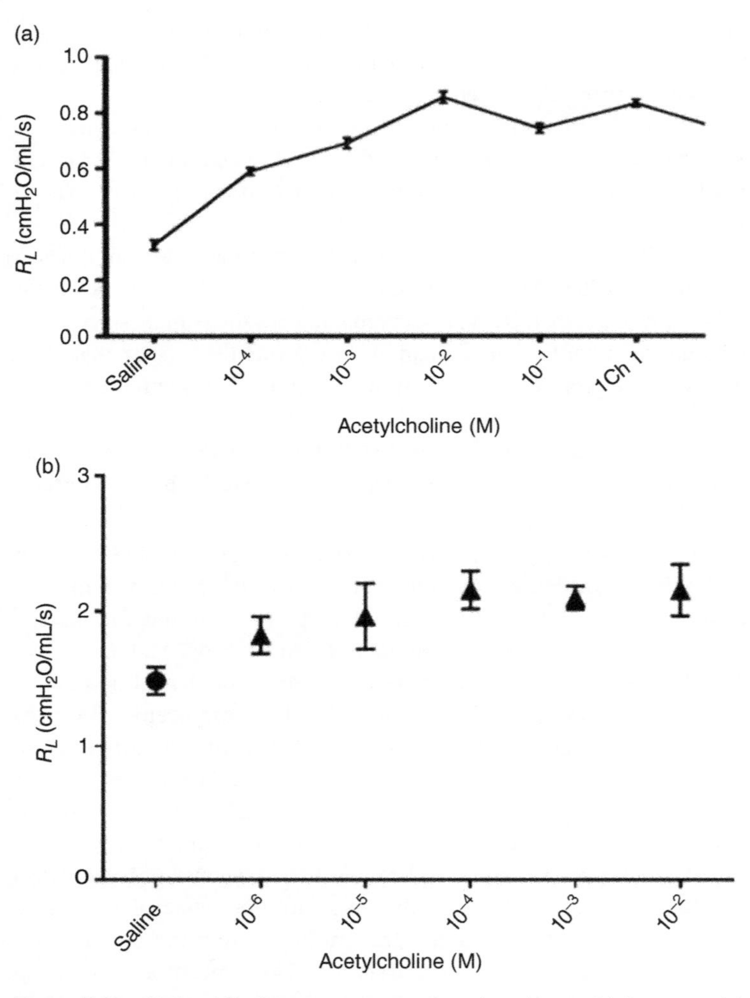

Figure 7.19 R_L for ACh dose–response asthmatic subjects: (a) Acute model $n = 6$; (b) Chronic $n = 5$.

7.9.3 ELISA (IgE)

Chemical evaluation is an essential part in asthma assessment. In asthmatic models, the levels of the antibodies IgE and IgG are normally increased. To confirm that our model was asthmatic, the levels of these two antibodies were tested using a direct sandwich enzyme-linked immunosorbent assay (ELISA) method. OVA-specific IgE levels were determined in duplicate. Blood samples from the mice heart were obtained right after the experimental protocols, and using coagulation and centrifugation, samples of serum were prepared from it and stored at $^{-}80\,°C$. Then IgE levels were quantified using the kit Mouse OVA-IgE ELISA from mdbioproducts (a division of mdbiosciences) and compared between controls and sensitized.

In order to analyze the data obtained from the healthy and sensitized mice, a standard curve with known concentrations for OVA IgE was determined. The standard

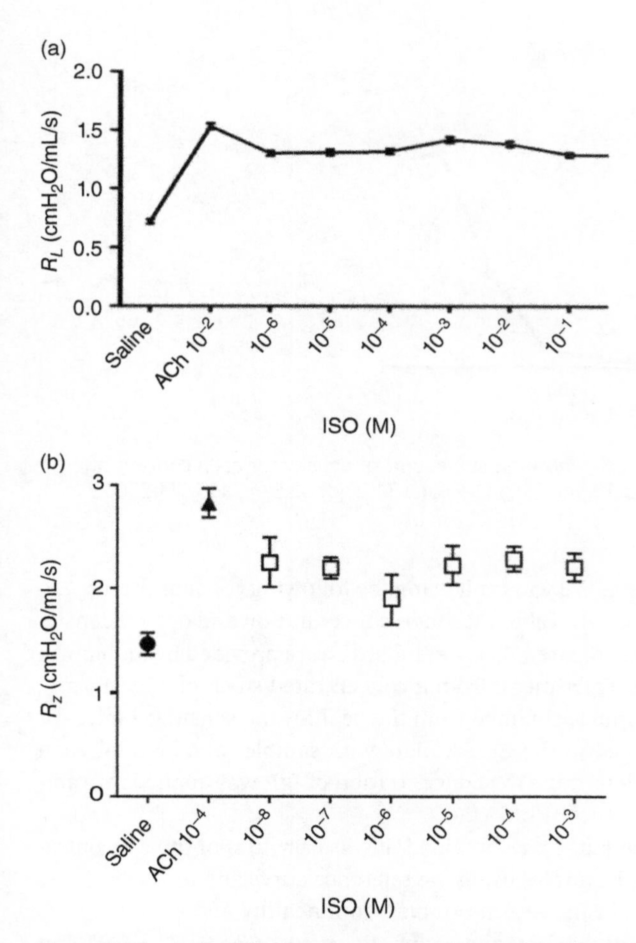

Figure 7.20 R_L for ISO dose–response: (a) Acute Model n = 6; (b) Chronic Model n = 5.

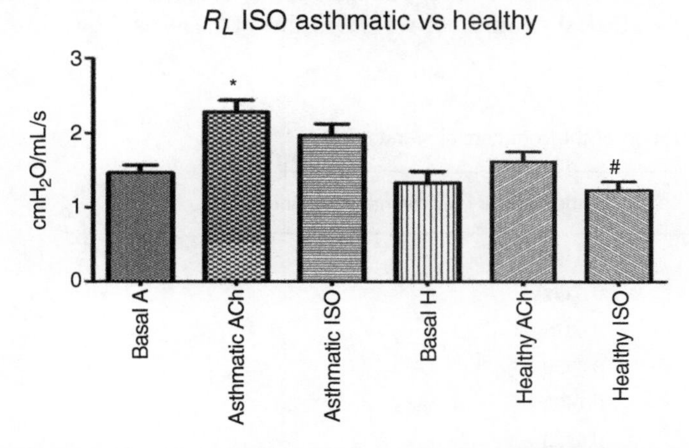

Figure 7.21 Comparison of R_L vertical axis between asthmatic and healthy subjects. (n = 6) and its statistical significance (* *or* # = p-value < 0.05).

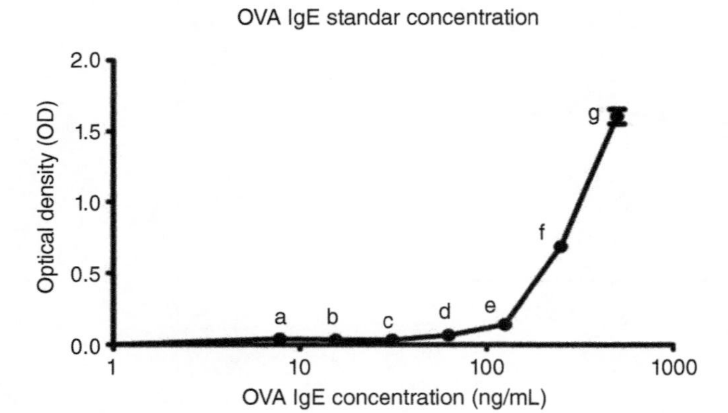

Figure 7.22 ELISA OVA IgE standard concentration curve: optical density for each concentration versus the concentration of each standard. (a) 7.8, (b) 15.6, (c) 31.2, (d) 62.5, (e) 125, (f) 250, (g) 500 ng/mL.

concentration curve shown in Figure 7.22 was built using the following concentrations: 7.8, 15.6, 31.2, 62.5, 125, 250, and 500 ng/mL. Table 7.2 shows concentration and optical density (OD) for each commercial standard prepared. These standards were provided by mdbioproducts and prepared just before the experiment, from a concentrated stock of 2500 ng/mL, prior to the testing of the serum samples obtained from the healthy and sensitized mice for the study. The standards were placed in the ELISA plate with samples obtained for each animal, and the same protocol to determine the concentration of IgE was applied for samples and standards (commercial protocol ELISA).

Once the OVA IgE concentration curve was obtained, it was easy to spot the concentration of the levels for OVA IgE in our samples using the reference curve and to compare the data with the absorbance obtained in the serum samples from healthy and asthmatics.

The absorbance obtained from the healthy and asthmatic groups was 0.032 ± 0.02 and 0.141 ± 0.03, respectively ($n = 11$). The absorbance for both groups was compared and analyzed with a t-test, obtaining a statistical significance with a p-value of 0.039, as shown in Figure 7.23.

Table 7.2 OD for each concentration of the commercial standards.

OVA IgE concentration (ng/mL)	Optical density for commercial standards
7.8	0.04085
15.5	0.0342
31.2	0.0338
62.5	0.06505
125	0.14025
250	0.6883
500	1.60535

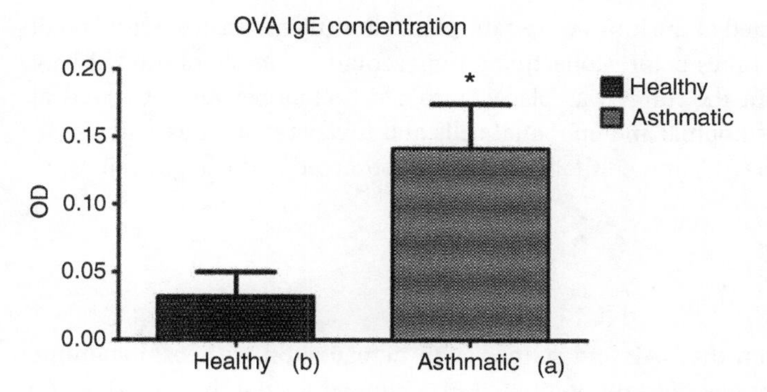

Figure 7.23 OVA IgE concentration: (a) sensitized animals and (b) control (healthy) animals (p-value < 0.05; $n = 11$).

7.9.4 BAL

BAL is a technique used to recover epithelial cells, proteins, and leukocytes, which are normally present in the airways under pathological conditions. This technique involves successive lavages of the airways with physiological solutions. The volume of the mouse lung is small, so BAL in mice is generally performed with 1-mL syringes to infuse smaller volumes of fluid. Multiple infusions are required to obtain enough recovered fluid for multiple analyses. Once the BAL is successfully obtained, the cells are counted in the Neubauer chamber (Hoymaan 2006), and slides stained with hematoxylin and eosin are prepared to differentiate the white cells (Figure 7.24).

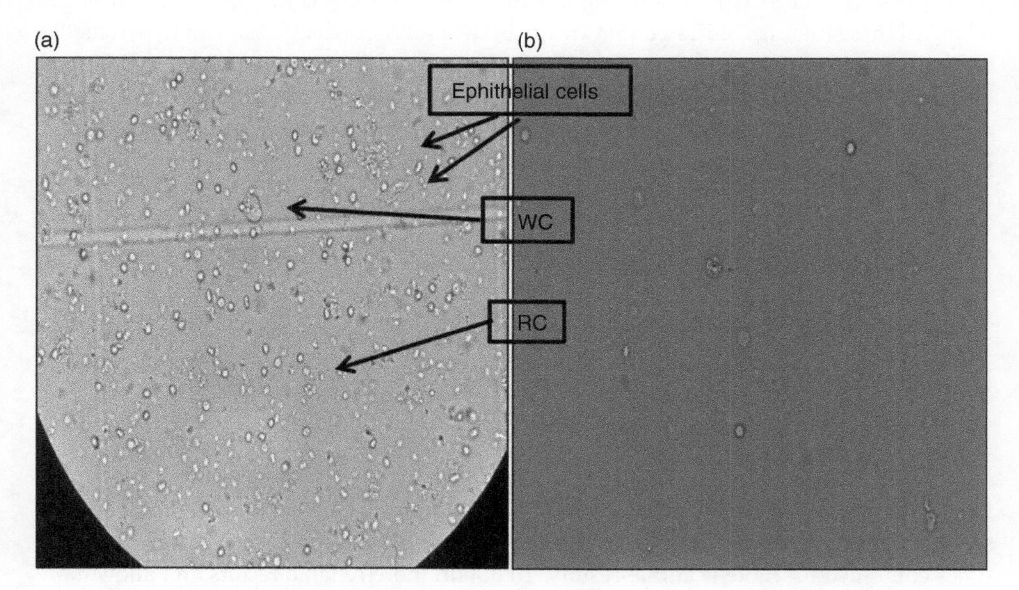

Figure 7.24 BAL images: (a) sensitized animals and (b) control (healthy) animals: epithelial cells, WC (white cells), and RC (red cells). *Source:* Based on Hoymaan (2006).

The main protocol consisted of the lungs were cannulated in situ and washed with 1ml of saline (NaCl 0.9%) several times before collecting a representative sample of the airways. The solution collected from the lungs was placed into a cell chamber and observed at 40× for the presence of eosinophils and epithelial cells and it was reported as follows:

The presence of red cells (RC) and epithelial cells were observed and categorized as:

+ = presence
++ = moderate presence
+++ = plenty
(–) = absence

Slides were prepared from the BAL and stained with hematoxylin and eosin staining. White cells (WCs) were counted and differentiated in a blinded fashion by counting 100 cells by light microscopy. The number of eosinophils was expressed as a % of the total WC.

When the BAL from the control mice was compared with the sensitized ones, it was found (as shown in Figure 7.22) that the BAL obtained from the sensitized airways showed increased presence of WC (white cells), RC (red cells), and epithelial cells in all the preparations, while in the slides prepared from BAL recovered from the healthy airways showed poor presence or complete absence of the same cellularity. A significant increase in eosinophils in the slides from sensitized mice ($21 \pm 6\%$) was also observed when compared with controls ($3 \pm 1\%$).

The first objective of this study establishment of an acute and a chronic asthmatic model using long-term DRA sensitization was achieved. Comparisons of asthmatic characteristics between healthy and sensitized subjects indicate markedly different responses to common stimuli. For sensitized subjects, increased AHR was demonstrated in response to ACh, and a reduced response to ISO was also observed. IgE levels were higher in the sensitized group of subjects compared to levels in the control group. Finally, BAL revealed that changes in cellularity occurred due to the sensitization, with higher white cell count present in sensitized mice. All these findings indicated that we are in the presence of subjects with hyperreactive airways. Our sensitization protocol to develop an acute asthmatic model was working as expected, exhibiting some typical features of asthma. Further, the assessments of these asthmatic characteristics also led to the conclusion that a chronic model was established for use in the study of pulmonary responses to applied pressure oscillations.

7.10 Superimposed Pressure Oscillation

To assess in vivo pulmonary functions in mice, different techniques are available. These usually are separated into two categories: invasive and noninvasive. This section summarizes and discusses the selected methodology for our in vivo experimental protocols.

7.10.1 Experimental Layout

Invasive and noninvasive techniques are available to study respiratory physiology, both with some advantages and disadvantages. These methods are sometimes used alone, and other times combined for more accurate results. To obtain more reliable results and allow more control over the overall manipulations, it was decided for this study to use an invasive approach.

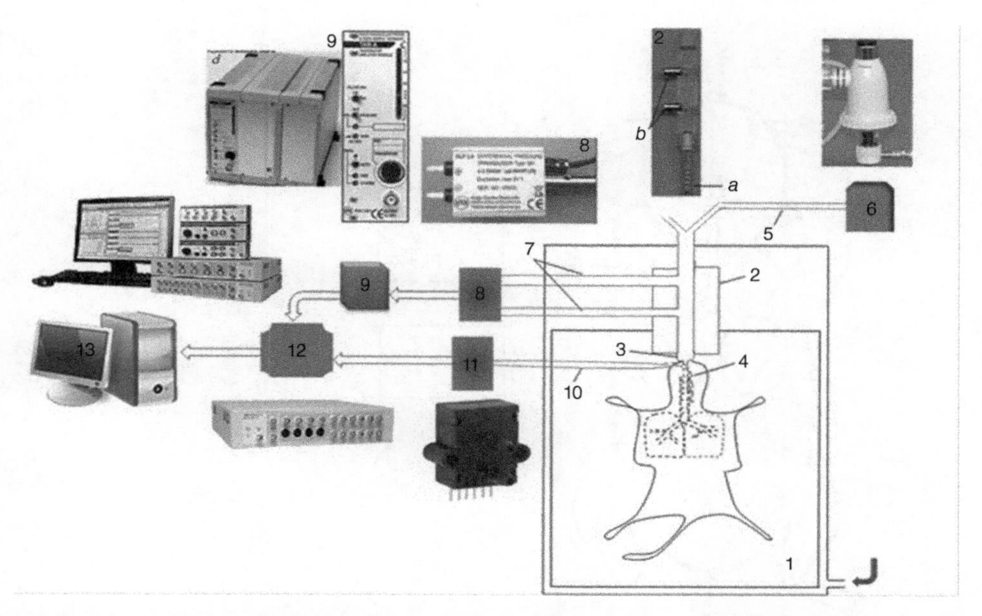

Figure 7.25 (1) Thermostated plexiglas chamber; (2) pneumotachometer; (3) tracheal cannula; (4) esophageal catheter; (5) tube supplying the nebulized solution; (6) nebulizer; (7) connection pipes; (8) differential pressure transducer; (9) Tam-A transducer amplifier (c) and PLUGSYS amplifier module (d); (10) connection pipe; (11) Honeywell pressure transducer; (12) 16 channels data acquisition board LabChart; (13) computer with data acquisition LabChart.

To measure the bronchoconstriction level using the parameters of R_L and C_{dyn}, a plethysmograph custom built based on the setup used by Glaab et al. (Glaab et al. 2004) was developed and built. The experimental layout presented in Figure 7.25 consists of two units: animal containment and measuring components. The animal containment unit consists of a testing chamber custom built and a thermoregulated bath. On the other hand, the measuring unit components consist of pressure transducers and amplifiers; drugs delivery system: Jet Nebulizer for the allergens and drugs; and a data acquisition system (LabChart).

The measurement components are different elements used to determine T_v, *flow*, P_{tp}, and T_p. These parameters are used to calculate R_L and C_{dyn} (as shown in Figure 7.26). The elements used to obtain these respiratory parameters were (a) pneumotachograph and tracheal cannula; (b) differential pressure transducer for T_p; (c) differential pressure transducer for P_{tp}, and (d) esophageal cannula.

The pneumotachometer and tracheal cannula are used to cannulate the trachea during the experimental procedures by keeping the airways open. Through the connection with the DLP 2.5 differential pressure transducer, the pneumotachometer also allows the measurement of pressure changes to occur at the level of the mice trachea and to measure the airflow and tidal volume occurring during the breathing process. This pressure transducer is used to register the T_p, T_v, and *flow*. To gain helpful readings from the pressure transducer outputs, the signals were needed to be amplified using a TAM-A transducer amplifier module (P/N 73-0065).

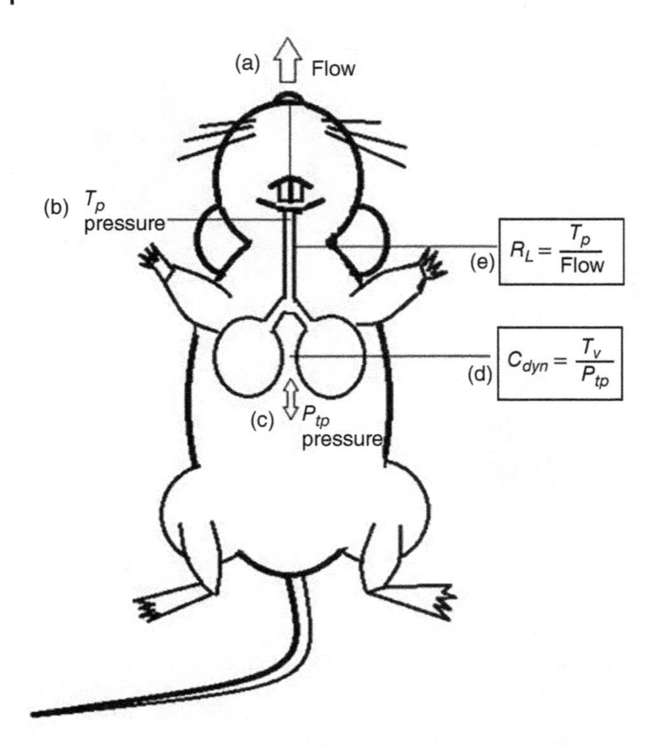

Figure 7.26 Respiratory parameters: (a) flow; (b) T_p pressure (tracheal); (c) P_{tp} pressure (transpulmonary); (d) R_L; C_{dyn}. *Source:* Dvorkin et al. (2010).

The esophageal catheter was required to measure the P_{tp}. The catheter was made of PE tubing (PE 90) and a blunt 20-g needle. The needle body was covered by silicone tubing with ID = 3 mm and OD = 5 mm. The PE tubing had a length of 120 mm, and the tip was beveled at 45°. Two oval holes were incised into the PE tubing at about 3 and 8 mm from the tip using a micro-spring scissor or a scalpel. The holes were on opposite sides and had a size of about 2 mm by 1 mm. The PE tubing was mounted on the 20 g blunt needle with the silicone tubing sleeve. The silicone tubing was used to maintain the needle into the esophagus, and the blunt 20 g needle was connected to the differential pressure transducer. The catheter was filled with saline solution, and changes in the position of the column of saline were registered by the pressure transducer due to changes of pressure at mid-thorax (Kumar and Foster 2001). This pressure transducer is required to measure the P_{tp} once that it is connected to the filled esophageal catheter.

All these measuring elements mentioned earlier correspond to the whole measuring setup that integrates the plethysmograph, but for the project it was also required to develop an oscillation device to deliver the oscillations into the airways in vivo. The system selected to record the data and process was the PowerLab 16/30. This is a high-performance data acquisition system suitable for a wide range of research applications that require up to 16 input channels. The unit is capable of recording at speeds of up to 400 000 samples per second continuously to disk and is compatible with instruments, signal conditioners, and transducers supplied by ADInstruments, as well as many other brands. In addition to standard single-ended BNC inputs, the PowerLab 16/30 features four differential Pod ports that allow for direct connection of Pod signal conditioners and appropriate transducers. Both pressure transducers are attached to the board using coaxial connectors into ports 1 and 7.

The PowerLab 16/30 works with the ADInstruments software LabChart, a flexible and easy tool to work with. This software brings several templates which could be used for different physiological experiences. The template for spirometry was used to develop the final program for the experiments of this study. This template has the option of automatically setting channels to register flow (L/s), tidal volume (L), and minute ventilation (L/min). This template was modified for purposes of the study and nine channels were activated and set Ch1) voltage; Ch2) Tp; Ch3) flow; Ch4) volume; Ch5) minute ventilation; Ch6) tidal volume; Ch7) P_{tp}; Ch8) RL, and Ch9) Cdyn. To obtain these two last parameters, specific equations for each of them were added into their respective channel settings. The pressure transducer was physically connected to the board using coaxial cables/connectors: DLP 2.5 differential pressure transducer to Ch1 and sensor techniques model 26PC0050D6A to Ch7 to register the voltage changes signals, which were converted into data pressure changes. These readings were useful to determine parameters such as T_v, *flow*, T_p, and P_{tp}, and indirectly to measure R_L and C_{dyn} in real time.

7.10.2 Nebulization System for the Drugs and Allergen

To sensitize and deliver drugs into the airways of the animal models, a nebulizer was required. For the purposes of this study, a jet nebulizer from Harvard apparatus was acquired (P/N 73-1963). This aerosol jet nebulizer required an operating pressure of approximately 1.5 bar (22 psi) from a compressed air source. This pressure was supplied using an AMPRO air brush compressor with a working range of 15–50 psi, which was fixed to the pressured required by the jet nebulizer. The particles generated by the jet nebulizer were 10 μm or less in size with 60% of the particles being 2.5 μm or less.

7.10.3 Pressure Oscillation Setup

For this study, a device based on pressure–volume oscillations was built. It was considered that this type of oscillations is the most direct approach to deliver the changes to be tested and also the less invasive of all those mentioned before. This device involved the use of an arbitrary waveform generator, power amplifier, electromagnetic shaker, a plastic chamber adapter with one inlet and one outlet, a piston built-in with silicone connected to the shaker and placed into the chamber, and a sturdy base to maintain the device in a fixed position. The device was capable of generating pressure changes, which were induced using a waveform generator that created the waveform desired and which was sent to the shaker and the piston, generating displacement of the piston inside the sealed chamber.

7.11 In Vivo Test

All devices and components discussed earlier and used in these protocols were calibrated and tested before use. To investigate the significance of superimposed oscillation in the treatment of asthmatics, the next step was to conduct in vivo investigation on the developed models, acute and chronic.

During the plethysmography, the animals were anaesthetized and intubated with a tracheal cannula and esophageal catheter. Unique doses of bronchoconstrictor (Ach 10^{-2} M) and bronchodilator (ISO 10^{-6} M) were used for the protocol to induce contraction and relaxation of the airways (respectively). These drugs were administered as a mist using the jet nebulizer. The jet nebulizer was indirectly connected to the mouse lungs via the duo pneumotachometer-tracheal cannula. A 5 mL of testing solution (standard saline solution, or a specific acetylcholine and ISO) was placed in the nebulizer tank (depending on the protocol). The following protocol was used to determine relaxation:

- Each mouse is initially challenged for 5 minutes with saline solution to obtain a basal reading, and after 5 minutes of nebulization, the nebulizer will be switched off before recording the respiratory parameters for 10 minutes. The recording time also corresponds to the animal's resting time, since it will not undergo any additional treatment or manipulation during this time.
- Then a unique dose of acetylcholine is administered to induce bronchoconstriction according to the following sequence: nebulization of acetylcholine 10^{-2} M for two minutes.
- Then the bronchorelaxation will be tested by administration of the unique dose of ISO for two minutes followed by five minutes of rest.
- After these steps have been recorded, the sequence will be repeated from the baseline with saline, and SIPOs will replace ISO.

Four sets of superimposed oscillations were tested after the contraction was induced using ACh 10^{-2} M:

- First set with a frequency of 5 Hz and amplitudes of 100, 200, 300, and 400 mV with 2 minutes per amplitude and after that 5 minutes of rest and saline.
- After these steps have been recorded, the sequence would be repeated from the baseline with saline.
- Second set with a frequency of 10 Hz and amplitudes of 100, 200, 300, and 400 mV with 2 minutes per amplitude and after that 5 minutes of rest and saline.
- After these steps have been recorded, the sequence would be repeated from the baseline with saline.
- Third set with a frequency of 15 Hz and amplitudes of 100, 200, 300, and 400 mV with 2 minutes per amplitude and after that 5 minutes of rest and saline.
- After these steps have been recorded, the sequence would be repeated from the baseline with saline.
- Finally, a fourth set with a frequency of 20 Hz and amplitudes of 100, 200, 300, and 400 mV with 2 minutes per amplitude and after that 5 minutes of rest and saline.
- After these steps have been recorded, the sequence would be repeated from the baseline with saline.

7.11.1 Relaxation

The percentage of relaxation ($\%R$) was calculated as the percentage of total force reduction (F_2) observed after five minutes after the application of ISO and/or oscillations relative to the force observed prior to these agents, which corresponds to the plateau (F_1), namely:

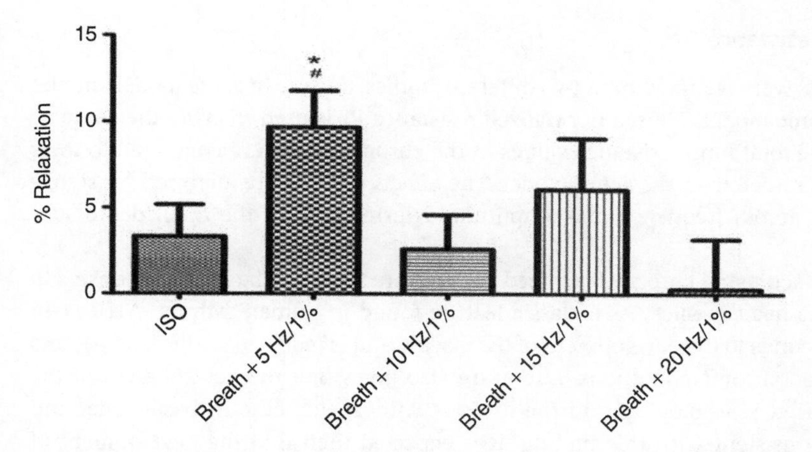

Figure 7.27 Effect of ISO compared with mechanical oscillations equivalent to breathing + superimposed length oscillations (1% of amplitude) on the contractile force. * indicates statistical significance for the mean of SILO at 5 Hz compared to the mean to the effect of ISO alone ($n = 7$).

$$\%R = \frac{(F1 - F2)}{F1} \tag{7.9}$$

When SILO was compared with ISO alone on asthmatic airways, they showed similar observed patterns to those compared with breathing alone. When ISO was compared with SILO of 1% amplitudes, only 5 Hz showed statistically significant increased relaxation (Figure 7.27).

When SILO at 1.5% were compared with ISO alone, similar increased relaxation patterns were observed when compared to breathing alone, with statistical significance for all the frequencies tested as presented in Figure 7.28.

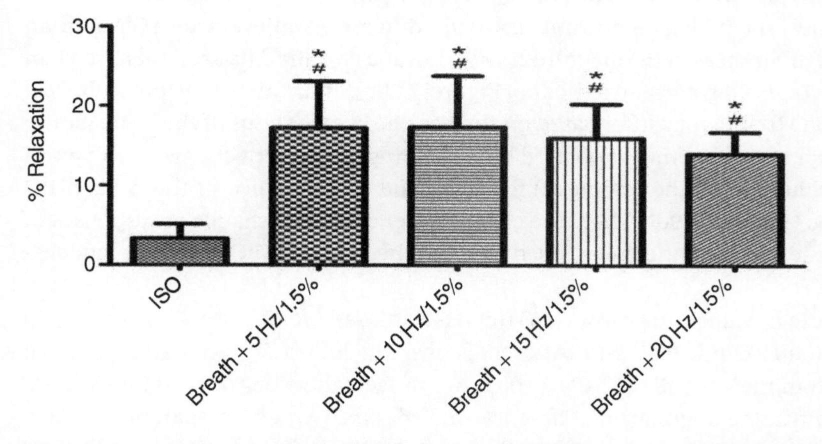

Figure 7.28 Effect of ISO compared with mechanical oscillations equivalent to breathing + superimposed length oscillations (1.5% of amplitude) on contractile force. * indicates statistical significance for the means of the effect of SILO when combined with breathing and compared with the effect of ISO alone using Wilcoxon; # indicates statistical significance between the means of the effect of SILO when combined with breathing and compared with the effect of ISO alone using *t*-test on asthmatic airways ($n = 7$).

7.11.2 Lung Resistance

The in vivo results were obtained from two different studies, one on an acute model and the second on a chronic model. Thus, a normalized resistance $\check{R}_L$ is used to make the comparison relevant. The total lung resistance values in the chronic model are compared to their corresponding parameters in the acute model. The effects of SIPO are analyzed for significant differences across frequency and amplitude determinants of the applied pressure oscillations.

The airways of acute and chronic sensitized subjects are shown to be hyperresponsive to ACh compared to healthy mice. As a classic feature found in human asthma, AHR is an effect that corresponds to the development of the disease and is modeled in this study across healthy, acute and chronic conditions. Literature also notes that the development of the disease is evidenced physiologically in the airways with an increase in basal tone, and our results are consistent with this finding. It is expected then that the development of the disease from healthy to chronic conditions is evidenced with greater AHR parameter values in long-term asthmatic ASM compared to short-term asthmatic ASM, as the tissues respond to a bronchoconstrictive stimulus.

Comparing one part of the AHR parameters is accomplished by using ŘL for acute and chronic data sets. $\check{R}_L$ values then serve as relative indicators of airway bronchoconstriction or bronchorelaxation. The established values of $\check{R}_L$ for ACh challenges and ISO treatments serve as standards to gauge the efficacy of applied pressure oscillations across the tested acute chronic mice. Figure 7.29a illustrates the 5-Hz acute $\check{R}_L$ values, indicating significant ($p < 0.05$) differences relevant to ACh with an asterisk (*). Chronic group $\check{R}_L$ values range from 1.05 $((\text{cmH}_2\text{O}\cdot\text{mL})/\text{s})^{-1}$ for 200-mV treatments up to a high of 1.64 $((\text{cmH}_2\text{O}\cdot\text{mL})/\text{s})^{-1}$ with ACh challenge. The lowest $\check{R}_L$ value for chronic mice was achieved from the 1.6 cmH_2O SIPO group and ISO treatment, which were statistically equivalent. All data for the chronic group were significantly lower than ACh ($p < 0.05$). Figure 7.29b illustrates the 5-Hz chronic $\check{R}_L$ values, indicating significant ($p < 0.05$) differences relevant to ACh with an asterisk (*). The differences in the magnitudes of acute and chronic data are apparent when viewed in the context of normalized data (Figure 7.29c), indicating that physiological responses to SIPO treatments differ based on the health/disease status of the same mouse strain. This conclusion is further supported by comparing the percent change in $\check{R}_L$ values (against ACh challenge as the reference) for acute and chronic mice in this 5-Hz SIPO treatment group (Figure 7.29d). There are clear differences in the change in responses of short- and long-term asthmatic mice, based on all of the applied SIPO and their modeled disease state.

The 10-Hz acute $\check{R}_L$ values range from 0.63 $((\text{cmH}_2\text{O}\cdot\text{mL})/\text{s})^{-1}$ for 200-mV treatment up to a high of 0.79 $((\text{cmH}_2\text{O}\cdot\text{mL})/\text{s})^{-1}$ with ACh challenge. The lowest $\check{R}_L$ value for acute mice was achieved from the 1.5 cmH_2O SIPO group, but in fact, all values other than ACh are statistically comparable. Significant differences are observed when comparing the ACh standard against the entire set. Figure 7.30a illustrates the 10-Hz acute $\check{R}_L$ values, indicating significant ($p < 0.05$) differences relevant to ACh with an asterisk (*). Chronic group $\check{R}_L$ values range from 0.92 $((\text{cmH}_2\text{O}\cdot\text{mL})/\text{s})^{-1}$ for 100-mV treatments up to a high of 1.64 $((\text{cmH}_2\text{O}\cdot\text{mL})/\text{s})^{-1}$ with ACh challenge. The lowest $\check{R}_L$ value for chronic mice was achieved from the 0.7 cmH_2O SIPO group, with all data for the chronic group significantly

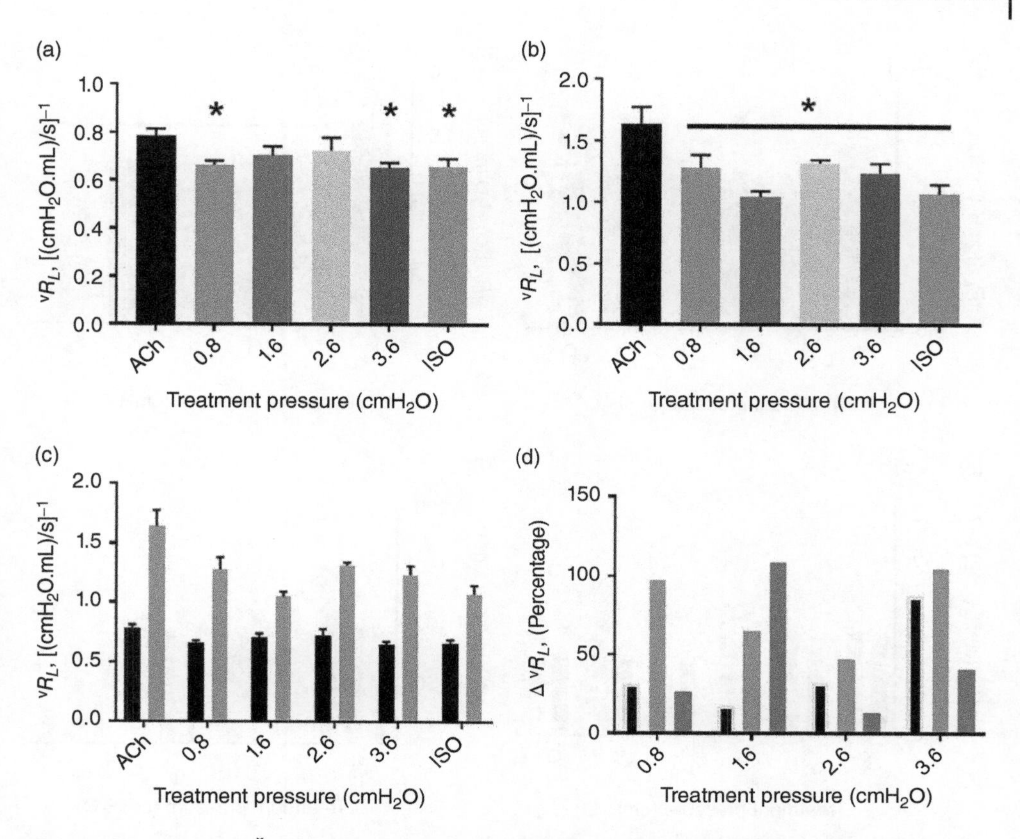

Figure 7.29 The 5-Hz $\check{R}_L$ values: acute (a) and chronic (b) groups with SIPO amplitudes of 0.8, 1.6, 2.6, and 3.6 cmH_2O are presented ($n = 7$, except for chronic SIPO, $n = 5$). (c) Magnitudes of acute (black) and chronic (gray) $\check{R}_L$ values; and (d) percent change in $\check{R}_L$ (compared to ACh) of acute (light gray) and chronic (dark gray) values; healthy changes are indicated (black) for comparison.

lower than ACh ($p < 0.05$). Figure 7.30b illustrates the 10-Hz chronic $\check{R}_L$ values, indicating significant ($p < 0.05$) differences relevant to ACh with an asterisk (*).

The differences in the magnitudes of acute and chronic 10-Hz groups are apparent when viewed in the context of normalized data (Figure 7.30c). They indicate that physiological responses to SIPO treatments differ based on the health/disease status of the same mouse strain. This conclusion is further supported by comparing the percent change in $\check{R}_L$ values for acute and chronic subjects in this 10-Hz SIPO treatment group (Figure 7.30d). There are apparent differences in the change in responses of short- and long-term asthmatic mice, based on the applied SIPO (except for 2.4 cmH_2O), and their modeled disease state.

The 15-Hz acute $\check{R}_L$ values range from 0.64 $((cmH_2O{\cdot}mL)/s)^{-1}$ for 200-mV treatment up to a high of 0.79 $((cmH_2O{\cdot}mL)/s)^{-1}$ with ACh challenge. The lowest $\check{R}_L$ value for acute mice was achieved from the 1.1 cmH_2O SIPO group, but it is not significantly different from other SIPO values for the group. When comparing the ACh standard with the group's SIPO or ISO data points, no significant differences are observed. Figure 7.31a illustrates the 15-Hz acute $\check{R}_L$ values, indicating the lack of significant ($p > 0.05$) differences relevant to ACh. The

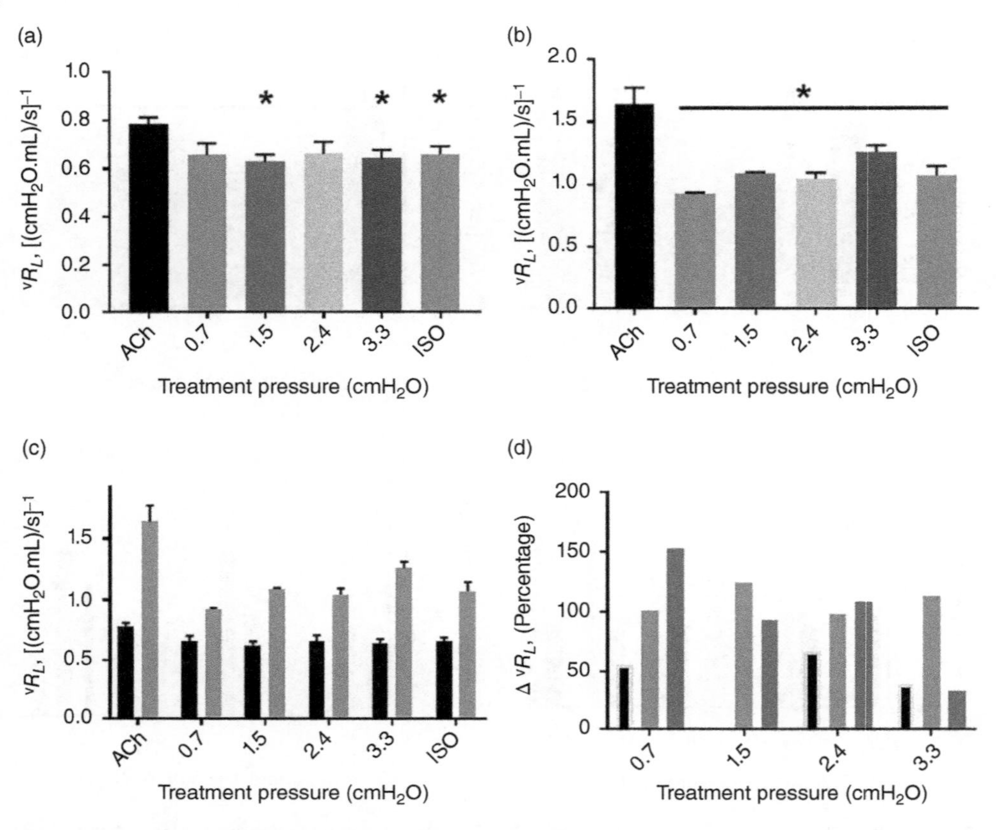

Figure 7.30 The 10-Hz $\check{R}_L$ values. Acute (a) and chronic (b) groups with SIPO amplitudes of 100–400 mV (0.7, 1.5, 2.4, and 3.3 cmH_2O) are presented ($n = 7$, except for chronic SIPO, $n = 5$). (c) Magnitudes of acute (black) and chronic (gray) $\check{R}_L$ values; and (d) percent change in $\check{R}_L$ (compared to ACh) of acute (light gray) and chronic (dark gray) values; healthy changes are indicated (black) for comparison.

chronic group $\check{R}_L$ values range from 0.85 $((cmH_2O{\cdot}mL)/s)^{-1}$ for 300-mV treatments up to a high of 1.64 $((cmH_2O{\cdot}mL)/s)^{-1}$ with ACh challenge. The lowest $\check{R}_L$ value for chronic mice was achieved from the 1.7 cmH_2O SIPO group. All data for the chronic group except for the 0.4 cmH_2O treatment was significantly lower than ACh ($p < 0.05$). Figure 7.31b illustrates the 15-Hz chronic $\check{R}_L$ values, indicating significant ($p < 0.05$) differences relevant to ACh with an asterisk (*).

The 20-Hz acute $\check{R}_L$ values range from 0.66 $((cmH_2O{\cdot}mL)/s)^{-1}$ for ISO treatment up to a high of 0.85 $((cmH_2O{\cdot}ml)/s)^{-1}$ with the 300-mV treatment. The lowest $\check{R}_L$ value for acute subjects treated with ISO was comparable to 400-mV (2.3 cmH_2O) treatments. No significant differences are observed when comparing the ACh standard with the group's data points, though notably, the 1.7 cmH_2O treatment is great than ACh. Figure 7.32a illustrates the 20-Hz acute $\check{R}_L$ values, indicating no significant ($p > 0.05$) differences relevant to ACh with an asterisk. Chronic group $\check{R}_L$ values range from 0.65 $((cmH_2O{\cdot}mL)/s)^{-1}$ for 300-mV treatments up to a high of 1.64 $((cmH_2O{\cdot}ml)/s)^{-1}$ with ACh challenge. The lowest $\check{R}_L$ value for chronic mice was achieved from the 1.7 cmH_2O SIPO group. All data for the chronic

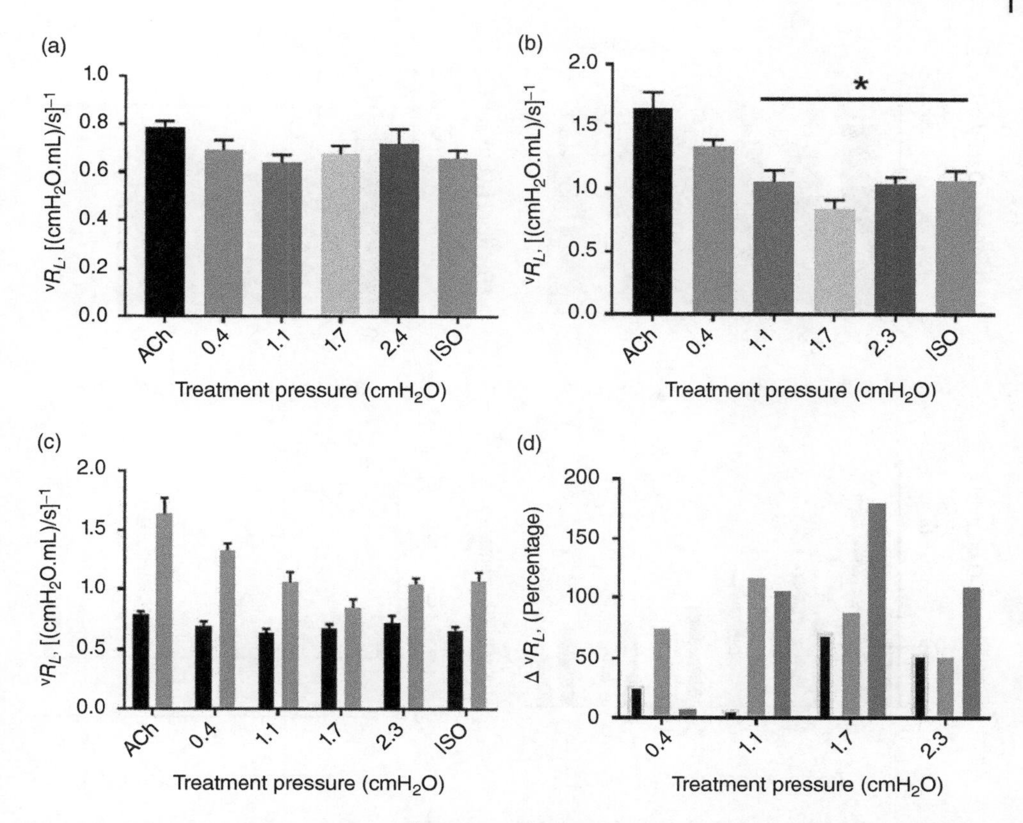

Figure 7.31 The 15-Hz R_L values to $\check{R}_L$. Acute (a) and chronic (b) groups with SIPO amplitudes of 100–400 mV (0.4, 1.1, 1.7, and 2.4 cmH$_2$O) are presented (n = 7, except for chronic SIPO, n = 5). (c) Magnitudes of acute (black) and chronic (gray) $\check{R}_L$ values; and (d) percent change in $\check{R}_L$ (compared to ACh) of acute (light gray) and chronic (dark gray) values; healthy changes are indicated (black) for comparison.

group were significantly lower than ACh ($p < 0.05$). Figure 7.32b illustrates the 20-Hz chronic $\check{R}_L$ values, indicating significant differences relevant to ACh with an asterisk (*).

The differences in the magnitudes of acute and chronic data are apparent when viewed in the context of normalized data (Figure 7.32c). They indicate that physiological responses to SIPO treatments differ based on the health/disease status of the same mouse strain. This conclusion is further supported by comparing the percent change in $\check{R}_L$ values for acute and chronic subjects in this 20-Hz SIPO treatment group (Figure 7.32d). There are apparent differences in the change in responses of short- and long-term asthmatic mice, based on the applied SIPOs of 1.1 and 1.7 cmH$_2$O and their modeled disease state.

7.11.3 Compliance

Similar to the resistance, a normalized compliance is used for comparison. The 5-Hz acute $\check{C}_{dyn}$ values range from 0.39 (cmH$_2$O·mL) for ACh challenge up to a high of 0.48 (cmH$_2$O·mL) with ISO and 3.6 cmH$_2$O SIPO treatment. No significant differences

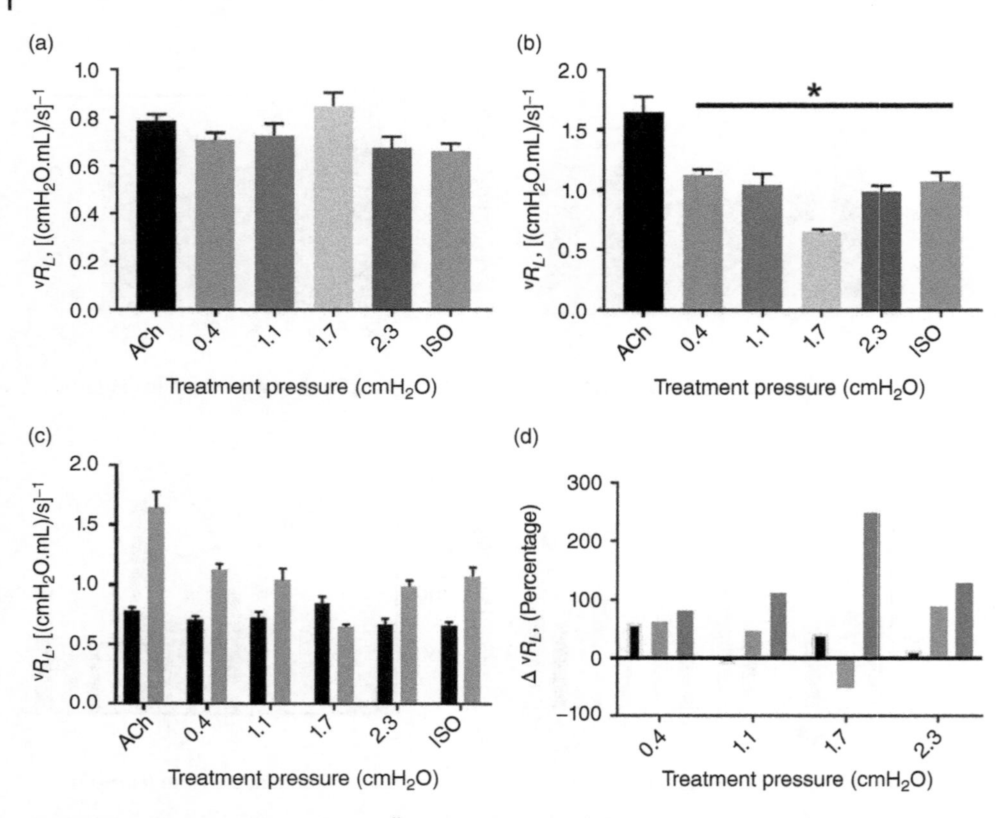

Figure 7.32 The 20-Hz R_L values to $\check{R}_L$. Acute (a) and chronic (b) groups with SIPO amplitudes of 100–400 mV (0.4, 1.1, 1.7, and 2.3 cmH_2O) are presented (n = 7, except for chronic SIPO, n = 5). (c) Magnitudes of acute (black) and chronic (gray) $\check{R}_L$ values; and (d) percent change in $\check{R}_L$ (compared to ACh) of acute (light gray) and chronic (dark gray) values; healthy changes are indicated (black) for comparison.

are observed when comparing the ACh standard with the groups' data points. Figure 7.33a illustrates the 5-Hz acute $\check{C}_{dyn}$ values. Chronic group $\check{C}_{dy}$ values range from 0.23 ($cmH_2O{\cdot}mL$) for 300-mV treatments up to a high of 0.31 ($cmH_2O{\cdot}mL$) with ISO treatment. The highest $\check{C}_{dyn}$ value for chronic mice was achieved from the 1.6 cmH_2O SIPO group and ISO treatment, which were statistically equivalent as well as significantly different from the ACh challenge. Figure 7.33b illustrates the 5-Hz chronic $\check{C}_{dyn}$ values, indicating significant ($p < 0.05$) differences relevant to ACh with an asterisk (*).

The differences in the magnitudes of acute and chronic data are apparent when viewed in the context of normalized data (Figure 7.33c), indicating that physiological responses to SIPO treatments differ based on the health/disease status of the same mouse strain. This conclusion is further supported by comparing the percent change in $\check{C}_{dyn}$ values for acute and chronic subjects in this 5-Hz SIPO treatment group (Figure 7.33d). There are apparent differences in short- and long-term asthmatic mice (except for the applied SIPO at 1.6 cmH_2O) and their modeled disease state.

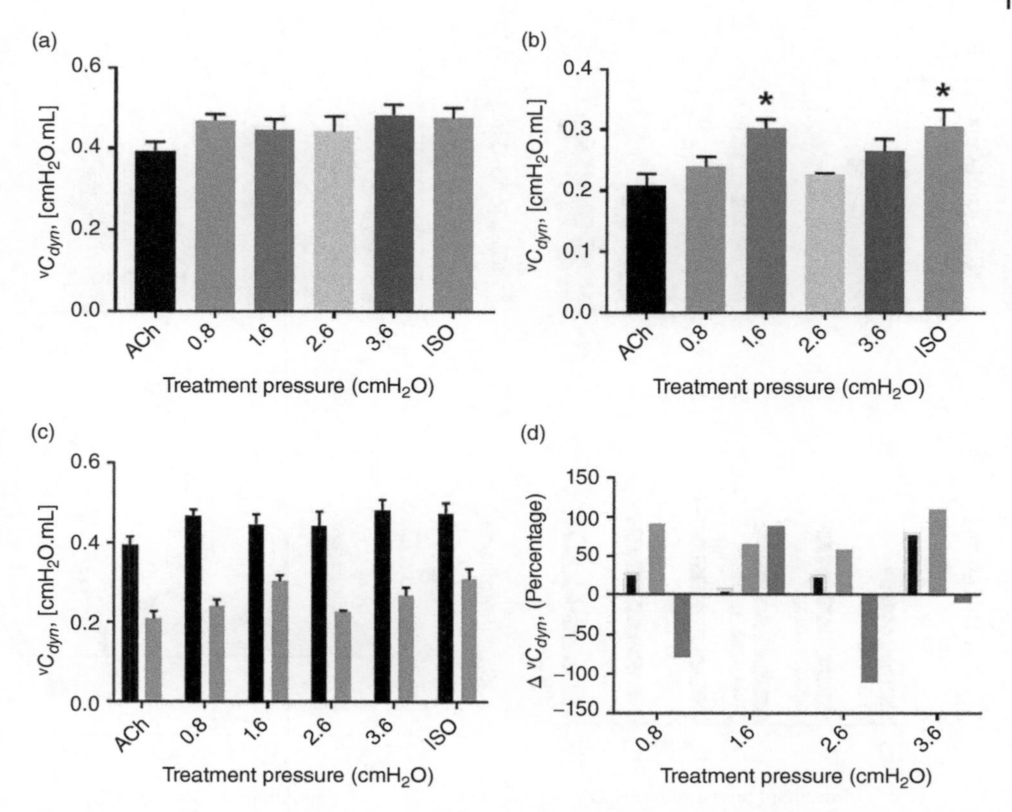

Figure 7.33 The 5-Hz C_{dyn} values to $\check{C}_{dyn}$. Acute (a) and chronic (b) groups with SIPO amplitudes of 100–400 mV (0.8, 1.6, 2.6, and 3.6 cmH_2O) are presented (n = 7, except for chronic SIPO, n = 5). (c) Magnitudes of acute (black) and chronic (gray) $\check{C}_{dyn}$ values; and (d) percent change in $\check{C}_{dyn}$ (compared to ACh) of acute (light gray) and chronic (dark gray) values; healthy changes are indicated (black) for comparison.

The 10-Hz acute $\check{C}_{dyn}$ values range from 0.39 ($cmH_2O{\cdot}mL$) for ACh challenge up to a high of 0.51 ($cmH_2O{\cdot}mL$) with 400-mV treatment. All values other than ACh are statistically comparable, and no significant differences are observed when comparing the ACh standard against the entire set. Figure 7.34a illustrates the 10-Hz acute $\check{C}_{dyn}$ values. Chronic group $\check{C}_{dyn}$ values range from 0.25 ($cmH_2O{\cdot}mL$) for 400-mV treatments up to a high of 0.34 ($cmH_2O{\cdot}mL$) with 100-mV treatments. Figure 7.34b illustrates the 10-Hz chronic $\check{C}_{dyn}$ values, indicating significant ($p < 0.05$) differences relevant to ACh with an asterisk (*). The differences in the magnitudes of acute and chronic 10-Hz groups are apparent when viewed in the context of normalized data (Figure 7.34c). They indicate that physiological responses to SIPO treatments differ based on the health/disease status of the same mouse strain. This conclusion is further supported by comparing the percent change in Čdyn values for acute and chronic mice in this 10-Hz SIPO treatment group (Figure 7.34d). There are clear differences in the change in responses of short- and long-term asthmatic mice, based on all applied SIPO of this group and the two modeled disease states.

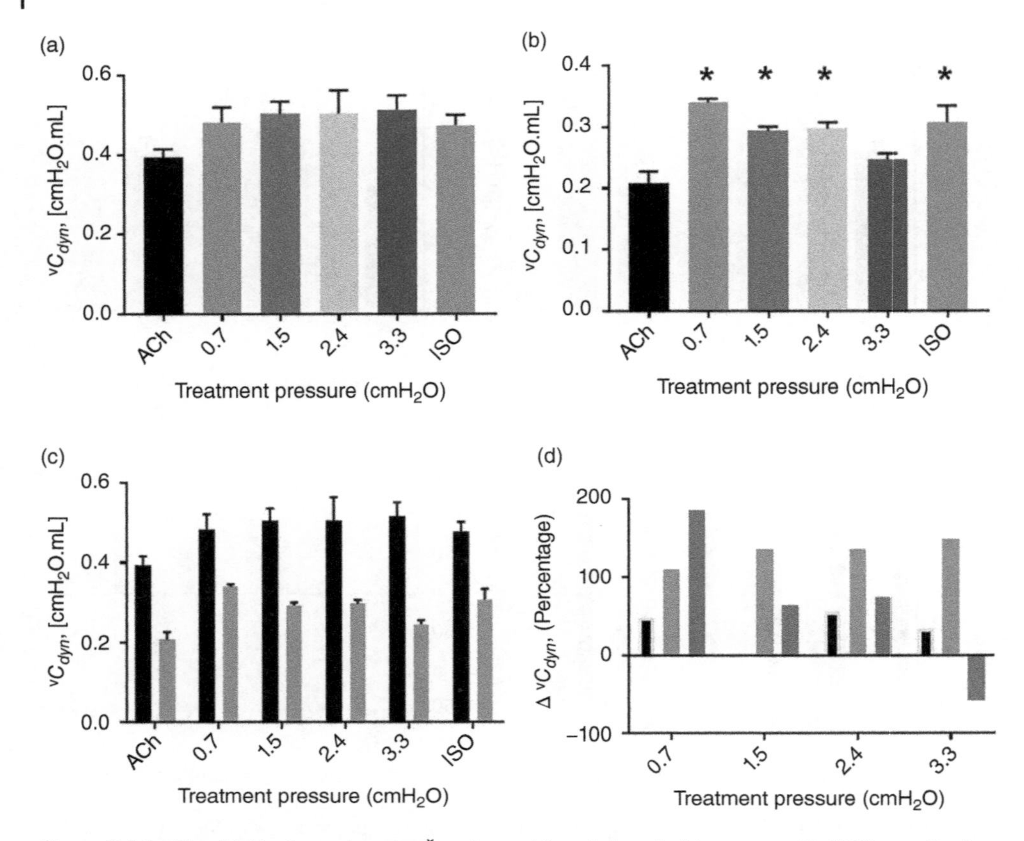

Figure 7.34 The 10-Hz C_{dyn} values to $\check{C}_{dyn}$. Acute (a) and chronic (b) groups with SIPO amplitudes of 100–400 mV (0.7, 1.5, 2.4, and 3.3 cmH_2O) are presented (n = 7, except for chronic SIPO, n = 5). (c) Magnitudes of acute (black) and chronic (gray) $\check{C}_{dyn}$ values; and (d) percent change in $\check{C}_{dyn}$ (compared to ACh) of acute (light gray) and chronic (dark gray) values; healthy changes are indicated (black) for comparison.

The 15-Hz acute $\check{C}_{dyn}$ values range from 0.39 ($cmH_2O{\cdot}mL$) for ACh challenge up to a high of 0.51 ($cmH_2O{\cdot}mL$) with 200-mV treatment. The highest $\check{C}_{dyn}$ value for acute mice was achieved from the 1.1 cmH_2O SIPO group, but it is not significantly different from other SIPO values for the group. No significant differences are observed when comparing the ACh standard with the group's SIPO or ISO data points. Figure 7.35a illustrates the 15-Hz acute $\check{C}_{dyn}$ values, indicating the lack of significant ($p > 0.05$) differences relevant to ACh. The chronic group $\check{C}_{dyn}$ values range from 0.23 ($cmH_2O{\cdot}mL$) for 100-mV treatments up to a high of 0.38 ($cmH_2O{\cdot}mL$) with 300-mV treatment. The highest $\check{C}_{dyn}$ value for chronic mice was achieved from the 1.7 cmH_2O SIPO group. All data for the chronic group except for the 0.4 cmH_2O treatment were significantly higher than ACh ($p < 0.05$). Figure 7.35b illustrates the 15-Hz chronic $\check{C}_{dyn}$ values, indicating significant ($p < 0.05$) differences relevant to ACh with an asterisk (*).

The differences in the magnitudes of acute and chronic data are apparent when viewed in the context of normalized data (Figure 7.35c), indicating that physiological responses to

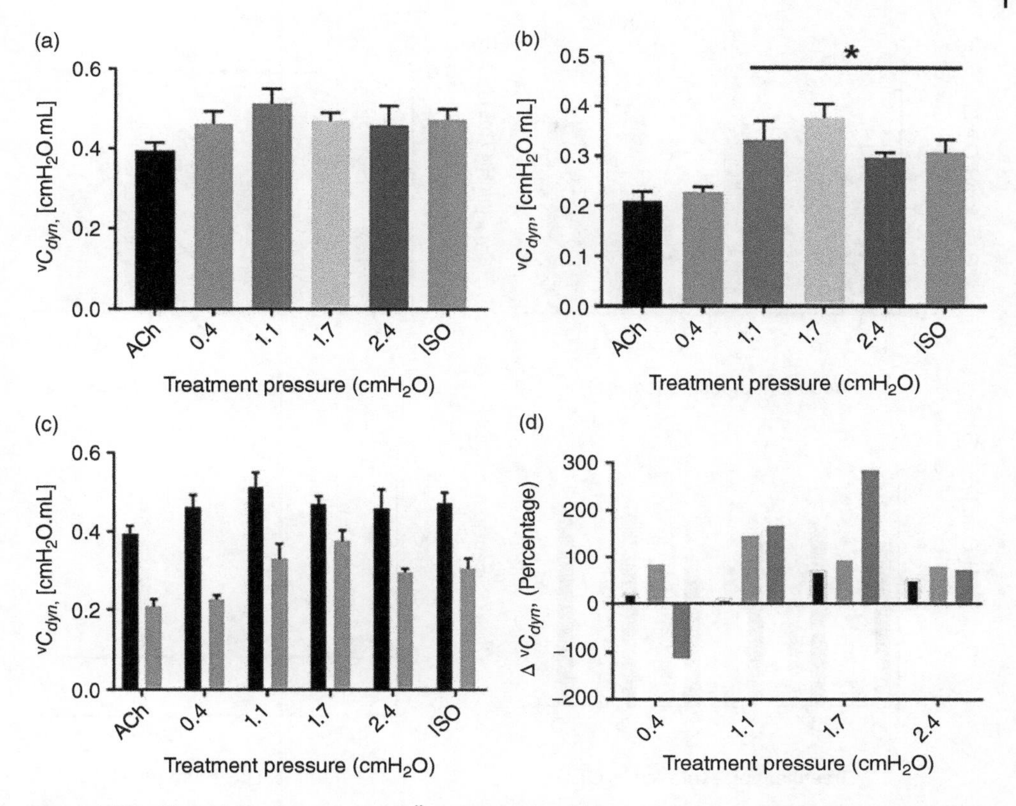

Figure 7.35 The 15-Hz C_{dyn} values to $\check{C}_{dyn}$. Acute (a) and chronic (b) groups with SIPO amplitudes of 100–400 mV (0.4, 1.1, 1.7, and 2.4 cmH$_2$O) are presented (n = 7, except for chronic SIPO, n = 5). (c) Magnitudes of acute (black) and chronic (gray) $\check{C}_{dyn}$ values; and (d) percent change in $\check{C}_{dyn}$ (compared to ACh) of acute (light gray) and chronic (dark gray) values; healthy changes are indicated (black) for comparison.

SIPO treatments differ based on the health/disease status of the same mouse strain. This conclusion is further supported by comparing the percent change in Čdyn values for acute and chronic mice in this 15-Hz SIPO treatment group (Figure 7.35d). There are apparent differences in the change in responses of short- and long-term asthmatic mice, based on the applied SIPOs of 0.4 and 1.7 cmH$_2$O, and their modeled disease state.

The 20-Hz acute $\check{C}_{dyn}$ values range from 0.38 (cmH$_2$O·mL) for 300-mV treatment up to a high of 0.48 (cmH$_2$O·mL) with the 400 mV ISO treatments. The highest $\check{C}_{dyn}$ value for acute mice treated with ISO was also comparable to 2.3 cmH$_2$O treatments. No significant differences are observed when comparing the ACh standard with the group's data points, though notably, the 1.7 cmH$_2$O treatment is equivalent to ACh. Figure 7.36a illustrates the 20-Hz acute $\check{C}_{dyn}$ values, indicating no significant ($p > 0.05$) differences relevant to ACh. Chronic group $\check{C}_{dyn}$ values range from 0.27 (cmH$_2$O·mL) for ACh challenge up to a high of 0.48 (cmH$_2$O·mL) with 300-mV treatment. The highest Čdyn value for chronic mice was achieved from the 1.7 cmH$_2$O SIPO group; this same group in the acute data was the lowest value of the groups' analyses. All data for the chronic group were significantly higher than

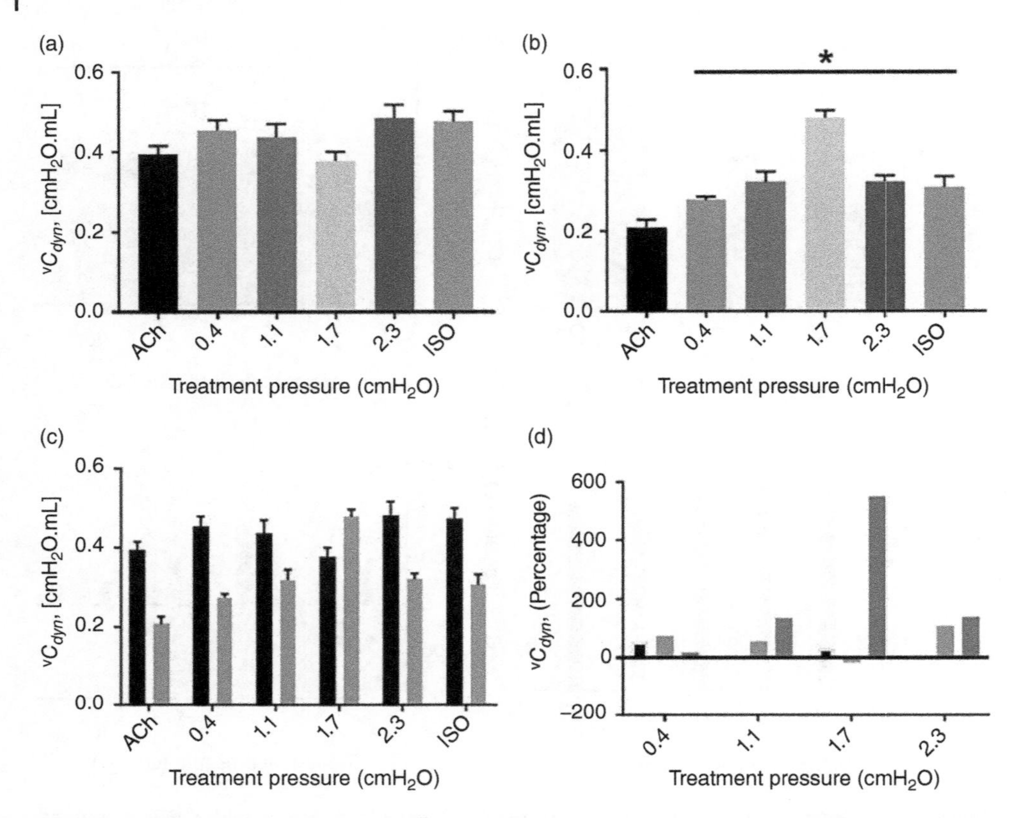

Figure 7.36 The 20-Hz C_{dyn} values to $\check{C}_{dyn}$. Acute (a) and chronic (b) groups with SIPO amplitudes of 100–400 mV (0.4, 1.1, 1.7, and 2.3 cmH_2O) are presented (n = 7, except for chronic SIPO, n = 5). (c) Magnitudes of acute (black) and chronic (gray) $\check{C}_{dyn}$ values; and d) percent change in $\check{C}_{dyn}$ (compared to ACh) of acute (light gray) and chronic (dark gray) values; healthy changes are indicated (black) for comparison.

ACh ($p < 0.05$). Figure 7.36b illustrates the 20-Hz chronic $\check{C}_{dyn}$ values, indicating significant differences relevant to ACh with an asterisk (*). The differences in the magnitudes of acute and chronic data are apparent when viewed in the context of normalized data (Figure 7.36c), indicating that physiological responses to SIPO treatments differ based on the health/disease status of the same mouse strain. This conclusion is further supported by comparing the percent change in $\check{C}_{dyn}$ values for acute and chronic subjects in this 20-Hz SIPO treatment group (Figure 7.36d). There is a clear difference in the change in responses of short- and long-term asthmatic mice, based on the applied SIPO of 1.7 cmH_2O, and their modeled disease state.

7.11.4 Concluding Remarks

This chapter builds on the theory that length oscillations of ASM during contraction have an effect on the interaction of actin and myosin, characterized as a disruption of the actinomycosis cross-bridge. The first part of the chapter focused on tissue testing from healthy subjects. It is concluded that superimposed length oscillation at effective frequency and amplitude induces relaxation on contracted ASMs. Oscillations of pre-constricted sensitized airways are therefore interpreted as being capable of inducing relaxation, in line with the hypothesis that the main components of the cross-bridge (actin, myosin, and the actino-myosin cross-bridge) do not change, yet can be disrupted or perturbed.

Changes in ASM occur as a result of asthma development. Length adaptation, change in the basal tone, and rearrangement of cellular structures are expected to present different tissue responses to a practical range of applied SIPO values. Indeed, comparisons of healthy, short-term, and long-term sensitization model responses indicate that this is the case for establishing a return to breathing patterns. SIPO applications vary in their ability to improve the work of the asthmatic airway based on the state of disease, as seen in the long-term sensitization treatment responses.

While healthy tissues (in vitro and in vivo) are influenced by mechanically applied (physiological value-based) tidal and deep inspiration oscillations, resulting in effective relaxation of the tissues tested, this is not the case for asthmatic tissues. The applied oscillations effects are believed to act on mechanical structures rather than chemical pathways. The mechanical systems involved in contraction, and relevant to this discussion, are the actin and myosin elements of the cross-bridge. Given that the identity of the cross-bridge is maintained regardless of healthy or diseased states, it is therefore plausible that cross-bridge activity changes with the state of the disease. Evidence for this differential tissue response is found in the application of similar oscillations to differing states of healthy and sensitized ASM. Adaptation of ASM over the course of disease progression is a valid characterization of changes to chemical and mechanical ASM identities. The adaptability of the tissues is consistent with the proposed conclusions of this comparison between healthy, acute and chronic asthma models. The observed effects of SIPO in a single strain of mice across different disease states allows for the assumption that the cross-bridge remains the same in terms of components but changes in its functional capabilities to effect force over the distance of its structures.

The results of this chapter are presented with the view that the cross-bridge cycle of ASM is a periodic function determined by the system's state of health/disease. The periodic cycle can be disrupted by specific interference patterns based on combinations of applied frequencies and amplitudes. The basal tone of the tissue is viewed as a physiological result of this periodic function of actinomyosin interactions within the cellular framework. Changes in the rate of cross-bridge cycling are then expected to change the basal tone of the tissue and are expected to influence the degree of AHR to external challenges – ACh in this study, for example. Relaxation of pre-contracted ASM due to applied SIPO can then be viewed as a disruption to the intrinsic disease-state cycling of the actinomyosin cross-bridge. Rates of myosin attachment and detachment are perturbed, in line with Fredberg's proposals. The consistency of the theory is found in the fact that in healthy tissues, the perturbation of the cross-bridge is possible with TO and DI applications (Kudo et al. 2002), while in acute or chronic asthmatic ASM, with a cycling rate expected to be different from the healthy rate, the successful "healthy" interference pattern from similar applications does not occur because the template cycle is no longer present in the cross-bridge of diseased tissues.

This chapter concludes with DRA-sensitized (long-term, chronic) asthmatic airways' responses to SIPO applications and analyzing the results with established short-term asthmatic modeling in our laboratory. Analysis of the collected respiratory parameters' data indicates results consistent with a model of cross-bridge perturbation. Further comparison of the chronic data with our laboratory's acute results also serves to augment the current body of knowledge relative to interference patterns applied to pre-constricted ASM. Based on data normalization, inferences also arise from this study regarding significant differences in power and work performed by short- and long-term asthmatic tissues, depending on the applied SIPO treatments.

Bibliography

Ali, F., Chin, L., Paré, P.D., and Seow, C.Y. (2007). Mechanism of partial adaptation in airway smooth muscle after a step change in length. *Journal of Applied Physiology* 103(2): 569–577.

Al-Jumaily, A. and Alizad, A. (ed.) (2008). *Biomedical Applications of Vibration and Acoustics in Therapy, Bioeffects and Modeling*, 215–244. ASME Press.

Al-Jumaily, A.M. and Du, Y. (2011). Fading memory model for airway smooth muscle dynamic response. *Journal of Theoretical Biology* 283(1): 10–13.

Al-Jumaily, A.M., Mbikou, P., and Redey, P.R. (2012). Effect of length oscillations on airway smooth muscle reactivity and cross-bridge cycling. *American Journal of Physiology—Lung Cellular and Molecular Physiology* 303(4): L286–L294.

Amrani, Y. and Panettieri, R.A. Jr. (2002). Modulation of calcium homeostasis as a mechanism for altering smooth muscle responsiveness in asthma. *Current Opinion in Allergy and Clinical Immunology* 2(1): 39–45.

An, S.S., Bai, T.R., Bates, J.H.T. et al. (2007). Airway smooth muscle dynamics: a common pathway of airway obstruction in asthma. *European Respiratory Journal* 29(5): 834–860.

Anafi, R.C. and Wilson, T.A. (2002). Empirical model for dynamic force–length behavior of airway smooth muscle contraction. *Journal of Applied Physiology* 98: 1356–1365.

Bai, T.R. (1990). Abnormalities in airway smooth muscle in fatal asthma 1–3. *American Review of Respiratory Disease* 141: 552–557.

Bai, T.R. (1991). Abnormalities in airway smooth muscle in fatal asthma: a comparison between trachea and bronchus. *American Review of Respiratory Disease* 143(2): 441–443.

Bai, T.R. (2004). On the terminology for describing the length–force relationship and its changes in airway smooth muscle. *Journal of Applied Physiology* 97: 2029–2034.

Bai, Y. and Sanderson, M.J. (2006). Airway smooth muscle relaxation results from a reduction in the frequency of Ca2+ oscillations induced by a cAMP-mediated inhibition of the IP3 receptor. *Respiratory Research* 7: 34.

Barnes, P.J., Rodger, I.W., and Thomson, N.C. (1992). *Asthma: Basic Mechanisms and Clinical Management*. London Academic Press.

Bateman, E.D., Hurd, S.S., Barnes, P.J. et al. (2008). Global strategy for asthma management and prevention: GINA executive summary. *European Respiratory Journal* 31(1): 143–178.

Bates, J.H. and Lauzon, A.M. (2005). Modelling the oscillation dynamics of activated airway smooth muscle strips. *American Journal of Physiology—Lung Cellular and Molecular Physiology* 289(5): L849–L855.

Bates, J., Bullimore, S., Politi, A. et al. (2009a). Transient oscillatory force–length behavior of activated airway smooth muscle. *American Journal of Physiology—Lung Cellular and Molecular Physiology* 297: L362–L372.

Bates, J.H.T., Rincon, M., and Irvin, C.G. (2009b). Animal models of asthma. *American Journal of Physiology—Lung Cellular and Molecular Physiology* 297: L401–L410.

Benayoun, L., Druilhe, A., Dombret, M.C. et al. (2003). Airway structural alterations selectively associated with severe asthma. *American Journal of Respiratory and Critical Care Medicine* 167(10): 1360–1368.

Benson, M.K. (1975). Bronchial hyperreactivity. *British Journal of Diseases of the Chest* 69(4): 227–239.

Berne, R.M. and Levy, M.N. (2008). *Physiology*, 6ee. Mosby: Maryland Heights, Missouri.

Berridge, M.J. (2009). Inositol trisphosphate and calcium signalling mechanisms. *Biochimica et Biophysica Acta (BBA)-Molecular Cell Research* 1793(6): 933–940.

Berry, R.B. and Fairshter, R.D. (1985). Partial and maximal expiratory flow-volume curves in normal and asthmatic subjects before and after inhalation of metaproterenol. *Chest* 88: 697–702.

Blanc, F.X., Coirault, C., Salmeron, S. et al. (2003). Mechanics and crossbridge kinetics of tracheal smooth muscle in two inbred trains. *European Respiratory Journal* 22: 227–234.

Bossé, Y., Sobieszek, A., Paré, P., and Seow, C. (2008). Length adaptation of airway smooth muscle. *Proceedings of the American Thoracic Society* 5: 62–67.

Bowler, S.D., Green, A., and Mitchell, C.A. (1998). Buteyko breathing techniques in asthma: a blinded randomised controlled trial. *Medical Journal of Australia* 169: 575–578.

Bramley, A.M., Thomson, R.J., Roberts, C.R., and Schellenberg, R.R. (1994). Hypothesis: excessive bronchoconstriction in asthma is due to decreased airway elastance. *European Respiratory Journal* 7(2): 337–341.

Brar, S.S., Kennedy, T.P., Sturrock, A.B. et al. (2002). NADPH oxidase promotes NF-κB activation and proliferation in human airway smooth muscle. *American Journal of Physiology—Lung Cellular and Molecular Physiology* 282(4): L782–L795.

Brown, R.H., Scichilone, N., Mudge, B. et al. (2001). High-resolution computed tomographic evaluation of airway distensibility and the effects of lung inflation on airway caliber in healthy subjects and individuals with asthma. *American Journal of Respiratory and Critical Care Medicine* 163: 994–1001.

Cerrina, J., Le Roy, L.M., Labat, C. et al. (1986). Comparison of human bronchial muscle responses to histamine in vivo with histamine and isoproterenol agonists in vitro. *The American Review of Respiratory Disease* 134(1): 57–61.

Cerrina, J., Labat, C., Haye-Legrande, I. et al. (1989). Human isolated bronchial muscle preparations from asthmatic patients: effects of indomethacin and contractile agonists. *Prostaglandins* 37(4): 457–469.

Chen, C., Kudo, M., Rutaganira, F. et al. (2002). Integrin α9β1 in airway smooth muscle suppresses exaggerated airway narrowing. *The Journal of Clinical Investigation* 122: 2916–2927.

Chin, L.Y., Bossé, Y., Pascoe, C. et al. (2012). Mechanical properties of asthmatic airway smooth muscle. *European Respiratory Journal* 40(1): 45–54.

Clifford, R.D., Radford, M., Howell, J.B., and Holgate, S.T. (1989). Prevalence of respiratory symptoms among 7 and 11 year old schoolchildren and association with asthma. *Archives of Disease in Childhood* 64(8): 1118–1125.

Cox, G., Miller, J.D., McWilliams, A. et al. (2006). Bronchial thermoplasty for asthma. *American Journal of Respiratory and Critical Care Medicine* 173(9): 965–969.

Craig, R. and Megerman, J. (1977). Assembly of smooth muscle myosin into side-polar filaments. *The Journal of Cell Biology* 75(3): 990–996.

Dahlen, S.E. (2006). Treatment of asthma with antileukotrienes: first line or last resort therapy? *European Journal of Pharmacology* 533: 40–56.

Dahlen, S.E., Dahlen, B., and Drazen, J.M. (2011). Asthma treatment guidelines meet the real world. *New England Journal of Medicine* 364(18): 1769–1770.

Donovan, G.M. (2013). Modelling airway smooth muscle passive length adaptation via thick filament length distributions. *Journal of Theoretical Biology* 333: 102–108.

Du, Y., Al-Jumaily, A.M., and Shukla, H. (2007). Smooth muscle stiffness variation due to external longitudinal oscillations. *Journal of Biomechanics* 40: 3207–3214.

Ducharme, F. and di Salvio, F. (2004). Anti-leukotriene agents compared to inhaled corticosteroids in the management of recurrent and/or chronic asthma in adults and children. *Cochrane Database of Systematic Reviews* 1: CD002314.

Dvorkin, M.A., D.P. Cardinali, and Lermoli, R., *Best &Taylor, Bases Fisiológicas de la Práctica Médica*. 14ª ed 2010: Editorial medica panamericana. 1164

Enarson, D.A., Vedal, S., Schulzer, M. et al. (1987). Asthma, asthmalike symptoms, chronic bronchitis, and the degree of bronchial hyperresponsiveness in epidemiologic surveys. *The American Review of Respiratory Disease* 136(3): 613–617.

Fabry, B. and Fredberg, J.J. (2003). Remodeling of the airway smooth muscle cell: are we built of glass? *Respiratory Physiology & Neurobiology* 137(2): 109–124.

Fanta, C.H. and Hockberger, B.S. (2021). Acute exacerbations of asthma in adults: emergency department and inpatient management. *UpToDate*.

Fish, J.E., Ankin, M.G., Kelly, J.F., and Peterman, V.I. (1981). Regulation of bronchomotor tone by lung inflation in asthmatic and nonasthmatic subjects. *Journal of Applied Physiology* 50: 1079–1086.

Forsythe, S.M., Kogut, P.C., McConville, J.F. et al. (2002). Structure and transcription of the human m3 muscarinic receptor gene. *American Journal of Respiratory Cell and Molecular Biology* 26(3): 298–305.

Fredberg, J.J. (2001). Airway obstruction in asthma: does the response to a deep inspiration matter? *Respiratory Research* 2: 273–275.

Fredberg, J.J., Inouye, D., Miller, B. et al. (1997). Airway smooth muscle, tidal stretches, and dynamically determined contractile states. *American Journal of Respiratory and Critical Care Medicine* 156(6): 1752–1759.

Fredberg, J.J., Inouye, D., Mijailovich, S. et al. (1999). Perturbed equilibrium of myosin binding in airway smooth muscle and its implications in bronchospasm. *American Journal of Respiratory and Critical Care Medicine* 159(3): 959–967.

Freedman, B.J. (1972). The functional geometry of the bronchi. The relationship between changes in external diameter and calibre, and a consideration of the passive role played by the mucosa in bronchoconstriction. *Bulletin de Physio-Pathologie Respiratoire* 8(3): 545–552.

Freedman, S., Lane, R., Gillett, M.K., and Guz, A. (1988). Abolition of methacholine induced bronchoconstriction by the hyperventilation of exercise or volition. *Thorax* 43: 631–636.

Ganong, W.F. (2001). *Review of Medical Physiology*. New York: McGraw-Hill.

Glaab, T., Mitzner, W., Braun, A. et al. (2004). Repetitive measurements of pulmonary mechanics to inhaled cholinergic challenge in spontaneously breathing mice. *Journal of Applied Physiology* 97: 1104–1111.

Glabb, T., Taube, C., Braun, A., and Mitzner, W. (2007). Review: invasive and noninvasive methods for studying pulmonary function in mice. *Respiratory Research* 8(1): 1–10.

Global Initiative for Asthma Report (2014). Global strategy for asthma management and prevention.

Grunstein, M.M., Hakonarson, H., Leiter, J. et al. (2002). IL-13-dependent autocrine signaling mediates altered responsiveness of IgE-sensitized airway smooth muscle. *American Journal of Physiology—Lung Cellular and Molecular Physiology* 282(3): L520–L528.

Gunst, S.J. (1983). Contractile force of canine airway smooth muscle during cyclical length changes. *Journal of Applied Physiology* 55: 759–769.

Gunst, S. and Stropp, J. (1988). Pressure-volume and length-stress relationships in canine bronchi in vitro. *Journal of Applied Physiology* 64: 2522–2531.

Gunst, S.J. and Tang, D.D. (2000). The contractile apparatus and mechanical properties of airway smooth muscle. *European Respiratory Journal* 15(3): 600–616.

Gunst, S.J., Stropp, J.Q., and Service, J. (1990). Mechanical modulation of pressurevolume characteristics of contracted canine airways in vitro. *Journal of Applied Physiology* 68: 2223–2229.

Gunst, S.J., Tang, D.D., and Opazo Saez, A. (2003). Cytoskeletal remodeling of the airway smooth muscle cell: a mechanism for adaptation to mechanical forces in the lung. *Respiratory Physiology & Neurobiology* 137(2): 151–168.

Guyton, A.C. and Hall, J.E. (2000). *Textbook of Medical Physiology*. Philadelphia: W. B. Saunders Co.

Hai, C.M. and Kim, H.R. (2005). An expanded latch-bridge model of protein kinase C-mediated smooth muscle contraction. *Journal of Applied Physiology* 98: 1356–1365.

Hai, C.M. and Murphy, R.A. (1998a). Cross-bridge phosphorylation and regulation of latch state in smooth muscle. *American Journal of Physiology—Cell Physiology* 254: C99–C106.

Hai, C.M. and Murphy, R.A. (1998b). Regulation of shortening velocity by cross-bridge phosphorylation in smooth muscle. *American Journal of Physiology—Cell Physiology* 255: L86–L94.

Halayko, A.J. and Solway, J. (2001). Molecular mechanisms of phenotypic plasticity in smooth muscle cells. *Journal of Applied Physiology* 90: 358–368.

Higgins, B.G., Britton, J.R., Chinn, S. et al. (1992). Comparison of bronchial reactivity and peak expiratory flow variability measurements for epidemiologic studies. *Am Rev Respir Dis* 145(3): 588–593.

Hoymaan, H. (2006). New developments in lung function measurements in rodents. *Experimental and Toxicologic Pathology* 57(S2): 5–11.

Hunter, P.J., Hunter, A.D., McCulloch, H.E., and Keurs, D.J. (1998). Modelling the mechanical properties of cardiac muscle. *Progress in Biophysics and Molecular Biology* 1998(69): 289–331.

Huxley, A.F. (1957). Muscle structure and theories of contraction. *Progress in Biophysics and Biophysical Chemistry* 7: 255–318.

Huxley, A.F. (1975). Muscle structure and theories of contraction. *Progress in Biophysics and Biophysical Chemistry* 7: 255–318.

James, A.L., Paré, P.D., and Hogg, J.C. (1989). The mechanics of airway narrowing in asthma. *American Review of Respiratory Disease* 139(1): 242–246.

Jenkins, M.A., Clarke, J.R., Carlin, J.B. et al. (1996). Validation of questionnaire and bronchial hyperresponsiveness against respiratory physician assessment in the diagnosis of asthma. *International Journal of Epidemiology* 25(3): 609–616.

Jo-Avila, M., Al-Jumaily, A.M., and Lu, J. (2014). Relaxant effect of superimposed length oscillation on sensitized airway smooth muscle. *American Journal of Physiology—Lung Cellular and Molecular Physiology* 308(5): L479–L484.

Kamm, R.D. (1999). Airway wall mechanics. *Annual Review of Biomedical Engineering* 01: 47–72.

Karol, M.H. (1994). Animal models of occupational asthma. *European Respiratory Journal* 7(3): 555–568.

Kazani, S., Ware, J.H., Drazen, J.H. et al. (2010). The safety of long-acting beta-agonists: more evidence is needed. *Respirology* 15(6): 881–886.

Knorr, B., Franchi, L.M., Bisgaard, H. et al. (2001). Montelukast, a leukotriene receptor antagonist, for the treatment of persistent asthma in children aged 2 to 5 years. *Pediatrics* 108 (3): e48–e58.

Knox, A.J., Corbett, L., Stocks, J. et al. (2001). Human airway smooth muscle cells secrete vascular endothelial growth factor: up-regulation by bradykinin via a protein kinase C and prostanoid-dependent mechanism. *The FASEB Journal* 15(13): 2480–2488.

Koning, P. (1981). Hidden asthma in childhood. *American Journal of Diseases of Children* 135: 1053–1055.

Kudo, M., Melton, A.C., Chen, C. et al. (2002). IL-17A produced by αβ T cells drives airway hyper-responsiveness in mice and enhances mouse and human airway smooth muscle contraction. *Nature Medicine* 18(4): 547.

Kudo, M., Ishigatsubo, Y., and Aoki, I. (2013). Pathology of asthma. *Frontiers in Microbiology* 4: 263.

Kumar, R.K. and Foster, P.S. (2001). Murine model of chronic human asthma. *Immunology and Cell Biology* 79: 141–144.

Kuo, K.H., Wang, L., Paré, P.D. et al. (2001). Myosin thick filament lability induced by mechanical strain in airway smooth muscle. *Journal of Applied Physiology* 90(5): 1811–1816.

Kuo, K.H., Herrera, A.M., and Seow, C.Y. (2003). Ultrastructure of airway smooth muscle. *Respiratory Physiology & Neurobiology* 137(2): 197–208.

Lakser, O.J., Lindeman, R.P., and Fredberg, J.J. (2002). Inhibition of the p38 MAP kinase pathway destabilizes smooth muscle length during physiological loading. *American Journal of Physiology. Lung Cellular and Molecular Physiology* 282: L1117–L1121.

Lalloo, U., Ainslie, G., Wong, M. et al. (2007). Guidelines for the management of chronic asthma in adolescents and adults. *S Afr Fam Pract.* 49(5): 19–31.

Lambert, L. (2013). Montelukast in the treatment of asthma. *SA Pharmaceutical Journal* 81(1): 22–24.

Lambert, R.K., Wiggs, B.R., Kuwano, K. et al. (1993). Functional significance of increased airway smooth muscle in asthma and COPD. *Journal of Applied Physiology* 74: 2771–2771.

Lambert, R.K., Pare, P.D., and Seow, C.Y. (2004). Mathematical description of geometric and kinematic aspects of smooth muscle plasticity and some related morphometrics. *Journal of Applied Physiology* 96(2): 469–476.

Lazaar, A.L. and Panettieri, R.A. Jr. (2001). Airway smooth muscle as an immunomodulatory cell: a new target for pharmacotherapy? *Current Opinion in Pharmacology* 1(3): 259–264.

Lecarpentier, Y., Blanc, F.X., Salmeron, S. et al. (2002). Myosin cross-bridge kinetics in airway smooth muscle: a comparative study of humans, rats, and rabbits. *American Journal of Physiology—Lung Cellular and Molecular Physiology* 282: L83–L90.

Lee, D.A., Winslow, N.R., Speight, A.N., and Hey, E.N. (1983). Prevalence and spectrum of asthma in childhood. *British Medical Journal (Clinical Research Ed.)* 286(6373): 1256.

Léguillette, R., Laviolette, M., Bergeron, C. et al. (2009). Myosin, transgelin, and myosin light chain kinase: expression and function in asthma. *American Journal of Respiratory and Critical Care Medicine* 179(3): 194–204.

Lim, T.K., Pride, N.B., and Ingram, R.H. Jr. (1987). Effects of volume history during spontaneous and acutely induced air-flow obstruction in asthma. *The American Review of Respiratory Disease* 135: 591–596.

Lim, T.K., Ang, S.M., Rossing, T.H. et al. (1989). The effects of deep inhalation on maximal expiratory flow during intensive treatment of spontaneous asthmatic episodes. *The American Review of Respiratory Disease* 140: 340–343.

Ma, X., Cheng, Z., Wang, Y. et al. (2002). Changes in biophysical and biochemical properties of single bronchial smooth muscle cells from asthmatic subjects. *American Journal of Physiology—Lung Cellular and Molecular Physiology* 283(6): L1181–L1189.

Manocha, R., Marks, G.B., Kenchington, P. et al. (2002). Sahaja yoga in the management of moderate to severe asthma: a randomised controlled trial. *Thorax* 57(2): 110–115.

Marieb, E. N. (2005). *Human Anatomy and Physiology Laboratory Manual*, cat version. 3(2), 82–86. Benjamin Cummings: Pearson.

Marieb, E.N., Mitchell, S.J., and Zao, P.Z. (2008). *Human Anatomy and Physiology Laboratory Manual*. Benjamin Cummings: Pearson.

Meiss, R.A. (1982). Transient responses and continuous behavior of active smooth muscle during controlled stretches. *American Journal of Physiology—Lung Cellular and Molecular Physiology* 242: L146–L158.

Mijailovich, S.M., Butler, J.P., and Fredberg, J.J. (2000). Perturbed equilibria of myosin binding in airway smooth muscle: bond-length distributions, mechanics, and ATP metabolism. *Biophysical Journal* 79: 2667–2681.

Montella, S., Maglione, M., De Stefano, S. et al. (2012). Update on leukotriene receptor antagonists in preschool children wheezing disorders. *Italian Journal of Pediatrics* 38(1): 29–46.

Moreno, R.H., Hogg, J.C., and Pare, P.D. (1986). Mechanics of airway narrowing. *American Review of Respiratory Disease* 133(6): 1171–1180.

National Institutes of Health. *Guidelines for the Diagnosis and Management of Asthma*. 2007.

New Zealand Guidelines Group (2002). *Diagnosis and Treatment of Adult Asthma*. Wellington: New Zealand Guidelines Group.

Oliver, M.N., Fabry, B., Marinkovic, A. et al. (2007). Airway hyperresponsiveness, remodeling, and smooth muscle mass: right answer, wrong reason? *American Journal of Respiratory Cell and Molecular Biology* 37(3): 264–272.

Panettieri, R.A. Jr. (1998). Cellular and molecular mechanisms regulating airway smooth muscle proliferation and cell adhesion molecule expression. *American Journal of Respiratory and Critical Care Medicine* 158: S133–S140.

Pathology, C. F. G. (2013). Respiratory system. http://ctrgenpath.net/static/atlas/mousehistology/Windows/respiratory/diagrams.html.

Pattemore, P.K., Asher, M.I., Harrison, A.C. et al. (1990). The interrelationship among bronchial hyperresponsiveness, the diagnosis of asthma, and asthma symptoms. *American Review of Respiratory Disease* 142(3): 549–554.

Peat, J.K., Salome, C.M., Toelle, B.G. et al. (1992). Reliability of a respiratory history questionnaire and effect of mode of administration on classification of asthma in children. *Chest Journal* 102(1): 153–157.

Pedersen, S., Agertoft, L., Williams-Herman, D. et al. (2007). Placebo-controlled study of montelukast and budesonide on short-term growth in prepubertal asthmatic children. *Pediatric Pulmonology* 42: 838–843.

Plopper, C.G. and Hyde, D.M. (2008). The non-human primate as a model for studying COPD and asthma. *Pulmonary Pharmacology & Therapeutics* 21: 755–766.

Rachelefsky, G. (2009). Inhaled corticosteroids and asthma control in children: assessing impairment and risk. *Pediatrics* 123(1): 353–366.

Rizzo, C.A., Yang, R., Greenfeder, S. et al. (2002). The IL-5 receptor on human bronchus selectively primes for hyperresponsiveness. *Journal of Allergy and Clinical Immunology* 109(3): 404–409.

Salome, C.M., Peat, J.K., Britton, W.J., and Woolcock, A.J. (1987). Bronchial hyperresponsiveness in two populations of Australian schoolchildren. I. Relation to respiratory symptoms and diagnosed asthma. *Clinical & Experimental Allergy* 17(4): 271–281.

Salpeter, S.R., Buckley, N.S., Ormiston, T.M., and Salpeter, E.E. (2006). Meta-analysis: effect of long-acting β-agonists on severe asthma exacerbations and asthma-related deaths. *Annals of Internal Medicine* 144(12): 904–912.

Schild, H.O., Hawkins, D.F., Mongar, J.L., and Herxheimer, H. (1951). Reactions of isolated human asthmatic lung and bronchial tissue to a specific antigen histamine release and muscular contraction. *The Lancet* 258(6679): 376–382.

Sears, M.R., Jones, D.T., Holdaway, M.D. et al. (1986). Prevalence of bronchial reactivity to inhaled methacholine in New Zealand children. *Thorax* 41(4): 283–289.

Seow, C.Y. (2000). Response of arterial smooth muscle to length perturbation. *Journal of Applied Physiology* 89: 2065–2070.

Seow, C.Y. (2005). Myosin filament assembly in an ever-changing myofilament lattice of smooth muscle. *American Journal of Physiology-Cell Physiology* 289(6): C1363–C1368.

Seow, C.Y. and Fredberg, J.J. (2001). Historical perspective on airway smooth muscle: the saga of a frustrated cell. *Journal of Applied Physiology* 91(2): 938–952.

Shen, X., Gunst, S.J., and Tepper, R.S. (1997a). Effect of tidal volume and frequency on airway responsiveness in mechanically ventilated rabbits. *Journal of Applied Physiology* 83: 1202–1208.

Shen, X., Wu, M.F., Tepper, R.S., and Gunst, S.J. (1997b). Mechanisms for the mechanical response of airway smooth muscle to length oscillation. *Journal of Applied Physiology* 83(3): 731–738.

Sieck, G.C. and Prakash, Y.S. (1997). Cross-bridge kinetics in respiratory muscles. *European Respiratory Journal* 10: 2147–2158.

Siersted, H.C., Mostgaard, G., Hyldebrandt, N. et al. (1996). Interrelationships between diagnosed asthma, asthma-like symptoms, and abnormal airway behaviour in adolescence: the Odense Schoolchild Study. *Thorax* 51(5): 503–509.

Silveira, P.S., Butler, J.P., and Fredberg, J.J. (2005). Length adaptation of airway smooth muscle: a stochastic model of cytoskeletal dynamics. *Journal of Applied Physiology* 99(6): 2087–2098.

Simon, A.R., Takahashi, S., Severgnini, M. et al. (2002). Role of the JAK-STAT pathway in PDGF-stimulated proliferation of human airway smooth muscle cells. *American Journal of Physiology—Lung Cellular and Molecular Physiology* 282(6): L1296–L1304.

Singh, V., Wisniewski, A., Britton, J., and Tattersfield, A. (1990). Effect of yoga breathing exercises (pranayama) on airway reactivity in subjects with asthma. *The Lancet* 335(8702): 1381–1383.

Skloot, G., Permutt, S., and Togias, A. (1995). Airway hyperresponsiveness in asthma: a problem of limited smooth muscle relaxation with inspiration. *The Journal of Clinical Investigation* 96: 2393–2403.

Smolensky, A.V., Ragozzino, J., Gilbert, S.H. et al. (2005). Length-dependent filament formation assessed from birefringence increases during activation of porcine tracheal muscle. *The Journal of Physiology* 563(2): 517–527.

Stephens, N.L., Li, W., Jiang, H. et al. (2003). The biophysics of asthmatic airway smooth muscle. *Respiratory Physiology & Neurobiology* 137: 125–140.

Sward-Comunelli, S.L., Mabry, S.M., Truog, W.E., and Thibeault, D.W. (1997). Airway muscle in preterm infants: changes during development. *The Journal of Pediatrics* 130(4): 570–576.

Takahashi, K., Yoshimoto, R., Fuchibe, K. et al. (2000). Regulation of shortening velocity by calponin in intact contracting smooth muscles. *Biochemical and Biophysical Research Communications* 279: 150–157.

Ulrich, K., Hincks, J.S., Walsh, R. et al. (2008). Anti-inflammatory modulation of chronic airway inflammation in the murine house dust mite model. *Pulmonary Pharmacology & Therapeutics* 21(4): 637–647.

Venables, K.M., Burge, P.S., Davison, A.G., and Newman Taylor, A.J. (1984). Peak flow rate records in surveys: reproducibility of observers' reports. *Thorax* 39: 828–832.

Vibert, P., Craig, R., and Lehman, W. (1993). Three-dimensional reconstruction of caldesmon-containing smooth muscle thin filaments. *The Journal of Cell Biology* 123(2): 313–321.

Wang, L. and Paré, P.D. (2003). Deep inspiration and airway smooth muscle adaptation to length change. *Respiratory Physiology & Neurobiology* 137(2): 169–178.

Wang, L., Pare, P.D., and Seow, C.Y. (2000). Effects of length oscillation on thesubsequent force development in swine tracheal smooth muscle. *Journal of Applied Physiology* 88: 2246–2250.

Wang, L., Pare, P.D., and Seow, C.Y. (2001). Selected contribution: effect of chronic passive length change on airway smooth muscle length-tension relationship. *Journal of Applied Physiology* 90: 734–740.

Wang, I., Politi, A.Z., Tania, N. et al. (2008). A mathematical model of airway and pulmonary arteriole smooth muscle. *Biophysical Journal* 94(6): 2053–2064.

Welker, L., Jörres, R.A., Costabel, U. et al. (2004). Predictive value of BAL cell differentials in the diagnosis of interstitial lung diseases. *European Respiratory Journal* 24: 1000–1006.

Whicker, S.D., Armour, C.L., and Black, J.L. (1988). Responsiveness of bronchial smooth muscle from asthmatic patients to relaxant and contractile agonists. *Pulmonary Pharmacology* 1(1): 25–31.

Whorwell, P.J., Lupton, E.W., Erduran, D., and Wilson, K. (1986). Bladder smooth muscle dysfunction in patients with irritable bowel syndrome. *Gut* 27(9): 1014–1017.

Wiggs, B.R., Bosken, C., Pare, P.D. et al. (1992). A model of airway narrowing in asthma and in chronic obstructive pulmonary disease. *American Review of Respiratory Disease* 145(6): 1251–1258.

Wolthers, O.D. (2009). Anti-inflammatory treatment of asthma: differentiation and trial-and-error. *Acta Paediatrica* 98: 1237–1241.

Wong, J.Z., Woodcock-Mitchell, J., Mitchell, J. et al. (1998). Smooth muscle actin and myosin expression in cultured airway smooth muscle cells. *American Journal of Physiology—Lung Cellular and Molecular Physiology* 274(5): L786–L792.

Woolcock, A.J., Salome, C.M., and Yan, K. (1984). The shape of the dose-response curve to histamine in asthmatic and normal subjects. *American Review of Respiratory Disease* 130(1): 71–75.

Xu, J.Q., Harder, B.A., Uman, P., and Craig, R. (1996). Myosin filament structure in vertebrate smooth muscle. *The Journal of Cell Biology* 134(1): 53–66.

Yim, S., Fredberg, J.J., and Malhotra, A. (2007). Continuous positive airway pressure for asthma: not a big stretch? *European Respiratory Journal* 29(2): 226–228.

Yu, S.N., Crago, P.E., and Chiel, H.J. (1997). A nonisometric kinetic model for smooth muscle. *American Journal of Physiology—Cell Physiology* 272: C1025–C1039.

Zahalak, G.I. (1981). A distribution-moment approximation for kinetic theories of muscle contraction. *Mathematical Biosciences* 55: 89–114.

Zosky, G.R. and Sly, P.D. (2007). Animal models of asthma. *Clinical and Experimental Allergy* 37: 973–988.

8

Pressure Oscillations in Neonatal Respiratory Diseases Treatment

8.1 Introduction

This chapter briefly introduces the common neonatal respiratory diseases, the traditional surfactant therapies, and the respiratory support devices currently used in practice. Various pressure oscillation techniques, including high-frequency ventilation (HFV), continuous positive airway pressure (CPAP), and "noisy" ventilation as effective and cheaper methods for respiratory support, are discussed. Clinical trials that describe the effectiveness of using such treatments are presented. The concept of stochastic resonance and its application to "noisy" ventilation is introduced. The potential advances in using pressure oscillations and "noisy" ventilation to treat both neonatal and adult diseases are presented.

8.2 Neonatal Respiratory Diseases

A neonate has a high larynx enabling the epiglottis to guide the larynx up behind the soft palate to produce a direct airway from the nasal cavity to the lungs. Neonates are obligated nose breathers who can almost simultaneously breathe and swallow until two to three months (Laitman and Crelin 1980). Compared to adults, neonates have less compliant lungs and more compliant chest walls (Kattan 1979; Gerhardt and Bancalari 1980). This difference leads to an increase in both airway resistance and obstruction. Airway obstruction can be caused due to the narrow airways of the infants and the absent cough reflex (Dunn and Lewis 1973). Neonates are more vulnerable to respiratory distress because of the structural and functional differences between children and adults (Crane 1981). Neonatal respiratory failure is not a single disease entity. The five most common disorders will be reviewed to emphasize the need for this disease-specific approach to treatment and understand how some of these diseases are tuned into physical therapy such as pressure oscillation rather than pharmacological treatment. Emphasis will focus on pressure oscillation implementation in the form of treatment and rehabilitation methods of various lung diseases and in particular, respiratory distress syndrome (RDS). Table 8.1 summarizes some of the abbreviations used in the remainder of the chapter.

8.2.1 Bronchopulmonary Dysplasia

Bronchopulmonary dysplasia (BPD) is the aftermath of underlying lung disease, and its treatment can lead to pulmonary fibrosis and reactive airway diseases (Northway et al.

Pressure Oscillation in Biomedical Diagnostics and Therapy, First Edition. Ahmed Al-Jumaily and Lulu Wang.

Table 8.1 Table of abbreviations.

BPD	**Bronchopulmonary Dysplasia**
CPAP	Continuous Positive Airway Pressure
CV	Conventional Ventilation
EIT	Electrical Impedance Tomography
HFJV	High-Frequency Jet Ventilation
HFOV	High-Frequency Oscillatory Ventilation
HFPPV	High-Frequency Positive Pressure Ventilation
HFV	High-Frequency Ventilation
I/E	Ratio of Inspiratory to Expiratory Time
IMV	Intermittent Mandatory Ventilation
MAP	Mean Airway Pressure
MV	Mechanical Ventilation
nCPAP	Nasal Continuous Positive Airway Pressure
OI	Oxygenation Index
$PaCO_2$	Pressures of Arterial Carbon Dioxide
PaO_2	Pressures of Arterial Oxygen
PCV	Pressure-Controlled Ventilation
PEEP	Positive End-Expiratory Pressure
PIP	Peak Inspiratory Pressure
RDS	Respiratory Distress Syndrome
RIP	Respiratory Impedance Plethysmography
SIMV	Synchronized Intermittent Mandatory Ventilation
VCV	Volume-Controlled Ventilation
VILI	Ventilator-Induced Lung Injury
WOB	Work Of Breath
ΔP	Oscillatory Pressure Amplitude
f	Frequency
FiO_2	Fraction Of Inspired Oxygen
P_{aw}	Airway Resistance
P_{mean}	Mean Airway Resistance
V_D	Dead Space Volume
V_{ts}	Tidal Volumes

1967). The pathogenesis of BPD is multifactorial. It includes the effects of positive pressure ventilation, over distention of the lungs by large tidal volume (V_{ts}) ventilation, the effects of repetitive opening and closing of lung units, and the effects of oxidant stress and inflammation. The ventilator-induced lung injury (VILI) has been used to describe these phenomena (Jobe and Ikegami 1998; Nelson et al. 1998; Attar and Donn 2002).

Several definitions of BPD including oxygen dependence at 28 days or 36 weeks postconceptional age have been proposed. Unfortunately, many infants with reasonably normal pulmonary function require oxygen for the treatment of episodic apnea and/or desaturation, making interpretation of the literature difficult. BPD treatments include the use of mechanical ventilation, augmented by bronchodilators, diuretics, and corticosteroids. To assess these methods, few clinical trials have been conducted. However, the advent of pulmonary mechanics testing may enable objectification of results in individual patients (Donn and Sinha 2003).

8.2.2 Pneumonia

Numerous microorganisms can cause congenital pneumonia with Group B β-hemolytic streptococci and Escherichia coli which are the most common cultured infectious agents used among infants with early-onset disease (first three to five days of life). Several other agents, such as chlamydia trachomatis and Ureaplasma Urealyticum, are still under extensive investigation. Infectious pneumonia leads to widespread inflammatory changes in the lung, including consolidation, edema, and both proteinaceous and hemorrhagic exudates, which can exacerbate difficulties with gas exchange.

Pneumonia can be complicated by systemic manifestations of sepsis, including hypotension and acidosis, causing diminished pulmonary blood flow and secondary persistent pulmonary hypertension of the newborn (PPHN), and disseminated intravascular coagulopathy resulting in pulmonary hemorrhage (Nelson et al. 1998; Faix 2000). Treatment of the underlying infection and its systemic complications augments the respiratory management. Unfortunately, some of these complications can alter the ventilator approach. For example, the impaired cardiac output that accompanies the sepsis syndrome may be a relative contraindication to high-frequency oscillatory ventilation (HFOV), a form of pressure oscillation, which itself may decrease cardiac output.

8.2.3 Persistent Pulmonary Hypertension of the Newborn

After birth, approximately 10% of infants have respiratory failure, which is PPHN (Steinhorn 2010). PPHN is associated with several symptoms such as asphyxia, tachypnea, low Apgar (a measure of a baby's condition after birth) scores, meconium staining, systemic hypotension, shock, poor cardiac function, and perfusion (Nelson et al. 1998). Normally, echocardiography is used to diagnose PPHN though it may be suspected from the evidence of ductal shunting, extreme lability, or a response to hyperoxia-hyperventilation. Severe parenchymal lung disease and cyanotic congenital heart disease with fixed right-to-left shunting have been diagnosed (Walsh and Stork 2001).

The treatment strategy for PPHN is to maintain good systemic blood pressure, decrease pulmonary vascular resistance, ensure oxygen release to tissues, and reduce lesions induced by high levels of inspired oxygen and ventilator high-pressure settings. General management principles include but are not limited to continuous monitoring of oxygenation, blood pressure, and perfusion; maintenance of normal body temperature; glucose abnormalities and metabolic acidosis; nutritional support; minimal stimulation/handling of the newborn and minimal use of invasive procedures.

8.2.4 Meconium Aspiration Syndrome

Meconium aspiration syndrome (MAS) occurs when meconium is presented in the amniotic fluid. It often occurs when the baby experiences stress that result from the reduction of oxygen available to the fetus. MAS can occur due to many reasons, including past due date pregnancy, difficult or long labor, and mother with health issues such as high blood pressure and diabetes. MAS may cause gas trapping and thoracic air leaks. Severe MAS can cause secondary PPHN with attendant intrapulmonary shunting, profound hypoxemia, and diminished pulmonary blood flow.

MAS can be diagnosed based on the newborn's symptoms and the presence of meconium in the amniotic fluid. Additional tests such as blood tests and chest X-rays are required to confirm the diagnosis (Wiswell and Fuloria 1999; Wiswell 2000). Several MAS treatments are available such as antibiotics, use of a ventilator to assist infant breathe, extracorporeal membrane oxygenation (ECMO) if the baby has persistent high blood pressure, oxygen therapy, and radiant warmer to keep normal body temperature (Wiswell 2001).

8.2.5 Respiratory Distress Syndrome

The most common neonatal respiratory illness is RDS. Its incidence increases with decreasing gestational age and is characterized by the lung's structural and biochemical immaturity. Very premature infants will have deficient alveolarization and therefore have diminished surface area for gas exchange and an increased distance from the alveolus to its adjacent capillary, which makes diffusion more difficult. Surfactant deficiency results in high alveolar surface tension and progressive atelectasis. Increased capillary permeability causes deposition of exudative debris in the air spaces and further inactivation of surfactant, which give rise to the original description of RDS as hyaline membrane disease. Radiography is a common tool for RDS detection. Typical findings include the reticulogranular, "ground glass" appearance, with air bronchograms and diminished lung volumes. Affected infants display the usual cardinal signs of respiratory distress: tachypnea, grunting, nasal flaring, and retracting (Nelson et al. 1998). Although surfactant replacement therapy has significantly altered RDS treatment, mortality has not been eradicated, and complications such as air leaks and BPD continue to occur with unacceptable frequency. Management styles differ widely, and there is a very limited evidence base on which to draw conclusions. In the following sections, we will elaborate on the therapeutical approaches to treat this disease, focusing on methods that implement pressure oscillation.

8.2.6 Neonatal RDS Treatments

In early years before the 1970s, CPAP and time-cycled, pressure-limited intermittent mandatory ventilation (IMV) were normally used to treat neonatal respiratory failure. There was no apparent reason why people get the neonatal pulmonary disease. Equipment was rudimentary, with monitoring limited to clinical assessment and intermittent radiography and blood gas assessment. Little thought was given to disease-specific strategies, and most newborns were treated similarly irrespective of the underlying disease.

Better appreciation of neonatal lung disease was achieved in the late 1970s by using newly developed monitoring devices such as transcutaneous oxygen sensors (Lunkenheimer et al.

1973; Bohn et al. 1980). The unique interdependence of the newborn's heart and lungs and the role of the ductus arteriosus in several disease states have been better described by echocardiography.

Proliferation of new techniques, widespread clinical trials, and the emergence of specific strategies were marked in the 1980s by aiming at the differences in pulmonary pathophysiology that characterize the array of neonatal lung disorders. HFV as a form of pressure oscillation was established by the late 1980s, and ECMO has been accepted as the ultimate rescue therapy for the term and near-term infants.

New technologies emerged by the closure of the twentieth century. Surfactant replacement therapy and antenatal corticosteroid treatment became accepted practices. Surfactant treatments improved oxygenation and lung mechanics as well as lowered the mean airway pressures (MAPs) and the percentage concentration of oxygen in the gas (fraction of inspired oxygen, FiO_2) delivered by ventilators to premature infants with RDS. A surfactant can be administered either prophylactically (within 10 minutes of birth), as an early treatment (within 2 hours of birth) or as a rescue treatment (when RDS has been established). Later treatment is less effective than early treatment until RDS has been established (Wung et al. 1979; Poets and Sens 1996; Nelson et al. 1998). Real-time pulmonary graphic monitoring achieved widespread clinical applicability. However, these methods require invasively adding foreign materials to the lung, which may have side effects.

Many new ventilation techniques were introduced, including ventilators operated based on different principles such as patient-triggered, volume-targeted and pressure support ventilation (PSV). The efficacy of inhaled nitric oxide as a selective pulmonary vasodilator was demonstrated. Extensive basic investigation into the mechanisms of lung injury added the concepts of oxidant stress, inflammatory mediators, and infectious agents to the previous associations of "barotrauma," "volutrauma," and "atelectrauma" with chronic lung disease. However, there is no consensus as to how best to manage neonatal respiratory failure.

8.2.7 Role of Pressure Oscillations

Bubble CPAP produces CPAP with superimposed low-pressure oscillations produced by a bubbler (more details in the following sections) that can be felt with hands on the chest. It has been postulated that this pressure oscillation facilitates gas transport and vibrates the lung and chest wall. The clinical research performed on the bubble CPAP system highlights the potential benefits of the pressure oscillations superimposed on the mean CPAP pressure (Avery et al. 1987; Gittermann et al. 1997; Lee et al. 1998; Van Marter et al. 2000; De Klerk and De Klerk 2001a; Jobe et al. 2002; Arold et al. 2003; Pillow et al. 2007).

Published results indicate that the geometry and mechanical properties of the lungs are important factors in the mechanical response of the lung (Fredberg and Hoeing 1978; Fredberg et al. 1978). Pressure oscillations introduce vibrations that can improve the elastic condition of the lung walls (Fredberg and Hoeing 1978; Fredberg et al. 1978), and the behavior of the lung is susceptible to the frequency of those pressure oscillations (Grimal et al. 2002; Manilal 2004; Manilal-Reddy 2004). However, the effect of pressure oscillations on respiratory parameters is not fully understood. Investigation of the mechanical response of the lung walls to high frequencies is a field of undergoing research. The remainder of this chapter will focus on implementing pressure oscillations in various techniques to treat RDS.

8.3 High-Frequency Ventilation

HFV is a process of implementing the pressure oscillation principle by supplying the lung with air at small amplitudes (small tidal volumes, V_{ts}) and certain frequencies (rapid respiratory rates, more than four times the natural breathing frequency). A continuous distending pressure is normally applied to maintain lung volume, and small tidal volumes are superimposed at a rapid rate. All high-frequency ventilators can deliver extremely rapid rates (300–1500 breaths/min, 5–25 Hz, or cycles per second). Patients with severe pulmonary failures have benefited from HFV because it uses small V_{ts}, maintains the lungs/alveoli open on the deflation limb of the pressure–volume curve at a relatively constant airway pressure, and improves ventilation/perfusion (V/Q) matching (Fort et al. 1997). Some explanations have been given to the attribution of HFV, such as it produces vibratory energy, which may facilitate gas exchange and help to mobilize pulmonary secretions; nevertheless, gas transport still cannot be explained by classic concepts of ventilation and lung mechanics.

8.3.1 Mechanics of High-Frequency Oscillation

Several variables need to be adjusted before HFV is applied. These are the frequency and amplitude of oscillation, MAP, bias flow, and inspiratory time. These must be adjusted to suit various patients' conditions and comfort.

The **frequency** affects the pressure cycles applied to the lung. A smaller percentage of the circuit change in pressure is transmitted at higher frequencies. Whereas normal lungs can be ventilated over a wide pressure range without inducing injury, the lung with poor compliance has a very limited zone of safety (Venegas and Fredberg 1994). There is no simple formula for estimating the ideal frequency for an individual patient; however, clinical judgment combined with arterial blood gas measurements are the two main factors considered in setting frequency value. A frequency of less than 3 Hz and not higher than 7 Hz in adults is not normally recommended to use due to the depth of oscillation increase and may increase the risk of barotrauma (Stawicki et al. 2009).

The **amplitude** is the maximum amount of pressure oscillation applied above a mean value. The way it is applied depends on the type of equipment used to generate the oscillation. For instance, it may represent the polarity voltage applied to an electrical coil, which determines the distance that a piston is driven toward/away from the patient's airway. The extent to which the amplitude increases depend on the resistance the piston encounters to forward movement. Generally, the starting amplitude should be 70–90 cmH_2O (Stawicki et al. 2009).

The **mean airway pressure (MAP)** is selected based on the lung resistance. A starting MAP value of 5 cmH_2O above the last plateau pressure noted during conventional ventilation (CV) with a maximal starting P_{aw} of 35 cmH_2O is recommended by many researchers (Fort et al. 1997).

Bias flow is the rate at which gas flows through the ventilator circuit. Normally, 40 L/min is the accepted starting bias flow rate.

The **inspiratory time** can be adjusted in conjunction with the amplitude, MAP, bias flow, and frequency control. Inspiratory time of 33% is optimal because it results in a drop in the mean intrapulmonary pressure because of higher flow-dependent ET tube resistance during inspiration (Pillow et al. 1999).

Frequency selection directly controls the speed of oscillation and the pressure cycles applied to the lung. The changes in frequency are inversely proportional to the amplitude and thus the delivered tidal volume. Alveoli with short time constants for airflow (high lung compliance and low airway resistance) are ventilated more effectively at higher frequencies than those that have longer time constants (low lung compliance or high airway resistance).

8.3.2 Modalities of High-Frequency Ventilation

There are several modes of operation of HFV, including but are not limited to high-frequency positive pressure ventilation (HFPPV), high-frequency jet ventilation (HFJV), and HFOV. These are summarized in Table 8.2. The physiological volume of tidal space V_{ts} in conventional ventilation is created by the significant pressure changes, and the exchange is dependent on bulk convection. The most widely used method of HFV in clinical practice is HFOV that relies on alternative mechanisms of gas exchange. In HFOV, lung recruitment is maintained by the application of relatively high MAP. At the same time, ventilation is achieved by superimposed sinusoidal pressure oscillations delivered by a motor-driven piston or diaphragm at a frequency of 3–15 Hz (Nelson et al. 1998; Wiswell 2000).

HFOV is the only HFV technique in which expiration is an active process, and it is decoupling of oxygenation and CO_2 elimination compared to other HFV techniques. The changes in power and frequency affect ventilation which can be improved by increasing the capacity to the maximum and reducing the frequency. However, these steps may cause larger V_{ts} and pressure swings at the alveoli and may potentially lead to negative impacts on lung protection (Fort et al. 1997; Hromi et al. 2000; Mehta et al. 2001).

Table 8.2 HFV modalities.

Ventilator type	Breath rate, f breath/min	Inspiration	Expiration	Description
HFPPV	60–100	Active	Passive	It is difficult to measure V_{ts} due to some gas flows through the expiratory conduit during inspiration. With high f, there is a risk of gas trapping with over distention of some lung regions and adverse circulatory effects.
HFJV	100–200	Active	Passive	During HFJV, high inspiratory airflow rates and the decompression of jet gas prevent optimal humidification and warming of inspired air, increasing the risk of airway obstruction with desiccated secretions and epithelial debris.
HFOV	Up to 2400	Active	Active	HFOV controls lung volumes better than HFPPV and HFJV. It reduces the risk of air trapping, over distention of airspaces, and circulatory depression.

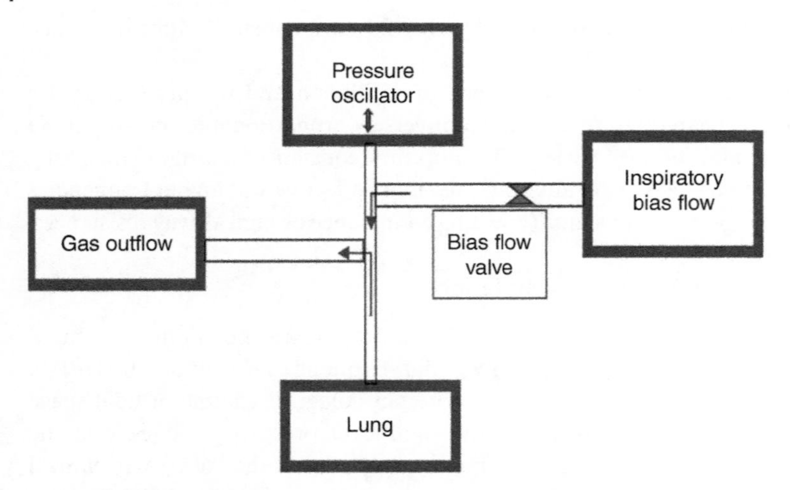

Figure 8.1 High-frequency oscillatory ventilation. *Source:* Based on Stawicki et al. (2009).

8.3.3 High-Frequency Oscillatory Ventilation System

To reduce lung injury or prevent further lung injury, HFOV has been applied to neonates and adults by Lunkenheimer et al. (Lunkenheimer et al. 1972). HFOV operates at a high respiratory range of 3.5–15 Hz (210–900 breaths/min) for inhalation and exhalation. The patient's size, age, and disease process affect the respiratory rates. In HFOV, the pressure oscillates around the constant distending pressure, which in effect is the same as positive end-expiratory pressure. The gas is pushed into the lung during inspiration while it is pulled out from the lung during expiration. HFOV generates very low V_{ts} that is normally less than the dead space of the lung. The value of V_{ts} depends on some factors such as endotracheal tube size, power, and frequency. In HFOV oxygenation can be separated from ventilation as they are not dependent on each other as is the case with conventional ventilation. Ventilation or CO_2 elimination is dependent on the amplitude and to a lesser degree frequency.

Figure 8.1 shows a typical HFOV system (Stawicki et al. 2009). It consists of a piston assembly that incorporates an electronic control circuit or square wave driver, which powers a linear drive motor. The inspiratory phase is created when the coil and the attached piston are driven toward the patient. The expiratory phase is generated when the electrical coil and the attached piston are driven away from the patient. The bias flow is oscillated through the airways if the coil and piston move rapidly. It controls and indicates the rate of the continuous flow of humidified blended gas through the patient circuit.

8.3.4 Gas Transport During HFOV

Gas flow distribution is one of the fundamental principles underlying the increased efficiency of HFOV (Slutsky et al. 1980). The major gas transport mechanisms operating during HFOV include convective ventilation, asymmetric velocity profiles, pendelluft, Taylor dispersion, molecular diffusion, and cardiogenic mixing (Slutsky and Drazen 2002). These mechanisms will be elaborated on as follows:

- **Convective ventilation** plays a relatively small role in gas transport during HFOV, although it is likely to contribute significantly to the ventilator exchange in the most proximal gas exchange units. Spahn et al. (Spahn et al. 1991) found that $PaCO_2$ is increased in an anesthetized dog model because the rebreathing volume is reduced to a level below the HFO circuit-related rebreathing volume. They suggested that the efficiency of CO_2 elimination is critically dependent on the net oscillatory volume, and that bulk convection has an essential role during HFOV. Turbulence in the large airways may also enhance gas mixing (Slutsky and Drazen 2002).
- **Asymmetric** velocity profiles result from the asymmetry between the inspiratory and expiratory velocity profiles (Taylor 1953). The velocity profile of the inhaled gas is initially parabolic, but on exhalation, it is flat. After a full cycle, gas particles initially near the centre of the flow are displaced to the right, while those near the wall are displaced to the left (Haselton and Scherer 1980; Scherer and Haselton 1982).
- **Pendelluft** occurs when regions of the lung have different dynamics of regional inflation and deflation. The alveoli are ventilated asynchronously, as opposed to synchronously in CV (Slutsky et al. 1981; Tanida 1990). This asynchronous ventilation occurs when small neighboring regions of the lung are different in compliance, air resistance, or the time constants of their filling or emptying. Variation in regional airway resistance and compliance cause some regions to fill and empty more rapidly than others. Some gas may flow between regions if these characteristics vary among regions that are in proximity.
- **Taylor dispersion** is one in which the longitudinal dispersion of tracer molecules in a diffusive process was proposed by Taylor, in which augmented diffusion occurs because of turbulent flow between the axial and radial gas concentrations in the airways. Fredberg et al subsequently used a semi-empiric analysis to predict that the combination of Taylor dispersion and molecular diffusion (augmented dispersion) accounts for almost all gas transport during HFOV (Fredberg 1980).
- **Molecular diffusion** is one of the major mechanisms for alveolar ventilation that is responsible for the gas exchange across the alveolar-capillary membrane and contributes to the transport of O_2 and CO_2 in the gas phase near the membrane. Chang et al. described the molecular diffusion occurring near the alveolo-capillary membrane as the random thermal oscillation of gas molecules. He suggested that molecular diffusion would always occur when the gas molecules have a temperature that is above absolute zero (Chang 1984).
- **Cardiogenic mixing** is one in which the heartbeat adds to the peripheral gas mixing. Spontaneous mixing of gas particles arising from Brownian motion contributes to the diffusion of gases in the respiratory tract. Gas velocities are approximately zero in the alveolar region because of the very high total cross-sectional area. The dominant mechanism for gas mixing in this zone is molecular diffusion, with net transport of gas best described by Fick's law. Theoretical studies in healthy animals (Fredberg 1980) and humans (Rossing et al. 1981; Slutsky et al. 1981; Jaeger et al. 1984) have demonstrated that V_{ts} has a greater effect on gas exchange than during HFOV. Ventilation efficiency during HFOV is expressed as (Pillow 2005):

$$Q = f^a V_{ts}^b \tag{8.1}$$

where f is the frequency, V_{ts} is the tidal volume, $a \approx 1$, and $b \approx 2$. The more dominant contribution of V_{ts} to ventilation during HFOV is the result of the oscillatory redistribution of gas from central to distal regions where molecular diffusion overcomes Taylor dispersion as the principal influence on gas transport (Kamm et al. 1984). The transition frequency varies in proportion to the ratio of metabolic rate to dead space and hence is lower in large animal species compared with small animal species and will be lower in adults compared with neonates (Venegas et al. 1986). It has been shown that HFV gas transport mechanisms come into play, whereas V_{ts} still exceeds airway dead space volume (V_D), and that the transition frequency occurs when alveolar ventilation/frequency = 20% of V_D and $V_{ts} = 1.20\ V_D$ (Weinmann et al. 1984).

8.3.5 Control of Gas Exchange

In HFOV, the operator controls the MAP, amplitude (ΔP), frequency (f), and the ratio of inspiratory to expiratory time (I/E). Previous studies indicated that the MAP and ΔP are significantly attenuated by the tracheal tube, where the alveolar pressure is inhomogeneously distributed during HFOV, and the ratio of I/E determines the alveolar pressure (deMello et al. 1994; Lawson and Reid 2000; Nogee et al., 2001; Crouch and Wright 2001). Experimental results showed that the expiratory time ($I/E = 1$) would promote alveolar gas trapping, especially at lower MAP, and $I/E \leq 0.5$ is suggested in applying of HFOV (LeVine et al. 1997; Wright 1997; Mason et al. 1998).

The polarity voltage applied to the electrical coil determines the driven distance between the piston and the patient's airway. The piston movement or amplitude increases with the high polarity voltage, and the more the volume is delivered to the patient, the greater the piston displacement.The extent to which the amplitude increases depend on the resistance the piston encounters to forward movement. The starting power is normally set at 70–90 cmH_2O.

There is no simple formula for estimating the ideal frequency for an individual patient, and the clinical judgment combined with arterial blood gas measurements is needed. The use of frequency lower than 3 Hz is not recommended because the depth of oscillation increases markedly, which may increase risk of barotrauma. In addition, the frequency should not be raised higher than 7 Hz in adults (Wheeler et al. 2009).

8.3.6 Ventilator

Figure 8.2 shows an example of an HFOV ventilator, where a bias flow of fresh, heated, humidified gas is provided across the proximal endotracheal tube (Ritacca and Stewart 2003). The bias flow is typically set at 20–40 L/min, and airway resistance (P_{aw}) at the proximal endotracheal tube is set at a relatively high level (25–35 cmH_2O). An oscillating piston pump akin to the woofer of a loudspeaker vibrates this pressurized, flowing gas at a generally set frequency of 3–10 Hz. A portion of this flow is pumped into and out of the patient by the oscillating piston. The P_{aw} is sensitive to the rate of bias flow but can be adjusted by varying the back pressure on the mushroom valve through which the bias flow vents into the room. The P_{aw} can thus be modified by either adjusting the bias flow rate or the back pressure. The clinics in practice approved HFOV ventilators include the SensorMedics 3100A for use in infants and children and the SensorMedics 3100B for use in adults (Healthcare, Yorba Linda, CA).

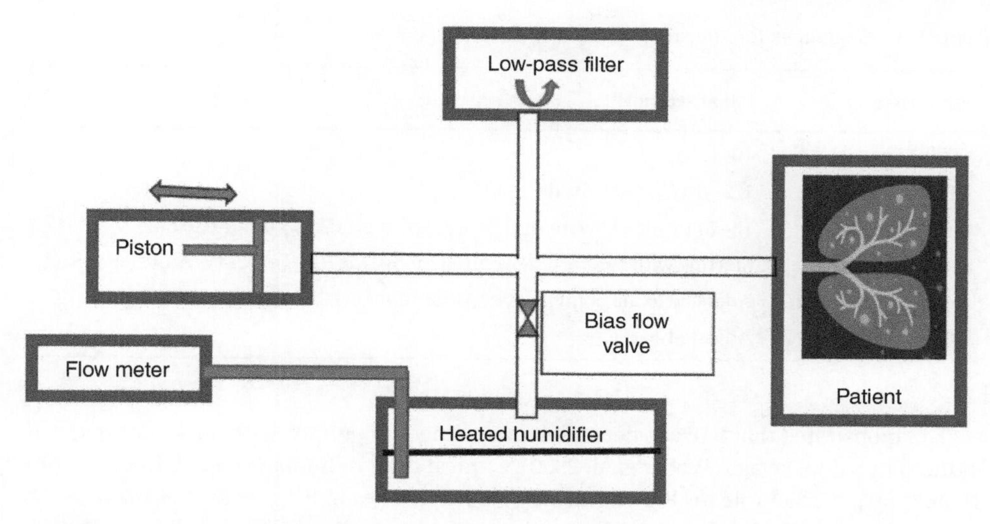

Figure 8.2 Major functioning parts of high-frequency oscillatory ventilation.

8.3.7 Adjusting Ventilatory Parameters

The set power on the ventilator controls the distance that the piston pump moves and, hence, controls the tidal volume V_{ts}. The result is a visible wiggle of the patient's body, which is typically titrated to achieve acceptable CO_2 elimination. The oscillatory pressure amplitude (ΔP) is measured in the ventilator circuit and is a surrogate of the actual pressure oscillations in the airways. These pressures are generally greatly attenuated through the endotracheal tube and larger airways, so the pressure swings in the alveoli are much less. The P_{aw} is similar in the ventilator circuit and the alveoli. The operator uses the parameters of power and frequency to manipulate the V_{ts}. It is important to note that the actual V_{ts} depends on several factors, including the size of the endotracheal tube, the airway resistance, and the compliance of the total respiratory system. Unfortunately, there is no predictable relationship between power and ΔP with the V_{ts} received by the patient. In addition, the V_{ts} can change on a breath-to-breath basis, and therefore ventilator settings are used with clinical factors such as the amount of wiggle in monitoring the patient. Table 8.3 displays the common recommendations applied for the HFOV during neonatal surgery.

8.3.8 Noninvasive Assessment of Lung Volume

Respiratory impedance plethysmography (RIP) and electrical impedance tomography (EIT) have emerged as two promising approaches by which pulmonary mechanics and alveolar recruitment can be assessed noninvasively at the bedside during HFOV. RIP is a monitoring technique capable of quantifying global lung volume by relating it to measurable changes in the cross-sectional area of the chest wall and the abdominal compartment. Brazelton et al. demonstrated that RIP-derived lung volumes correlated well with those that were obtained using a super syringe ($\gamma^2 = 0.78$), and that RIP is capable of tracking global changes in lung volume and creating a pressure–volume curve during HFOV (Brazelton et al. 2001). Weber

Table 8.3 Guidelines for adjusting ventilatory parameters.

Parameters	Initial settings
Frequency	10–15 Hz
I/E ratio	1/2 to 1/1 if not fixed
Gas bias flow	15–20 L/min (beware of flow influence on P_{mean}) if not fixed
P_{mean}	Starting with a P_{mean} of 2–4 cmH_2O above the P_{mean} on CMV. Adjusted to SaO_2
Amplitude pressure	Adjusted to the point that chest vibrations can be seen and to $PaCO_2$ or $TcCO_2$
FiO_2	Adjusted to SaO_2

et al. demonstrated that RIP can detect relative changes in pulmonary compliance that was induced by saline lavage (Weber et al. 2000). Clinical trials of human subjects have demonstrated the potential role for RIP in tracking global changes in lung volume at the bedside. Experience with RIP in human subjects is limited to investigations of its application in conventional phasic ventilation (Choong et al. 2003; Maggiore et al. 2003).

EIT is the most suitable method for regional heterogeneity detection at the bedside of people with the diffused alveolar disease. EIT has been applied in conjunction with CV and HFOV to describe regional lung characteristics in the laboratory. Investigations on CV in large animal models have validated EIT against the super-syringe method to determine regional pressure–volume curves (Kunst et al. 1999; Wolf and Arnold 2005) and have demonstrated a good correlation between EIT-derived regional changes in lung impedance and CT-derived regional variations in aeration (Frerichs et al. 2002). Van Genderingen et al. (van Genderingen et al. 2004) demonstrated that regional pressure–volume curves constructed using maneuvers on HFOV show less variation along the gravitational axis than using the super-syringe method. Clinical trials of human subjects with acute lung injury or RDS have correlated regional impedance changes induced by slow inflation maneuvers using the DAS-01P EIT system (Sheffield, UK) in conjunction with CT scanning (Victorino et al. 2004). Recently, investigators at Children's Hospital Boston have utilized EIT to detect regional changes in lung volume during a standardized suctioning maneuver in children with acute lung injury or RDS with HFOV (Wolf and Arnold 2005).

It is expected that EIT will soon facilitate the development of standard HFOV protocols. Theoretically, this technology can create opportunities for therapeutic intervention by dynamically tracking the regional differences in alveolar recruitment that make portions of the lung highly susceptible to VILI. One drawback of EIT is the substantial bias generated due to the tendency for electrical current to follow the lowest impedance path rather than the shortest distance between the transmitting and receiving electrodes (Hedenstierna 2004). The changes in baseline regional intrathoracic impedance may cause errors in the interpretation of EIT-derived data. Several investigators have reported that EIT reliably detects regional alterations in pulmonary blood flow (Kunst et al. 1998a) and extravascular lung water (Kunst et al. 1998b).

8.3.9 Weaning

Numerous studies have shown that limiting exposure to potentially harmful strategies on CV may enhance outcome benefits attributable to HFOV among patients with severe lung injury. Large trials in the neonatal and pediatric populations have demonstrated favorable outcomes when HFOV is applied early in disease status. It seems logical to expect that timing the transition back to conventional ventilation may be of significant importance.

Weaning a patient from HFOV may be considered when the clinician determines that gas exchange and pulmonary mechanics are suitable for transition to acceptable settings on conventional ventilation. HFOV has been successfully applied to infants (Gerstmann et al. 1997; Courtney et al. 2002; Johnson et al. 2002). However, it is difficult to accomplish in older pediatric and the adult patient who may be less likely to tolerate a degree of sedation that would allow spontaneous respiration while on HFOV and in whom spontaneous breathing may significantly depressurize the circuit, resulting in recurrent alveolar derecruitment.

In general, it is appropriate to undertake a more detailed evaluation of the patient's response to phasic ventilation provided by conventional means when clinical improvement occurs to the point that $P_{aw} \leq 20$ cmH_2O, $FiO_2 \leq 0.4$, and the patient tolerates endotracheal suctioning without significant desaturation (Arnold 2000). This can be done by hand ventilating while noting the pressures, tidal volume, and inspiratory to expiratory time ratio necessary to sustain satisfactory oxygen saturation. It is common to know that patients on CV normally demonstrate satisfactory gas exchange on a MAP (P_{mean}) several cmH_2O below the last P_{aw} on HFOV.

8.4 Noisy Ventilation

Noisy ventilation or biologically variable ventilation uses a computer controller to mimic the normal variability in a spontaneously breathing lung by producing random variations in tidal volume and respiratory rate. Studies on porcine (Mutch et al. 2000a, 2000b) and rodent models have shown that such techniques improve the oxygenation of arterial blood and have enhanced the performance of mechanical ventilators (Arold et al. 2003). Mathematical models have suggested that mechanical ventilation accompanied by randomly varying breathing patterns improved alveolar recruitment by opening collapsed alveoli and increasing the net lung volume without causing increases in MAPs (Suki et al. 1998). The mechanisms of alveolar recruitment for this technique are not fully understood, although theories do exist on the mechanisms of gas transport. These are mostly associated with the benefits due to the better mixing of gases in noisy ventilation (Otis et al. 1956; High et al. 1991).

Various authors have suggested that noisy ventilation (like biologically variable ventilation and bubble CPAP) is an example of stochastic resonance (Suki et al. 1998; Mutch et al. 2000a, 2000b). Previous studies (Suki et al. 1998) found that parameters such as the timing and amplitude of the pressure fluctuations can optimize the effect of the small animals' respiratory system. However, this needs to be further proven on larger animals and humans.

8.5 Continuous Positive Airway Pressure

As previously mentioned in Chapter VI, the CPAP device is considered as the gold standard for the treatment of obstructive sleep apnea. However, several modifications have been made to use this therapy technique for other lung diseases.

8.5.1 Nasal CPAP

Nasal continuous positive airway pressure (nCPAP) is a noninvasive type of respiratory support compared to endotracheal ventilation. It applies continuous positive pressure to the alveoli throughout the respiratory cycle and has been increasingly used in hospitals to treat neonates with RDS (Locke et al. 1991; Gittermann et al. 1997). Variable resistance at exhalation creates the positive pressure applied to the lung. It produces a more regular breathing pattern when compared to mechanical ventilation, since it allows the infant to breathe spontaneously. The continuous distending pressure provided by nCPAP increases lung volume and promotes better gas exchange in the alveoli by keeping them open (Donn and Sinha 2003; Polin et al. 2011). Studies showed that nCPAP reduced the need for mechanical ventilation in neonates with moderate RDS and proved to be an adequate ventilation by improving oxygenation without posing any harmful side effects (Verder et al. 1999; De Klerk and De Klerk (2001a, 2001b); Mazzella et al. 2001).

8.5.2 Bubble CPAP System

Compared to HFV, the bubble CPAP system facilitates better gas exchange in the lungs, and it has been applied for RDS treatment in newborn infants (Wung et al. 1975). Researchers at Columbia Presbyterian Medical Centre applied the bubble CPAP bottle system. (Wung et al. 1979). The CPAP bottle produces bubbles in addition to the mean pressure. Clinical studies of neonates suffering from RDS showed that the respiratory rates and minute ventilation decrease, but alveolar ventilation does not decrease when applying bubble CPAP, which means the infants perform less muscle work to achieve adequate respiration.

The working principle of bubble CPAP is similar to ordinary CPAP, but it offers more accuracy in maintaining the mean pressures and flows and considers patients' comfort. A bubble CPAP system, Figure 8.3, typically contains a humidifier, a gas flow circuit, a nasal prong, and a water bottle, which offers respiratory support to spontaneously breathing

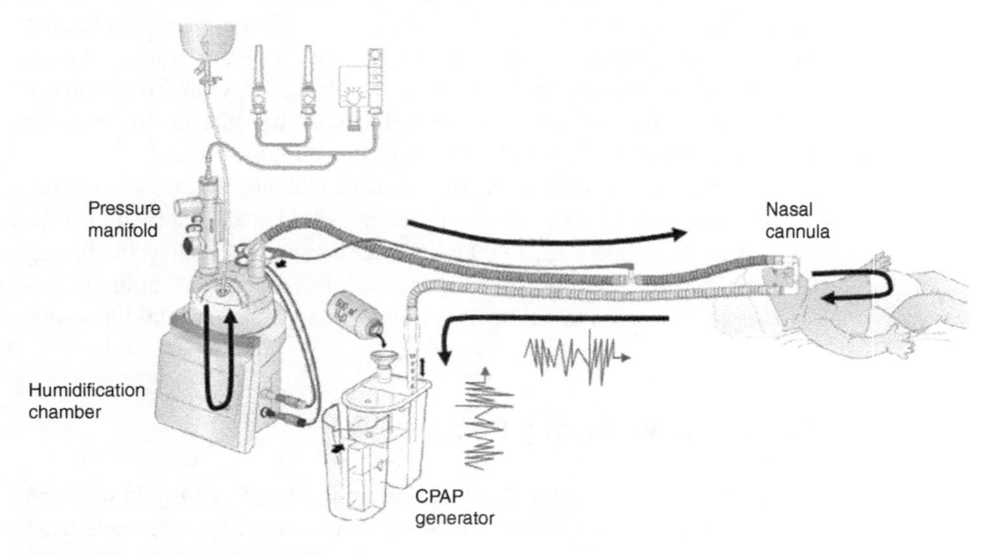

Figure 8.3 Bubble CPAP system.

neonates suffering from RDS. The system applies a mean CPAP accompanied by pressure oscillations to the airways and alveoli throughout the respiratory cycle. This maintains a degree of lung inflation during expiration, thus preventing lung collapse and making it easier for the neonate to breathe, since a partially inflated alveolus is more accessible to expand than a fully collapsed one. The bubble CPAP includes a delivery system (consists of a humidification chamber, a single-heated breathing circuit, CPAP generator, and pressure manifold) and a patient interface (consists of the nasal tubing, nasal prongs, and infant bonnet). Headgear can be used in place of the bonnet, and chin straps can be added to the patient interface to optimize the effect of the bubble CPAP (Manilal-Reddy 2009; Manilal-Reddy and Al-Jumaily 2009).

The **humidification chamber** is the source of the humidity required by the neonate. Sterile water is fed into the chamber from a flexible bag, and a dual float mechanism controls the water level. The float prevents the chamber from flooding and provides a closed system and a constant mean CPAP pressure. The breathing circuit (consists of the inspiratory and expiratory lines) links the flow source, humidification chamber, patient interface, and CPAP generator. The inspiratory line connects the humidifier to the patient interface, and the transparent expiratory line connects the patient interface to the CPAP generator.

The **pressure manifold** is a pressure relief mechanism situated upstream of the neonate that limits the maximum pressure in the system to 17 cmH_2O at a flow rate of 8 L/min in the event of an occlusion in the delivery system. It also has measurement ports for pressure and air/oxygen analysis.

The **CPAP generator** is the innovative version of the bubble CPAP system used at the Columbia Presbyterian Medical Centre since 1974. Lowering or raising the adjustable probe in the water container creates the required level of positive pressure in the circuit. Achievable mean pressures range from 3 to 10 cmH_2O. Excess water spills out into an overflow container, ensuring that the mean CPAP pressure is constant. As air enters the water through the CPAP probe, it creates bubbles, creating pressure oscillations about a mean CPAP level.

The **nasal tubing and interface** deliver the air into the neonate and allow a balanced flow into the nares. To address the variety of nares diameters and septum gaps, nasal prongs of various sizes can be attached to the end of the nasal tubing. The specially designed infant bonnet allows the nasal tubing to be fixed in position and permits repositioning of the nasal prongs with the infant's movements. Headgear as an alternative to the infant bonnet is also available. Chinstraps are also available to prevent excessive leaks from the mouth, thereby preventing significant pressure loss.

Clinical studies showed that several parameters, such as minute ventilation, respiratory rate, ventilation requirement, and respiratory support period, dropped the use of the bubble CPAP. It has also been speculated that the vibrations produced by the pressure oscillations enhance gas exchange and respiratory mechanics.

8.6 Modeling of Bubble CPAP

A simplified mathematical model of bubble CPAP has been developed by Manilal-Reddy et al. (Manilal-Reddy and Al-Jumaily 2009) to demonstrate the dynamic behavior of the neonatal respiratory system when using the bubble CPAP. Experimental results showed

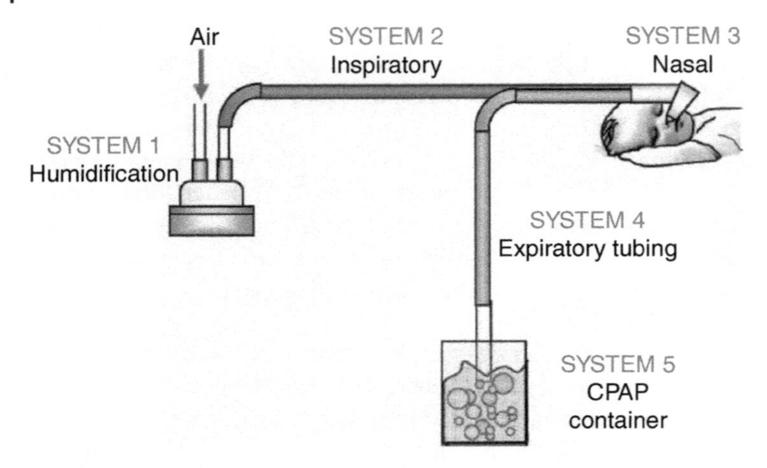

Figure 8.4 Schematic diagram of a bubble CPAP setup showing model subsystems. *Source:* Manilal-Reddy (2009).

that the bubble CPAP model can provide realistic values of pressure and flow under real operating conditions. The model was compatible with a simple viscoelastic neonatal respiratory model based on kinetic considerations. This section describes the development of the bubble CPAP model.

8.6.1 Model Formulation

Figure 8.4 shows the bubble CPAP system used in the study (Manilal-Reddy and Al-Jumaily 2009). The system contains five subsystems, and the dynamic model for each subsystem is formulated based on conservation of mass and/or energy, basic laws of physics. The subsystems in the form of a block diagram are represented in Figure 8.5.

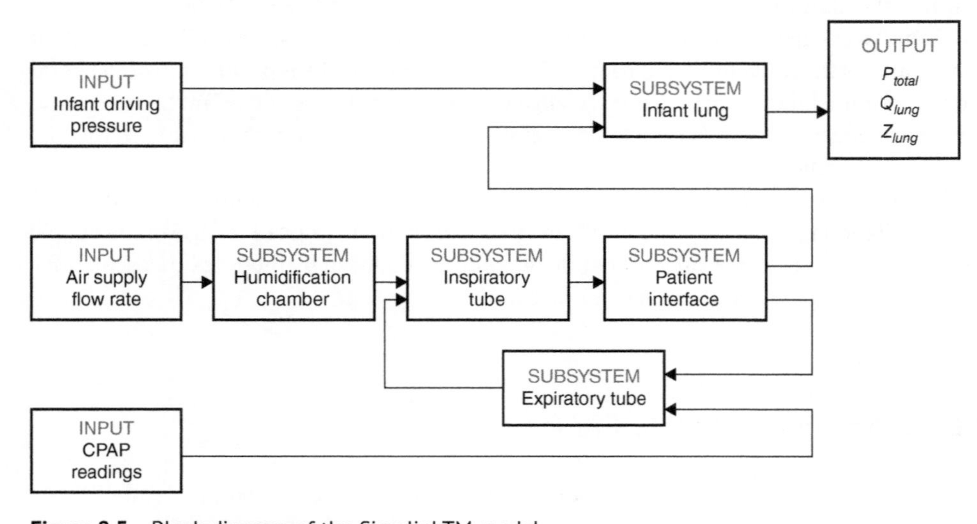

Figure 8.5 Block diagram of the SimulinkTM model.

In the study, air humidity is neglected, and flow restrictions and shock factors at circuit connections are disregarded. Conservation of mass at the patient interface model can be written as follows:

$$\dot{m}_{tube} - \dot{m}_{out} = C_{interface}\frac{dP_{interface}}{dt} \tag{8.2}$$

where $\dot{m}_{in}$ and $\dot{m}_{tube}$ are the mass inflow rate and mass outflow rate of the chamber, respectively. The interface capacitance $C_{interface}$ can be expressed as:

$$C_{interface} = \frac{V_{interface}}{R_a T_{interface}} \tag{8.3}$$

where $V_{interface}$ is the interface volume, $T_{interface}$ is the temperature, and R_a is the gas constant of air.

The lung volume flow rate (Q_{lung}) can be written as:

$$Q_{lung} = \frac{P_{total}A^2}{m_{wall}D^2 + B_{wall}D + k_{wall}} \tag{8.4}$$

where m_{wall}, k_{wall}, and B_{wall} are the lung wall mass, overall stiffness, and damping coefficient respectively. $P_{total} = P_{lungwall} + P_{interface}$, A is surface area, and $D = d/dt$.

The effective lung impedance can be obtained by:

$$Z_{effective} = \frac{P_{total}}{Q_{lung}} \tag{8.5}$$

8.6.2 Structural Correlation

Each airway branch is modeled as a thin-walled, elastic pressure vessel that is exposed to a pressure (p_O) and circumferential stress (σ). Figure 8.6 shows the airway branch cut along the longitudinal axis, exposing an element to investigate its dynamic characteristics. Applying Newton's second law and assuming that $d\theta$ has very small results in (Al-Jumaily and Du 2002):

$$\frac{1}{\omega_n^2}\frac{d^2r}{dt^2} + r = k_w P_{man} \tag{8.6}$$

where P_{man} is the MAP (airway pressure minus the pleural pressure), and the natural frequency is defined as:

$$\omega_n = \sqrt{\frac{Et_w - p_O r_O}{\rho_w t_w r_O^2}} \tag{8.7}$$

where E is the lung wall elastance, t_w is the wall thickness, p_O is the fluid pressure, r_O is the airway radius, and ρ_w is the wall mass density.

The wall stiffness can be obtained by:

$$k_{wall} = \frac{r_O^2}{Et_w - p_O r_O} \tag{8.8}$$

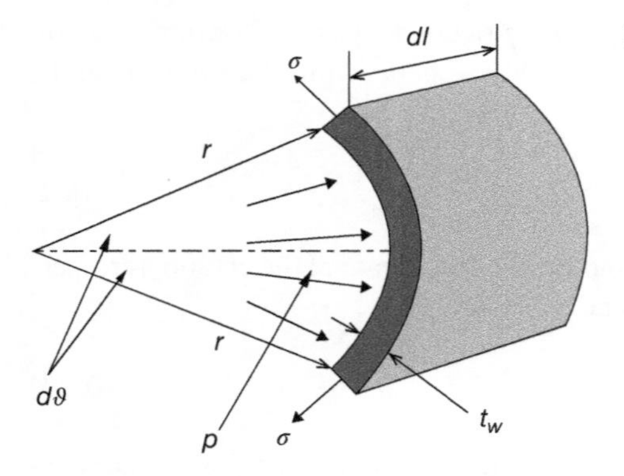

Figure 8.6 An element of an airway passage.

Equation (8.7) is used to calculate the natural frequencies of the upper airway branches. Due to the lack of data on the properties of neonatal lungs, values are obtained by finding approximate values and adult-to-infant ratios from Weibel (1963).

8.6.3 Results and Discussion

A computer model is developed to demonstrate the effect of pressure oscillations generated at different flow rates. Figure 8.7a and b show the simulation results of the total lung pressure and lung flow rate for one breath cycle of a healthy lung ($k_{wall} = 30$ MPa), respectively. The pressure and flow rate values for a healthy neonatal lung with no bubble CPAP are the representatives of published values (De Jongh 1995).

The normalized impedance is considered to investigate the lung performance as exposed to bubble CPAP. Figure 8.8 shows the effect of the frequencies of pressure oscillations produced by a different flow rate at 5 cmH_2O. Pressure data obtained from experiments on respiratory support devices that produce pressure oscillations at flow rates of 2, 4, 8, 12 L/min are applied to different lung models by varying the lung stiffness (k_{wall}) values (Manilal-Reddy 2004).

The healthier lungs ($k_{wall} = 10 - 60$ MPa) experience the lowest effective impedance values at 2 L/min and show a minimal change in lung performance as the flow rate increases. The moderately unhealthy lung ($k_{wall} = 70$ MPa) shows the best lung performance at 8 L/min, where there are strong pressure oscillation strengths at 30 and 60 Hz. The extremely unhealthy lungs ($k_{wall} = 90 - 110$ MPa) show a different trend in that maximum improvement in lung performance at 12 L/min, where there are strong powers at frequencies around 30 Hz as well as around 60–80 Hz.

The presence of strong powers of pressure oscillation at frequencies 30–60 Hz reflects on the flow setting range of 8–12 L/min for unhealthy lungs, which are most beneficial to lung performance. Before applying oscillatory CPAP, the differences in practical impedance values are large across all k_{wall} values, and these differences reduced after applying oscillatory CPAP. The steep impedance gradients of the unhealthier lungs are easily seen here and show that on the application of CPAP with pressure oscillations, the unhealthier the lungs, the greater the benefit gained.

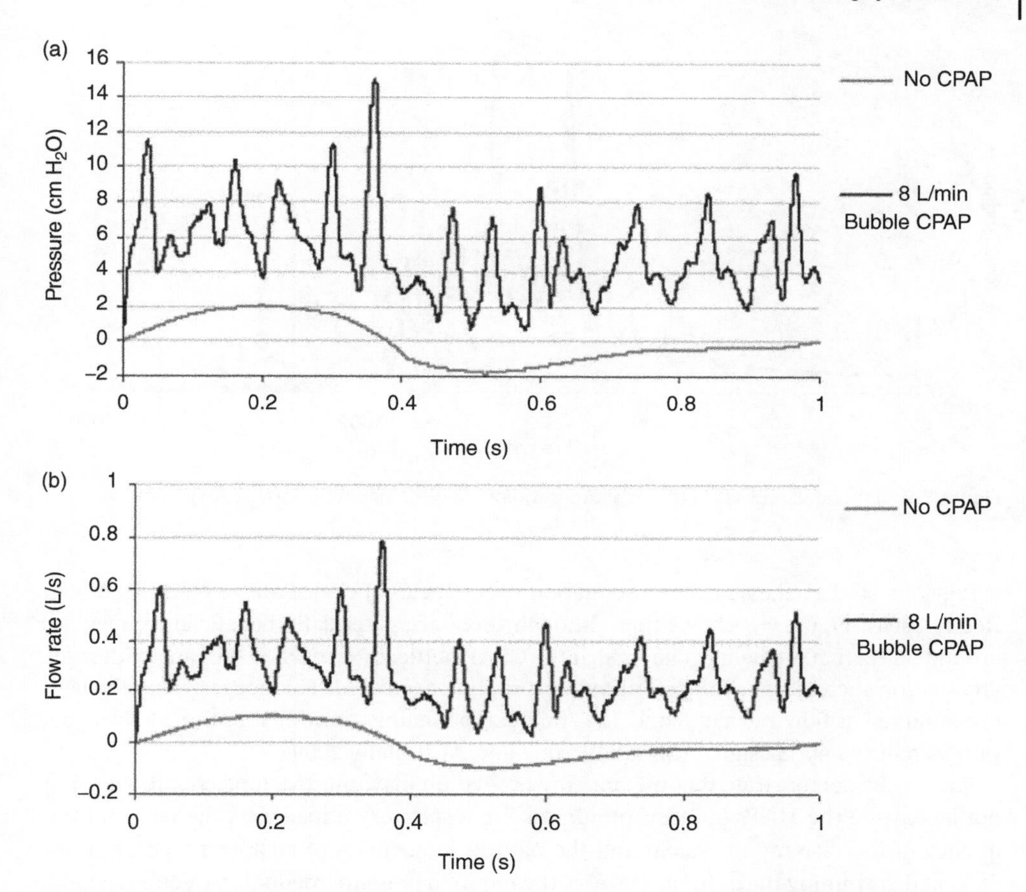

Figure 8.7 Simulated lung for k_{wall} = 30 MPa: (a) pressure for k_{wall} = 30 MPa and (b) flow rate.

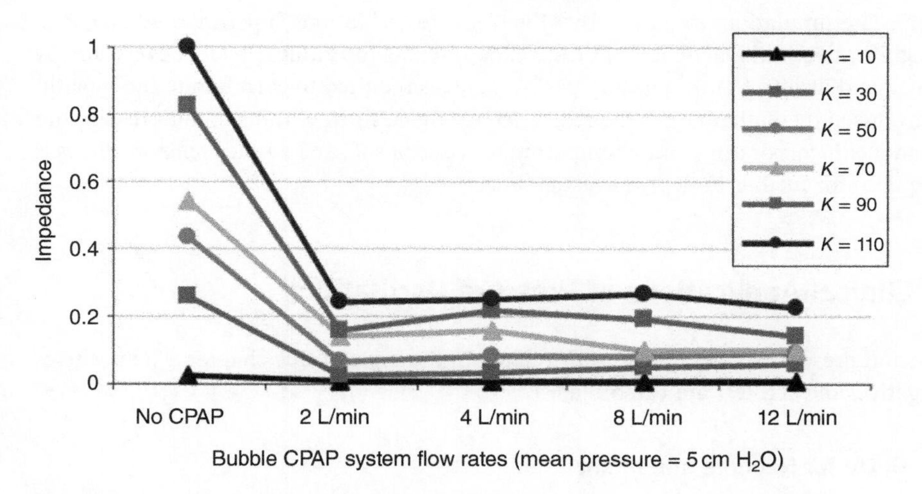

Figure 8.8 Lung performance at different bubble CPAP system settings.

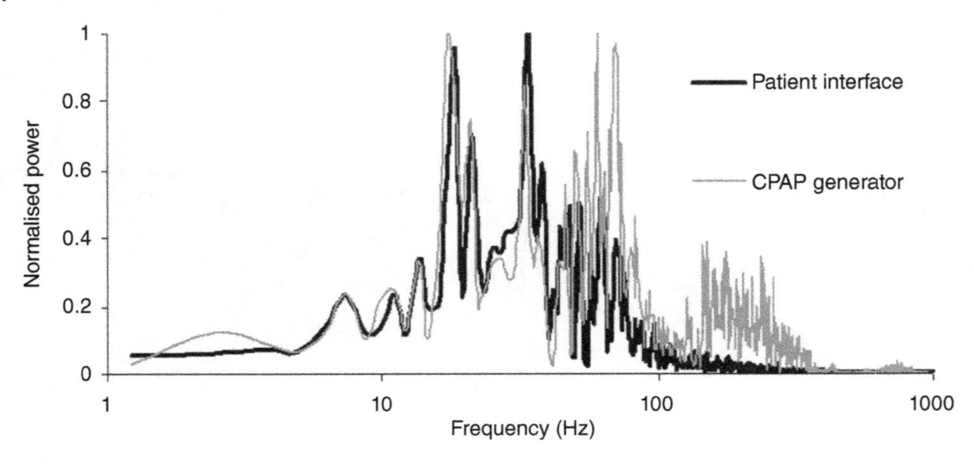

Figure 8.9 Frequency "Loss" in the expiratory tubing. *Source:* Manilal-Reddy (2004).

Previous studies showed that attenuation occurs at frequencies above 45 Hz (Manilal-Reddy 2004). Figure 8.9 shows that although there are potentially beneficial frequencies around 60 Hz that are being generated at the CPAP bottle, they reached the lung with some attenuation such that the lung recruitment is slightly decreased. This suggests that the benefit observed at flow rates of 8 and 12 L/min can be mainly attributed to the 30-Hz signal rather than the 60-Hz signal (Manilal-Reddy and Al-Jumaily 2009).

The results demonstrate that the pressure oscillation may improve lung conditions without increasing the MAPs in terms of minimizing respiratory impedance; the resonant frequency of the respiratory system and the natural frequencies of airways play significant roles in determining the optimum level of the required pressure oscillation frequency. Only lung stiffness is considered in this study. In the actual case, it should consider mass and lung structure. A more complex model is required to determine the dynamic characteristics of the lungs. The simulations suggested that the pressure oscillations superimposed on CPAP had a significant additional benefit. Further experimental and clinical evidence into airway resonance and its effect on respiratory performance is required to corroborate the hypotheses introduced by mathematical modeling. An extended form of the neonatal respiratory system model incorporating multi-compartment, viscoelastic, and surface tension effects is also required for further in-depth investigations.

8.7 Clinical Applications of Pressure Oscillations

This section presents some clinical trials related to the topic of this chapter. The positives and negatives of each test are elaborated.

8.7.1 HFOV for Neonate and Infant

Marchak et al. reported the use of HFOV in neonates who are suffering from RDS (Marchak et al. 1981). Various studies have suggested that infants who had severe hyaline membrane disease could achieve adequate gas exchange with HFOV at MAP significantly lower than that required with conventional mechanical ventilation (CMV) (Bhuta et al. 2001; Henderson-Smart et al. 2001).

Based on preliminary findings (Riphagen and Bonn 1999; Bhuta et al. 2001; Henderson-Smart et al. 2001), the HIFI study group (HiFi Study Group 1989) conducted a comparison between HFOV and CMV in a large number of premature infants ($n = 673$). Infants were selected with the weight of 750–2000 g who had been supported less than 12 hours on the CV for respiratory failure in the first 24 hours of life. Significant differences were not found in the incidence of BPD or death in both groups. HFOV did not produce more effectiveness or reduce the need for ventilator support compared to CMV.

Courtney et al. (Courtney et al. 2002) studied 500 infants with a strategy in the CMV that targeted a V_{ts} of 5–6 mL/kg body weight and the HFOV at a frequency in the range of 10–15 Hz. This study successfully observed a significant decrease in the need for supplemental oxygen at 36 weeks postmenstrual age. By defining a disease threshold in the study of infants, adhering to lung-protective protocols for mechanical ventilation, and extubating from the assigned ventilator according to specific criteria, this study identified a set of circumstances in which HFOV may be used with clear benefits in preterm infants with RDS.

Johnson et al. (Johnson et al. 2002) investigated 797 preterm infants with a gestation age of 23–28 weeks to either HFOV or CMV. No significant difference was found between HFOV and CMV for early treatment of RDS. Several investigators have studied HFOV in infants with a congenital diaphragmatic hernia (CDH) and found that the HFOV has been associated with an improvement in survival, dramatic reductions in $PaCO_2$, and concurrent improvements in oxygenation (Williams 1977; Crouch et al. 1991; Vorhout et al. 1992).

8.7.2 HFOV for Children

The HFOV has been applied for children with acute severe respiratory failure attributable to diffused alveolar disease and/or air leak syndromes. Investigations suggested that HFOV could be applied as a rescue therapy for pediatric patients with severe hypoxic lung injury (Venegas and Fredberg 1994; Stawicki et al. 2009).

Arnold et al. evaluated the effect of HFOV on respiratory outcomes in pediatric patients. They randomized 70 patients to receive CV using a strategy to limit peak inspiratory pressure, or HFOV at a frequency of 5–10 Hz, using an open lung strategy in which the lung volume optimal oxygenation occurred was defined $SaO_2 \geq 90\%$ and $FiO_2 < 0.6$ (Arnold et al. 1994). In patients with air leaks, airway pressure was limited while accepting preferential increases in FiO_2 to achieve saturation of $\geq 85\%$ and pH ≥ 7.25 until it resolved (Fort et al. 1997). No difference in survival or duration of CV between the two groups was observed, but significantly fewer patients randomized to receive HFOV remained dependent on supplemental oxygen at 30 days, compared with those who were randomized to receive a CV, despite the use of significantly higher P_{aw} in the HFOV group (Arnold et al. 1994).

Using HFOV for the treatment of lower airways disease in older pediatric patients is limited. HFOV has been applied to older pediatric patients with status asthmatics (Duval and van Vught 2000). Duval et al. achieved optimal CO_2 clearance using an open lung strategy with P_{aw} (20 cmH_2O), low frequency (6 Hz), I/E (0.33), and relatively high ΔP without apparent air leak. However, the use of HFOV in obstructive lung diseases must be considered thoroughly.

8.7.3 HFOV for Adolescent and Adult

The 3100B HFOV (SensorMedics, Yorba Linda, CA) is available for larger pediatric and adult patients (Slutsky et al. 1981; Lunkenheimer et al. 1994). Compared to the 3100A model, 3100B has a higher maximal bias flow that allows for higher MAP delivery. It also has a more powerful electromagnet that produces faster acceleration to maximal ΔP (Mehta et al. 2001).

Early investigations of HFOV in adolescent and adult patients with hypoxic respiratory failure were published in (Fort et al. 1997; Mehta et al. 2001). People with RDS who failed CV were given rescue therapy with low-frequency (maximum 5–6 Hz) HFOV. These studies included people with severe disease, including mean values for PaO_2/FiO_2 in the range of 60 at the time of enrollment. Although these studies failed to measure significant differences in outcomes such as mortality, the majority of patients in the studies demonstrated an improvement in short-term physiologic variables such as FiO_2, PaO_2/FiO_2, and OI. Non-survivors were exposed to significantly larger extended periods of CV, suggesting once again the importance of instituting HFOV early in the course of the disease.

To evaluate the safety and effectiveness of HFOV and CMV, clinical studies of adult patients with RDS have been conducted (Derdak et al. 2002). However, the study failed to assess the mortality differences between HFOV and CMV groups, but there was a clear trend toward increased 30-day mortality among the patients randomized to receive CMV (52%) compared to those who received HFOV (37%).

It is unclear whether low-frequency HFOV can be used as protective as high-frequency HFOV. The conducted experimental results by using the 3100B HFOV model demonstrated that tidal volumes approaching those used in conventional ventilation are produced under the conditions of low-frequency and high-pressure amplitude (Sedeek et al. 2003).

8.7.4 Clinical Benefits and Disadvantages of HFOV

HFOV is ideally suited as a lung-protective strategy because of its theoretical ability to minimize many of potential adverse effects. Clinical trials suggested that HFOV can be a safe alternative to conventional ventilation in neonates and pediatric populations. However, whether HFOV offers benefits over the current best conventional strategies is still unclear. In addition, many of the theoretical benefits of HFOV are unproven, and the lung volumes achieved while using high MAPs and various frequencies are unknown.

The difficulty in proving significant clinical outcome benefits in pediatric and adult patients may be largely due to the diverse potential etiologies of respiratory failure in these populations and a wide range of approaches to their medical management applied over a relatively long period of mechanical ventilatory support. The use of low-frequency HFOV in adult patients may not be as successful as the high-frequency HFOV used in small animals and human infants.

Heavy sedation and/or neuromuscular blockade are often required for using HFOV, which may be problematic, particularly, given the supporting evidence of a benefit in daily wakening of sedated mechanically ventilated patients. Such an approach is often not possible in patients requiring HFOV. Suctioning patients on HFOV can be achieved using a closed inline system that does not require the patient to be disconnected from the oscillator. The extent to which this prevents derecruitment is not clear. In addition, a higher P_{aw} may explain the reductions in cardiac preload that are occasionally seen with HFOV. Consequently, a fluid balance needs to be carefully monitored as hypoxemia can, at times, be exacerbated by relative hypovolemia. Transportation out of the intensive care unit on the oscillator is currently not possible. Procedures like bronchoscopy may lead to loss of P_{aw}. Other limitations include loss of the ability to auscultate the lung, the heart, and the abdomen, and the difficulty in recognizing pneumothorax, right mainstem bronchus intubation, and endotracheal tube dislodgement.

8.7.5 Clinical Applications of CPAP

The bubble CPAP has several benefits for managing RDS in neonates. Clinical studies on preterm and deficient birth weight infants concluded that the bubble CPAP reduces the possibility of lung injury by reducing the need and number of days required for mechanical ventilation (Gittermann et al. 1997; Narendran et al. 2003). BPD rate has been reduced in preterm infants with RDS when they used bubble CPAP compared to those who did not use it.

The bubble CPAP improves arterial oxygen levels, stabilizes lung volumes at low pressure, and enhances lung mechanics. Studies on preterm lambs have shown that bubble CPAP improves the compliance of the lung by preventing repetitive inflammatory stresses on the preterm lung and preserving surfactant function better than mechanical ventilation (Jobe et al. 2002). It has also been speculated that the noisy pressure signal may also promote surfactant secretion in the lung (Arold et al. 2003). The improvements noted in ventilatory parameters during bubble CPAP could be due to several physiological, chemical, and mechanical mechanisms. These include the avoidance of aggressive initiation of ventilation with high tidal volumes and inadvertent hyperventilation of the lung (Narendran et al. 2003), similar to those seen in HFV (Lee et al. 1998), and that of stochastic resonance resulting from the superimposed noisy pressure signal from the bubbling action. However, the relative contributions of the different proposed mechanisms are yet to be established.

The essence of stochastic resonance in relation to the bubble CPAP system can be explained as follows. The response is improved when "noise" is added to the input into a nonlinear system (like the lung). It is possible that if the timing or the pressure amplitude of the noise is increased further, it can produce a detrimental effect. Published reports (Arold et al. 2002) suggested that "noise" can be tuned to provide optimal ventilation. This improvement has been illustrated in terms of the net volume recruited due to the nonlinearity of the pressure–volume curves of the lung.

Bibliography

Al-Jumaily, A.M. and Du, Y. (2002). Obstruction identification in a compliant tube with application to airway passages. *Journal of Vibration and Control* 8(5): 643–657.

Arnold, J.H. (2000). High-frequency ventilation in the pediatric intensive care unit. *Pediatric Critical Care Medicine* 1(2): 93–99.

Arnold, J.H., Hanson, J.H., Toro-Figuero, L.O. et al. (1994). Prospective, randomized comparison of high-frequency oscillatory ventilation and conventional mechanical ventilation in pediatric respiratory failure. *Critical Care Medicine* 22(10): 1530–1539.

Arold, S.P., Mora, R., Lutchen, K.R. et al. (2002). Variable tidal volume ventilation improves lung mechanics and gas exchange in a rodent model of acute lung injury. *American Journal of Respiratory and Critical Care Medicine* 165(3): 366–371.

Arold, S.P., Suki, B., Alencar, A.M. et al. (2003). Variable ventilation induces endogenous surfactant release in normal guinea pigs. *American Journal of Physiology. Lung Cellular and Molecular Physiology* 285(2): L370–L375.

Attar, M.A. and Donn, S.M. (2002). Mechanisms of ventilator-induced lung injury in premature infants. *Seminars in Neonatology* 7(5): 353–360.

Avery, M.E., Tooley, W.H., Keller, J.B. et al. (1987). Is chronic lung disease in low birth weight infants preventable? A survey of eight centers. *Pediatrics* 79(1): 26–30.

Bhuta, T., Clark, R.H., and Henderson-Smart, D.J. (2001). Rescue high frequency oscillatory ventilation vs conventional ventilation for infants with severe pulmonary dysfunction born at or near term. *Cochrane Database of Systematic Reviews* 1(CD002974).

Bohn, D.J., Miyasaka, K., Marchak, B.E. et al. (1980). Ventilation by high-frequency oscillation. *Journal of Applied Physiology* 48: 710–716.

Brazelton, T.B. 3rd, Watson, K.F., Murphy, M. et al. (2001). Identifi cation of optimal lung volume during high-frequency oscillatory ventilation using respiratory inductive plethysmography. *Critical Care Medicine* 29(12): 2349–2359.

Chang, H.K. (1984). Mechanisms of gas transport during ventilation by high frequency oscillation. *Journal of Applied Physiology* 56(3): 553–563.

Choong, K., Chatrkaw, P., Frndova, H., and Cox, P.N. (2003). Comparison of loss in lung volume with open versus in-line catheter endotracheal suctioning. *Pediatric Critical Care Medicine* 4 (1): 69–73.

Clark, R.H., Gerstmann, D.R., Null, D.M. et al. (1992). Prospective randomized comparison of highfrequency oscillatory and conventional ventilation in respiratory distress syndrome. *Pediatrics* 89: 5–12.

Courtney, S.E., Durand, D.J., Asselin, J.M. et al. (2002). High-frequency oscillatory ventilation versus conventional mechanical ventilation for very-low-birth-weight infants. *New England Journal of Medicine* 347(9): 643–652.

Crane, L. (1981). Physical therapy for neonates with respiratory dysfunction. *Physical Therapy* 61 (12): 1764–1773.

Crouch, E. and Wright, J.R. (2001). Surfactant proteins A and D and pulmonary host defense. *Annual Review of Physiology* 63: 521–554.

Crouch, E., Rust, K., Marienchek, W. et al. (1991). Developmental expression of pulmonary surfactant protein D (SP-D). *American Journal of Respiratory Cell and Molecular Biology* 5: 13–18.

De Jongh, F. (1995). Ventilation modelling of the human lung. Faculty of Aerospace Engineering. *Delft University of Technology: Delft* 197.

De Klerk, A.M. and De Klerk, R.K. (2001a). Nasal continuous positive airway pressure and outcomes of preterm infants. *Journal of Paediatrics and Child Health* 37(2): 161–167.

De Klerk, A.M. and de Klerk, R.K. (2001b). Use of continuous positive airway pressure in preterm infants: comments and experience from New Zealand. *Pediatrics* 108(3): 761–763.

deMello, D.E., Nogee, L.M., Heyman, S. et al. (1994). Molecular and phenotypic variability in the congenital alveolar proteinosis syndrome associated with inherited surfactant protein B defi ciency. *The Journal of Pediatrics* 125: 43–50.

Derdak, S., Mehta, S., Stewart, T.E. et al. (2002). High-frequency oscillatory ventilation for acute respiratory distress syndrome in adults: a randomized, controlled trial. *American Journal of Respiratory and Critical Care Medicine* 166(6): 801–808.

Donn, S.M. and Sinha, S.K. (2003). Invasive and noninvasive neonatal mechanical ventilation. *Respiratory Care* 48(4): 426–441.

Dunn, D. and Lewis, A.T. (1973). Some important aspects of neonatal nursing related to pulmonary disease and family involvement. *Pediatric Clinics of North America* 20(2): 481–498.

Duval, E.L. and van Vught, A.J. (2000). Status asthmaticus treated by high-frequency oscillatory ventilation. *Pediatric Pulmonology* 30(4): 350–353.

Faix, R.G. (2000). Neonatal pneumonia. In: *Manual of Neonatal Respiratory Care* (ed. S.K. Sinha and S.M. Donn), 273–280. Armonk, NY: Futura Publishing Co.

Fort, P., Farmer, C., Westerman, J. et al. (1997). High-frequency oscillatory ventilation for adult respiratory distress syndrome-a pilot study. *Critical Care Medicine* 25(6): 937–947.

Fort, P., Farmer, C., Westerman, J. et al. (1997). High-frequency oscillatory ventilation for adult respiratory distress syndrome—a pilot study [comment]. *Critical Care Medicine* 25(6): 937–947.

Fredberg, J.J. (1980). Augmented diffusion in the airways can support pulmonary gas exchange. *Journal of Applied Physiology* 49: 232–238.

Fredberg, J.J. and Hoenig, A. (1978). Mechanical response of the lungs at high frequencies. *Journal of Biomechanical Engineering* 100(2): 57–66.

Fredberg, J.J., Sidell, R.S., Wohl, M.E., and DeJong, R.G. (1978). Canine pulmonary input impedance measured by transient forced oscillations. *Journal of Biomechanical Engineering* 100(2): 67–71.

Frerichs, I., Hinz, J., Herrmann, P. et al. (2002). Detection of local lung air content by electrical impedance tomography compared with electron beam CT. *Journal of Applied Physiology* 93(2): 660–666.

van Genderingen, H.R., van Vught, A.J., and Jansen, J.R. (2004). Regional lung volume during high-frequency oscillatory ventilation by electrical impedance tomography. *Critical Care Medicine* 32(3): 787–794.

Gerhardt, T. and Bancalari, E. (1980). Chestwall compliance in full-term and premature infants. *Acta Paediatrica Scandinavica* 69: 359–364.

Gerstmann, D.R., Minton, S.D., Stoddard, R.A. et al. (1996). The provo multicenter early highfrequency oscillatory ventilation trial: improved pulmonary and clinical outcome in respiratory distress syndrome. *Pediatrics* 98: 1044–1057.

Gittermann, M.K., Fusch, C., Gittermann, A.R. et al. (1997). Early nasal continuous positive airway pressure treatment reduces the need for intubation in very low birth weight infants. *European Journal of Pediatrics* 156(5): 384–388.

Grimal, Q., Watzky, A., and Naili, S. (2002). A one-dimensional model for the propagation of transient pressure waves through the lung. *Journal of Biomechanics* 35(8): 1081–1089.

Haselton, F.R. and Scherer, P.W. (1980). Bronchial bifurcations and respiratory mass transport. *Science* 208: 69–71.

Hedenstierna, G. (2004). Using electric impedance tomography to assess regional ventilation at the bedside. *American Journal of Respiratory and Critical Care Medicine* 169(7): 777–778.

Henderson-Smart, D.J., Bhuta, T., Cools, F., and Offringa, M. (2001). Elective high frequency oscillatory ventilation vs conventional ventilation for acute pulmonary dysfunction in preterm infants. *Cochrane Database of Systematic Reviews* 3: CD000104.

HiFi Study Group (1989). High-frequency oscillatory ventilation compared with conventional mechanical ventilation in the treatment of respiratory failure in preterm infants. *The New England Journal of Medicine* 320: 88–93.

HIFO Study Group (1993). Randomized study of high-frequency oscillatory ventilation in infants with severe respiratory distress syndrome. *The Journal of Pediatrics* 122: 609–619.

High, K.C., Ultman, J.S., and Karl, S.R. (1991). Mechanically induced pendelluft flow in a model airway bifurcation during high frequency oscillation. *Journal of Biomechanical Engineering* 113(3): 342–347.

Hromi, J.M., Tekeuchi, M., and Godden, S. (2000). Kacmarek: tidal volumes during high-frequency oscillatory partial liquid ventilation in an ovine model of adult ARDS [abstract]. *American Journal of Respiratory and Critical Care Medicine* 161: A388.

Jaeger, M.J., Kurzweg, U.H., and Banner, M.J. (1984). Transport of gases in high-frequency ventilation. *Critical Care Medicine* 12: 708–710.

Jobe, A.H. and Ikegami, M. (1998). Mechanisms initiating lung injury in the preterm. *Early Human Development* 53(1): 81–94.

Jobe, A.H., Kramer, B.W., Moss, T.J. et al. (2002). Decreased indicators of lung injury with continuous positive expiratory pressure in preterm lambs. *Pediatric Research* 52(3): 387–392.

Johnson, A.H., Peacock, J.L., Greenough, A. et al. (2002). High-frequency oscillatory ventilation for the prevention of chronic lung disease of prematurity. *New England Journal of Medicine* 347(9): 633–642.

Kamm, R.D., Collins, J., Whang, J. et al. (1984). Gas transport during oscillatory flow in a network of branching tubes. *Journal of Biomechanical Engineering* 106(4): 315–320.

Kattan, M. (1979). Long-term sequelae of respiratory illness in infancy and childhood. *Pediatric Clinics of North America* 26(3): 525–535.

Kunst, P.W., Vonk Noordegraaf, A., Hoekstra, O.S. et al. (1998a). Ventilation and perfusion imaging by electrical impedance tomography: a comparison with radionuclide scanning. *Physiological Measurement* 19(4): 481–490.

Kunst, P.W., Vonk Noordegraaf, A., Straver, B. et al. (1998b). Infl uences of lung parenchyma density and thoracic fl uid on ventilatory EIT measurements. *Physiological Measurement* 19(1): 27–34.

Kunst, P.W., de Vries, P.M., Postmus, P.E., and Bakker, J. (1999). Evaluation of electrical impedance tomography in the measurement of PEEP-induced changes in lung volume. *Chest* 115(4): 1102–1106.

Laitman, J.T. and Crelin, E.S. (1980). Developmental change in the upper respiratory system of human infants. *Perinatal/Neonatal* 4: 15–21.

Lawson, P.R. and Reid, K.B.M. (2000). The roles of surfactant proteins A and D in innate immunity. *Immunological Reviews* 173: 66–78.

Lee, K.S., Dunn, M.S., Fenwick, M., and Shennan, A.T. (1998). A comparison of underwater bubble continuous positive airway pressure with ventilator-derived continuous positive airway pressure in premature neonates ready for extubation. *Neonatology* 73(2): 69–75.

LeVine, A.M., Bruno, M.D., Huelsman, K.M. et al. (1997). Surfactant protein A defi cient mice are susceptible to group B streptococcal infection. *Journal of Immunology* 158: 4336–4340.

Locke, R., Greenspan, J.S., Shaffer, T.H. et al. (1991). Effect of nasal CPAP on thoracoabdominal motion in neonates with respiratory insufficiency. *Pediatric Pulmonology* 11(3): 259–264.

Lunkenheimer, P.P., Rafflenebell, W., Keller, H. et al. (1972). Application of transtracheal pressure oscillations as a modification of "diffusion respiration". *British Journal of Anaesthesia* 44(6): 627–627.

Lunkenheimer, P.P., Frank, I., Ising, H., and Keller, D.H.H. (1973). Intrapulmonary gas exchange during simulated apnea due to transtracheal periodic intrathoracic pressure changes. *Anaesthesist* 22: 232–238.

Lunkenheimer, P.P., Redmann, K., Stroh, N. et al. (1994). High-frequency oscillation in an adult porcine model. *Critical Care Medicine* 22(9 Suppl): S37–S48.

Maggiore, S.M., Lellouche, F., Pigeot, J. et al. (2003). Prevention of endotracheal suctioning-induced alveolar derecruitment in acute lung injury. *American Journal of Respiratory and Critical Care Medicine* 167(9): 1215–1224.

Manilal, P.I., Al-Jumail, A.M., and Prime, N. (2004). Effect of pressure oscillations on neonatal breathing. In: *ASME 2004 International Mechanical Engineering Congress and Exposition*, 103–105. ASME.

Manilal-Reddy, P.I. (2004). The effect of pressure oscillations on neonatal breathing. Master of Engineering Thesis, Auckland University of Technology, p. 134.

Manilal-Reddy, P.I. (2009). The effect of pressure oscillations on respiratory performance. PhD thesis, AUT University.

Manilal-Reddy, P.I. and Al-Jumaily, A.M. (2009). Understanding the use of continuous oscillating positive airway pressure (bubble CPAP) to treat neonatal respiratory disease: an engineering approach. *Journal of Medical Engineering & Technology* 33(3): 214–222.

Marchak, B.E., Thompson, W.K., Duffty, P. et al. (1981). Treatment of RDS by high-frequency oscillatory ventilation: a preliminary report. *The Journal of Pediatrics* 99: 287–292.

Mason, R.J., Greene, K., and Voelker, D.R. (1998). Surfactant protein A and surfactant protein D in health and disease. *The American Journal of Physiology* 275: L1–L13.

Mazzella, M., Bellini, C., Calevo, M.G. et al. (2001). A randomised control study comparing the Infant Flow Driver with nasal continuous positive airway pressure in preterm infants. *Archives of Disease in Childhood. Fetal and Neonatal Edition* 85(2): F86–F90.

Mehta, S., Lapinsky, S.E., Hallett, D.C. et al. (2001). Prospective trial of high-frequency oscillation in adults with acute respiratory distress syndrome. *Critical Care Medicine* 29(7): 1360–1369.

Mutch, W.A.C., Harms, S., Graham, M.R. et al. (2000a). BVV increases arterial oxygenation over that seen with positive end-expiratory prssure alone in a porcine model of acute respiratory distress syndrome. *Critical Care Medicine* 28: 2457–2463.

Mutch, W.A.C., Harms, S., Ruth Graham, M. et al. (2000b). Biologically variable or naturally noisy mechanical ventilation recruits atelectatic lung. *American Journal of Respiratory and Critical Care Medicine* 162(1): 319–323.

Narendran, V., Donovan, E.F., Hoath, S.B. et al. (2003). Early bubble CPAP and outcomes in ELBW preterm infants. *Journal of Perinatology* 23(3): 195–199.

Nelson, M., Becker, M.A., and Donn, S.M. (1998). Basic neonatal respiratory disorders. In: *Neonatal and Pediatric Pulmonary Graphics: Principles and Clinical Applications* (ed. S.M. Donn), 253–278. Armonk, NY: Futura Publishing Co.

Nogee, L.M., Dunbar, A.E., Wert, S.E. et al. (2001). A mutation in the surfactant protein C gene associated with familial interstitial lung disease. *The New England Journal of Medicine* 344: 573–579.

Northway, W.H. Jr., Rosan, R.C., and Porter, D.Y. (1967). Pulmonary disease following respirator therapy of hyaline membrane disease: bronchopulmonary dysplasia. *The New England Journal of Medicine* 276(7): 357–368.

Ogawa, Y., Miyasaka, Y., Kawano, T. et al. (1993). A multicenter randomized trial of high frequency oscillatory ventilation as compared with conventional mechanical ventilation in preterm infants with respiratory failure. *Early Human Development* 32: 1–10.

Otis, A.B., McHerrow, C.B., Bartlett, R.A. et al. (1956). Mechanical factors in distribution of pulmonasry ventilation. *Journal of Applied Physiology* 8: 427–443.

Pillow, J.J. (2005). High-frequency oscillatory ventilation: mechanisms of gas exchange and lung mechanics. *Critical Care Medicine* 33(3): S135–S141.

Pillow, J.J., Neil, H., Wilkinson, M.H., and Ramsden, C.A. (1999). Effect of I/E ratio on mean alveolar pressure during highfrequency oscillatory ventilation. *Journal of Applied Physiology* 87(1): 407–414.

Pillow, J.J., Hillman, N., Moss, T.J. et al. (2007). Bubble continuous positive airway pressure enhances lung volume and gas exchange in preterm lambs. *American Journal of Respiratory and Critical Care Medicine* 176(1): 63–69.

Poets, C.F. and Sens, B. (1996). Changes in intubation rates and outcome of very low birth weight infants: a population-based study. *Pediatrics* 98(1): 24–27.

Polin, R.A., Fox, W.W., and Abman, S.H. (2011). *Fetal and Neonatal Physiology: Expert Consult (2)*. Elsevier Health Sciences.

Rettwitz-Volk, W., Veldman, A., Roth, B. et al. (1998). A prospective, randomized, multicenter trial of high-frequency oscillatory ventilation compared with conventional ventilation in preterm infants with respiratory distress syndrome receiving surfactant. *The Journal of Pediatrics* 132(2): 249–254.

Riphagen, S. and Bohn, D. (1999). High frequency oscillatory ventilation. *Intensive Care Medicine* 25(12): 1459–1462.

Ritacca, F.V. and Stewart, T.E. (2003). Clinical review: high-frequency oscillatory ventilation in adults–a review of the literature and practical applications. *Critical Care* 7(5): 385.

Rossing, T.H., Slutsky, A.S., Lehr, J.L. et al. (1981). Tidal volume and frequency dependence of carbon dioxide elimination by high-frequency ventilation. *The New England Journal of Medicine* 305: 1375–1379.

Scherer, P.W. and Haselton, F.R. (1982). Convective exchange in oscillatory flow through bronchial- tree models. *Journal of Applied Physiology* 53: 1023–1033.

Sedeek, K.A., Takeuchi, M., Suchodolski, K., and Kacmarek, R.M. (2003). Determinants of tidal volume during high-frequency oscillation. *Critical Care Medicine* 31(1): 227–231.

Slutsky, A.S. (1981). Gas mixing by cardiogenic oscillations: a theoretical quantitative analysis. *Journal of Applied Physiology* 51: 1287–1293.

Slutsky, A.S. and Drazen, J.M. (2002). Ventilation with small tidal volumes. *The New England Journal of Medicine* 347(9): 630.

Slutsky, A.S., Drazen, F.M., Ingram, R.H. et al. (1980). Effective pulmonary ventilation with small-volume oscillations at high frequency. *Science* 209(4456): 609–671.

Slutsky, A.S., Kamm, R.D., Rossing, T.H. et al. (1981). Effects of frequency, tidal volume, and lung volume on CO2 elimination in dogs by high frequency (2–30 Hz), low tidal volume ventilation. *The Journal of Clinical Investigation* 68(6): 1475–1484.

Spahn, D.R., Leuthold, R., Schmid, E.R., and Niederer, P.F. (1991). Significance of bulk convection during high-frequency oscillation. *Respiration Physiology* 84(1): 1–11.

Stawicki, S.P., Goyal, M., and Sarani, B. (2009). Analytic reviews: high-frequency oscillatory ventilation (HFOV) and airway pressure release ventilation (APRV): a practical guide. *Journal of Intensive Care Medicine* 24(4): 215–229.

Steinhorn, R.H. (2010). Neonatal pulmonary hypertension. Pediatric critical care medicine. *A journal of the Society of Critical Care Medicine and the World Federation of Pediatric Intensive and Critical Care Societies* 11(2 Suppl): S79.

Suki, B., Alencar, A.M., Sujeer, M.K. et al. (1998). Life-support system benefits from noise. *Nature* 393(6681): 127–128.

Tanida, Y. (1990). Analysis of gas transport in highfrequency ventilation. *Frontiers of Medical and Biological Engineering* 2: 181–185.

Taylor, G.I. (1953). Diffusion and mass transport in tubes. *Proceedings of the Royal Society A* 216: 186.

Van Marter, L.J., Allred, E.N., Pagano, M. et al. (2000). Do clinical markers of barotrauma and oxygen toxicity explain interhospital variation in rates of chronic lung disease? *Pediatrics* 105 (6): 1194–1201.

Venegas, J.G. and Fredberg, J.J. (1994). Understanding the pressure cost of ventilation: why does high-frequency ventilation work? *Critical Care Medicine* 22(9 suppl): S49–S57.

Venegas, J.G., Hales, C.A., and Strieder, D.J. (1986). A general dimensionless equation of gas transport by high-frequency ventilation. *Journal of Applied Physiology* 60: 1025–1030.

Verder, H., Albertsen, P., Ebbesen, F. et al. (1999). Nasal continuous positive airway pressure and early surfactant therapy for respiratory distress syndrome in newborns of less than 30 weeks' gestation. *Pediatrics* 103(2): e24–e24.

Victorino, J.A., Borges, J.B., Okamoto, V.N. et al. (2004). Imbalances in regional lung ventilation: a validation study on electrical impedance tomography. *American Journal of Respiratory and Critical Care Medicine* 169(7): 791–800.

Vorhout, W.F., Veenendaal, T., Kuroki, Y. et al. (1992). Immunocytochemical localization of surfactant protein D (SP-D) in type II cell, Clara cells, and alveolar macrophages of rat lung. *The Journal of Histochemistry and Cytochemistry* 40: 1589–1597.

Walsh, M.C. and Stork, E.K. (2001). Persistent pulmonary hypertension of the newborn: rational therapy based on pathophysiology. *Clinics in Perinatology* 28(3): 609–627.

Weber, K., Courtney, S.E., Pyon, K.H. et al. (2000). Detecting lung overdistention in newborns treated with high-frequency oscillatory ventilation. *Journal of Applied Physiology* 89(1): 364–372.

Weibel, E.R. (1963). *Morphometry of the Human Lung*. New York: Academic Press and Springer-Verlag.

Weinmann, G.G., Mitzner, W., and Permutt, S. (1984). Physiological dead space during high-frequency ventilation in dogs. *Journal of Applied Physiology* 57: 881–887.

Wheeler, D.S., Wong, H.R., and Shanley, T.P. (2009). *The Respiratory Tract in Pediatric Critical Illness and Injury*. Springer.

Williams, M.C. (1977). Conversion of lamellar body membranes into tubular myelin in alveoli of fetal rat lungs. *The Journal of Cell Biology* 72: 260–277.

Wiswell, T.E. (2000). Meconium aspiration syndrome. In: *Manual of Neonatal Respiratory Care* (ed. S.K. Sinha and S.M. Donn), 266–272. Armonk, NY: Futura Publishing Co.

Wiswell, T.E. (2001). Expanded uses of surfactant therapy. *Clinics in Perinatology* 28(3): 695–711.

Wiswell, T.E. and Fuloria, M. (1999). Management of meconium-stained amniotic fluid. *Clinics in Perinatology* 26(6(3)): 59–668.

Wolf, G.K. and Arnold, J.H. (2005). Non-invasive assessment of lung volume: respiratory inductance plethysmography and electric impedance tomography. In: *Yearbook of Intensive Care and Emergency Medicine*, 116–128. Springer-Verlag, Berlin.

Wright, J.R. (1997). Immunomodulatory functions of surfactant. *Physiological Reviews* 77: 931–962.

Wung, J.T., Koons, A.H., Driscoll, J.M. Jr. et al. (1975). A new device for CPAP by nasal route. *Critical Care Medicine* 3(2): 76–78.

Wung, J.T., Koons, A.H., Driscoll, J.M. Jr., and James, L.S. (1979). Changing incidence of bronchopulmonary dysplasia. *The Journal of Pediatrics* 95(5 Pt 2): 845–847.

Index

Page numbers in italics refer to figures and in bold to tables.

Pressure Oscillation in Biomedical Diagnostics and Therapy, First Edition. Ahmed Al-Jumaily and Lulu Wang.

b

m

t